1·11·99

Asymmetry, Developmental Stability, and Evolution

Oxford Series in Ecology and Evolution
Edited by Robert M. May and Paul H. Harvey

The Comparative Method in Evolutionary Biology
Paul H. Harvey and Mark D. Pagel

The Causes of Molecular Evolution
John H. Gillespie

Dunnock Behaviour and Social Evolution
N. B. Davies

Natural Selection: Domains, Levels, and Challenges
George C. Williams

Behaviour and Social Evolution of Wasps:
The Communal Aggregation Hypothesis
Yosiaki Itô

Life History Invariants: Some Explorations of Symmetry
in Evolutionary Ecology
Eric L. Charnov

Quantitative Ecology and the Brown Trout
J. M. Elliott

Sexual Selection and the Barn Swallow
Anders Pape Møller

Ecology and Evolution in Anoxic Worlds
Tom Fenchel and Bland J. Finlay

Anolis Lizards of the Caribbean: Ecology, Evolution, and Plate Tectonics
Jonathan Roughgarden

From Individual Behaviour to Population Ecology
William J. Sutherland

Evolution of Social Insect Colonies: Sex Allocation and Kin Selection
Ross H. Crozier and Pekka Pamilo

Biological Invasions: Theory and Practice
Nanako Shigesada and Kohkichi Kawasaki

Cooperation Among Animals: An Evolutionary Perspective
Lee Alan Dugatkin

Natural Hybridization and Evolution
Michael L. Arnold

The Evolution of Sibling Rivalry
Douglas W. Mock and Geoffrey A. Parker

Asymmetry, Developmental Stability, and Evolution
Anders Pape Møller and John P. Swaddle

Asymmetry, Developmental Stability, and Evolution

Anders Pape Møller
Université Pierre et Marie Curie

and

John P. Swaddle
University of Bristol

Oxford New York Tokyo
OXFORD UNIVERSITY PRESS
1997

Oxford University Press, Great Clarendon Street, Oxford OX2 6DP

Oxford New York

Athens Auckland Bangkok Bogota Bombay
Buenos Aires Calcutta Cape Town Dar es Salaam
Delhi Florence Hong Kong Istanbul Karachi
Kuala Lumpur Madras Madrid Melbourne
Mexico City Nairobi Paris Singapore
Taipei Tokyo Toronto Warsaw

and associated companies in
Berlin Ibadan

Oxford is a trade mark of Oxford University Press

Published in the United States
by Oxford University Press, Inc., New York

A catalogue record for this book is available from the British Library

Library of Congress Cataloging in Publication Data

Møller, A.P. (Anders Pape)
Asymmetry, developmental stability, and evolution / Anders Pape
Møller and John P. Swaddle
(Oxford series in ecology and evolution)
1. Evolution (Biology) 2. Symmetry (Biology) 3. Developmental
biology. I. Swaddle, John, P. II. Title. III. Series.
QH371.M77   1997   576.8–dc21   97-25866

ISBN 0 19 854895 8 (Hbk)
0 19 854894 X (Pbk)

Typeset by Hewer Text Composition Services, Edinburgh

Printed in Great Britain by Bookcraft (Bath) Ltd
Midsomer Norton, Avon

Preface

In this book we illustrate how developmental stability and asymmetry are intrinsically entwined and why both of these concepts are important to evolutionary biologists. The chapters that follow, and also information that is contained on the associated Web site (see later in this preface), are devoted to examinations of developmental instability in separate fields and areas of biology that we hope will be of general interest to most people interested in the life sciences.

The first six chapters describe how and why asymmetry (developmental instability) is linked to evolutionary biology and how these morphological asymmetries develop in particular organisms. More specifically, Chapter 1 explores the connections between developmental instability and asymmetry, and explains how primarily one specific kind of asymmetry (fluctuating asymmetry) relates to genetic stability and fitness. Chapter 2 describes how these asymmetries might be produced during developmental processes. Chapters 3 and 4 investigate the theoretical and empirical links between developmental asymmetry, developmental stress, and evolutionary selection processes. The fifth and sixth chapters discuss the genetic and environmental variables that cause increased asymmetry.

The final three chapters are dedicated to the consequences of the asymmetries that have arisen during development. Chapter 7 explains the mechanical function of asymmetry. Bilateral structures require symmetry (i.e. zero asymmetry) for optimal performance and we discuss how asymmetry can be both directly and indirectly related to reduced locomotory ability, with a particular emphasis on bird flight. The next chapter, Chapter 8, proposes that asymmetry is not only mechanically functional but is also an intrinsic property of an individual's genome and so provides an 'uncheatable' method of signalling individual quality. Females and males may be able to use this honest signalling information when assessing conspecifics in social and sexual interactions. This link between fitness and asymmetry is developed in Chapter 9, with examples from both the animal and plant kingdoms. Symmetric individuals appear to have quantifiable and evolutionarily significant advantages over their asymmetric counterparts.

Each chapter can be treated as a self-contained unit and so could be referred to separately by anyone who is familiar with evolutionary concepts. The first six chapters essentially go in pairs; Chapters 1 and 2 cover the basis and the development of asymmetries; Chapters 3 and 4 explain asymmetry in terms of evolutionary processes; and Chapters 5 and 6 describe the possible causes of asymmetry. Chapters 7 to 9 (and Chapter W10 on the Web site) provide the real meat of the story and should provide stimulating reading for anyone interested in the life sciences. These chapters relate the theory and empirical evidence of the previous sections to real-life, evolutionarily important situations in humans, other animals, and plants. We hope that, before the end of the last chapter, the reader will be as intrigued with asymmetry as we are.

We have structured the book with three main priorities in mind. First, we wanted each chapter to be self-contained so that they could be read independently and information could be assimilated with ease. We hope that this structure will lend itself to student tutorials as each chapter could be discussed separately. Second, we intended there to be sufficient flow between the chapters for the dedicated reader to travel from cover to cover without encountering too much repetition. The chapters are organised into a logical scientific order so that individual ideas, references, and evidence can be found quickly and easily, whilst taking the reader through a chronological history of scientific discovery. We hope that this style will lend itself to both 'seminar' and 'bedtime' reading and will stimulate scientists to look at their world in a slightly different way.

The third structural element of this book is that we have developed a Web site in association with the writing of the manuscript, which we intend to up-date with recent findings and publications. The Web site also contains two chapters of text and numerous large tables of studies that review areas of the literature. We hope that readers who want to glean additional information about each topic will visit the Web site, whereas readers who are satisfied with the reviews printed in the book chapters need trek no further. If a chapter, table, or figure appears on the Web site and not in the printed book it is prefaced with a 'W'. For example, Table W2.1 reviews studies that have investigated individual and population asymmetry parameters and hence is relevant to Chapter 2 but appears on the Web site.

The two chapters of text that are published on the Web site (Chapters W1 and W10) provide, respectively, (i) a general introduction to the study of asymmetry in science and biology; and (ii) a review of possible applications of asymmetry and developmental stability in the life sciences. We recommend that everyone reads these two additional chapters, as they help to put asymmetry into its evolutionary context, introduce some interesting tales of early, pioneering research, and also propose some intriguing possible applications of asymmetry research to the world around us. What we also hope to convey in Chapter W1 is the enthusiasm for asymmetry research and indicate why symmetry/asymmetry is important in many fields of interest, including evolutionary biology. Hence, many issues are discussed from a broader standpoint and are not treated in depth. Chapter W10 is intended to encourage further diversification of the ever-growing interest in asymmetry and developmental stability and illustrate the general importance of much of the research reviewed in this book to many aspects of science. We pay particular attention to the role that asymmetry may play in human medicine, conservation biology, pollution monitoring, body condition, animal welfare, and forestry and farming policies.

Many people have helped us to produce this volume. Tim Birkhead, Jean Clobert, Denis Couvet, Innes Cuthill, Rowan Lockwood, Andrew Pomiankowski, Jacqui Shykoff, Randy Thornhill, Adam Wilkins, and Mark Witter have read all or part of various versions of the chapters and have improved the book's quality and content immeasurably with insightful and encouraging comment. Also, Paul Harvey must receive the credit for suggesting the associated Web site. Tim Birkhead kindly provided the cover illustration. We are gratefully indebted to them all. We would also like to thank the Natural Environment Research Council (UK) for financial support of JPS and the Swedish and Danish Natural Science Research Councils for financial support of APM during the preparation of this manuscript.

The Web site can be located at:

http://www.oup.co.uk/MS-asymmetry

Paris   A.P.M.
Glasgow   J.P.S.
December 1996

Acknowledgements

The following societies, publishers, and author kindly provided permission to reproduce figures and tables: Academic Press (Fig. 3.3), Alan R. Liss, Inc. (Table 5.1), Allen Press (Figs 5.2, 9.7), American Society of Mammalogists (Fig. 9.1), Annual Reviews Inc. (Table 1.3), J. M. Aparicio (Fig. 2.6), Birkhäuser Verlag (Figs 2.7, 9.6), Blackwell Science Ltd, (Fig. 9.5), Cambridge University Press (Fig. 3.2), Elsevier Science Ltd. (Fig. 8.3), Entomological Society of America (Figs 6.1, 6.4), Finnish Zoological Publishing Board (Fig. 4.1), B. Fuller (Fig. 9.9), Genetics (Fig. 3.7), Growth Publishing Co., Inc. (Fig. 5.1), Kluwer Academic Publishers (Figs 2.1, 2.2, 2.3, 2.5, 9.10), Munksgaard International Publishers (Fig 7.4), Oxford University Press (Fig. 4.2), Macmillan Magazines Ltd. (Figs 5.3, 8.2, 8.4, 8.5, 9.2, Table 7.1), The Royal Society (Figs 1.1, 3.4, 6.2, 7.1, 7.2, 7.3, 7.5, 8.7, 8.8, 9.3, 9.4, 9.8, Table 6.1), and US National Marine Fisheries Service (Fig. 6.3).

Contents

1   Asymmetries and developmental stability   1

1.1   Introduction   1
1.2   Developmental homeostasis, canalisation, and stability   3
1.3   Developmental stress   4
1.4   Frequency indices of developmental stability   5
1.5   Repeated-formation indices of developmental stability   9
1.6   Types of asymmetry   14
1.7   Which asymmetries estimate developmental instability?   17
1.8   Evolutionary transitions between asymmetries   20
1.9   Why is adaptive external asymmetry not more common?   21
1.10 The measurement and statistical analysis of fluctuating asymmetry   23
1.11 Summary   35

2   Ontogeny of asymmetry and phenodeviants   37

2.1   Introduction   37
2.2   The genetics of the formation of developmental stability   39
2.3   Developmental stability and chaos theory   42
2.4   Developmental biology and the formation of symmetry and
      handedness   55
2.5   Fitness, phenotypic quality, and reaction norms of asymmetry   58
2.6   Ontogeny of individual asymmetry and phenodeviants: empirical
      studies   60
2.7   Summary   66

3   Developmental stability and mode of selection   67

3.1   Introduction   67
3.2   Canalisation and the micro-evolutionary process   68
3.3   Phenotypic and genetic variance under different modes of selection   72
3.4   Phenotypic and genetic variance: lessons from selection experiments   76
3.5   Phenotypic and genetic variance: lessons from sexual selection and
      other directional selection phenomena   77
3.6   Summary   84

4   Adverse environmental conditions and evolution   85

4.1   Introduction   85
4.2   Consequences of adverse environmental conditions   86

4.3   Evolutionary consequences of adverse environments   92
4.4   Adverse environmental conditions and the distribution of species   103
4.5   Adverse environmental conditions and the speciation process   104
4.6   Adverse environmental conditions and major evolutionary
      innovations   108
4.7   Summary   109

5   Causes of developmental instability. I. Genetic factors   111

5.1   Introduction   111
5.2   Heritability of developmental stability   112
5.3   Genetic factors   117
5.4   Summary   133

6   Causes of developmental instability. II. Environmental factors 134

6.1   Introduction   134
6.2   Stress and energetic expenditure   135
6.3   Adverse temperatures   135
6.4   Nutritional stress   137
6.5   Chemical factors   138
6.6   Population density   142
6.7   Audiogenic stress   144
6.8   Trait-specific susceptibility to environmental factors   145
6.9   Stress-specific responses   147
6.10 Environmental stress   148
6.11 Environmental monitoring by fluctuating asymmetry   150
6.12 Summary   152

7   Developmental instability and performance   154

7.1   Introduction   154
7.2   Asymmetry and performance   155
7.3   Avian asymmetry and aerodynamics   159
7.4   Experimental investigations of asymmetry   168
7.5   Asymmetry and functional importance   172
7.6   Performance costs maintain asymmetry   175
7.7   Summary   177

8   Developmental stability and signalling   178

8.1. Introduction   178
8.2. General signalling theory   179
8.3. Origins of developmentally stable signals   182
8.4. Developmentally unstable signals   189
8.5. Sexual selection and developmental stability   194
8.6. Maintenance of developmental stability   203
8.7. Summary   205

9   **Developmental stability and fitness**   207

9.1   Introduction   207
9.2   Developmental stability, metabolism, and growth   209
9.3   Developmental stability and fecundity   210
9.4   Developmental stability and developmental selection   212
9.5   Developmental stability and competitive ability   217
9.6   Developmental stability and susceptibility to parasitism   220
9.7   Developmental stability, predation, and competition   223
9.8   Developmental stability and survival   226
9.9   Summary   228

References   231

Author index   273

Taxonomic index   283

Subject index   289

1

Asymmetries and developmental stability

1.1 Introduction

In this chapter we introduce the concepts of homeostasis, canalisation, and developmental stability and how this relates to developmental stress (sections 1.2–3). We then review the ways in which developmental stability has been assessed and measured. These methods include monitoring the frequency of deviant phenotypes (phenodeviants—section 1.4.1) and measuring variation in trait size (coefficient of variation—section 1.4.2) both among and within populations. Both of these methods are limited in that it is not possible to make *a priori* predictions concerning the true frequency of phenodeviants or the coefficient of variation within and among populations. Therefore the populations under investigation have to be thoroughly sampled to render accurate estimates of developmental stability. These two indices also tend not to control for differences in genotype or environmental variation during development, and so may not be good estimators of developmental stability (section 1.4.3).

Indices that are constructed from the comparison of structures that are repeatedly formed on the same individual provide estimates of developmental stability that do control for genetic and environmental differences (section 1.5). The difference in the two sides of a bilaterally symmetric trait (fluctuating asymmetry) may be a particularly good example of such an index. The use of asymmetry as an estimate of developmental stability, both among and within populations, is discussed in sections 1.5.1–3. Among its applications, population level fluctuating asymmetry could be important in tracing developmental stability over evolutionary time, or when comparing the response of populations to changing environmental conditions. At the individual level, asymmetry may reveal aspects of individual phenotypic and genotypic quality.

Frequency of asymmetric characters in a population, and within an individual, can also be employed as an indicator of developmental instability (section 1.5.2). Such an index could be useful as it can be employed when examining differences in developmental stability in studies with low sample sizes. However, large numbers of traits often have to be assessed and the discrimination between symmetry and asymmetry may not be straightforward in metric traits.

Developmental instability can be viewed as deviations from an intended phenotype and hence can be estimated by a number of additional techniques (section 1.5.4). Disruptions in the regular arrangement of leaves, deviations in radial symmetry and spiralling angles of snails' shells, and reduction in fractal dimension of morphology and behaviour have all been applied as indicators of developmental instability.

Generally, indices of developmental instability that are based on detecting dissimilarity between repeated-formation structures on the same individual appear to be the most accurate estimators of developmental instability, as genetic and environmental differences tend to be controlled for. Fluctuating asymmetry is most likely to control for these factors as the two sides of a bilateral trait are grown simultaneously, therefore minimising the probability that environmental conditions are different during the development of each element (section 1.5.5).

We describe the two forms of adaptive asymmetry: directional asymmetry and antisymmetry in sections 1.6.1 and 1.6.2, respectively. As both of these forms of asymmetry can be represented by an intended phenotype, they may also be employed in estimating developmental stability, although their use may be limited (section 1.7).

We also describe the evolutionary transitions between the three forms of morphological asymmetry in section 1.8. Under directional selection, it has been observed that fluctuating asymmetry can be transformed into directional asymmetry or antisymmetry; although a transition from directional asymmetry or antisymmetry to fluctuating asymmetry has not been reported.

The two forms of adaptive asymmetry may be relatively uncommon (especially in external characters), perhaps as a result of the increased costs associated with these forms of morphological and functional specialisation. Alternatively, adaptive asymmetry may be uncommon because directional asymmetry and antisymmetry could represent evolutionary dead-ends (section 1.9).

In sections 1.10.1–6 we review the recommendations for the statistical analysis of fluctuating asymmetry data. This review indicates several important considerations that should be taken into account. These are: (i) the statistical identification of fluctuating asymmetry; (ii) the magnitude of measurement error in relation to the asymmetry; (iii) the relationship between trait size and asymmetry, (iv) the statistical techniques employed in analysing asymmetry data; and (v) the problem of interpreting results from pooled and biased samples. Once these considerations have been made, we recommend that population fluctuating asymmetry data are analysed using variance asymmetry indices and individual asymmetry data are investigated using absolute or relative asymmetry indices. Composite asymmetry indices based on pooled measurements from several traits can often provide very powerful estimates of developmental instability.

1.2 Developmental homeostasis, canalisation, and stability

The concepts of symmetry and asymmetry have a distinguished and venerable past. In terms of modern-day biological theories, symmetry and asymmetry can be most clearly explained in terms of the control and stability of developmental processes.

A term that is often used in conjunction with the concept of controlled developmental processes is 'developmental homeostasis'. Homeostasis is a general expression indicating the tendency of the internal environment of an organism to be maintained constant. This process often involves self-regulating mechanisms in which the maintenance of a given level of development is initiated and controlled by the substance that is being regulated, most frequently by some sort of feedback mechanism. In other words, growth patterns occur that monitor and regulate themselves to maintain homeostasis. A self-correcting growth system. Maintaining developmental homeostasis is a synonym for stabilised growth patterns.

Homeostasis can be broken down into two elements, developmental stability and canalisation. Developmental canalisation is the production of consistent phenotypes under a range of environmental and genetic conditions. Therefore, highly canalised development reduces the effects of the environment and genes on developmental pathways. More specifically, it is the robust progression of a cell, or group of cells, down an evolved developmental pathway. In other words, canalisation is a fidelity-assurance mechanism that ensures that developmental pathways reach their intended end-point. Only at certain times of development, when the walls of canalisation are low, is the cell sufficiently vulnerable to be diverted onto another developmental trajectory.

One of the grandfathers of genetical research, C. H. Waddington, had a fondness for descriptive analogies, and he depicted canalisation as large balls travelling along valleys in an 'epigenetic landscape' (Waddington 1940). If the sides to these valleys are steep and the hills between them are mountainous in proportions, the balls will return to their original pathway in the valley very quickly and deviation will be minimal. Such a landscape has highly canalised developmental pathways. Development can be shunted to another pathway if genetic and environmental influences reach a threshold level, large enough to reduce the steepness of the valley sides and hence increase the probability of the ball wandering out of the intended pathway. Waddington proposed that this landscape is determined by the genome as a whole, and so may have heritable components. This view of developmental canalisation has changed little since these early days and the processes are still only crudely understood.

In a progression from the work and theorising of Waddington, more recent research has suggested that there are specific gene-complexes that determine canalisation (Zakharov 1989; Scharloo 1991). These gene complexes can be selected for and this type of selection is, logically, termed canalising selection (Waddington 1958). Hence, highly canalised strains of species can be produced through evolutionary processes.

Developmental stability, on the other hand, is the production of a phenotype, predetermined by adaptive design, under a given set of specified environmental and genetic conditions. It, therefore, refers to the capacity for developmental pathways to resist accidents and perturbations during growth processes. Developmental stability can be greatly affected by both the specified genetic and environmental influences, as genetically different strains of the same species and identical strains under different environmental conditions show variations in their levels of developmental stability. Developmental stability could be viewed as a balance between developmental homeostasis and developmental noise.

Highly stabilised development produces the ideal phenotype and any instability results in imperfect growth, which is caused by random developmental errors. There are certain genotypes and environments that increase the probability of these accidents occurring, by interrupting developmental processes.

The genetic basis of both canalisation and developmental stability appears to be complex and, as yet, unclear. Waddington performed early experiments in which he artificially selected for a high degree of canalisation in the development of fruit flies, *Drosophila*. Through selection of highly canalised family groups, he managed to produce populations of flies that were more canalised for an eye facet mutation (*Bar*) than others (Waddington 1960); however, there was no associated increase in developmental stability in the highly canalised flies. Waddington explains this apparent discrepancy by suggesting that the genes that determine overall developmental stability will not be the same as genes that determine the canalisation of any one particular trait (e.g. eye facet number). It is also worth pointing out that Waddington used mutant stocks of *Drosophila* for this experiment, which may have disrupted overall developmental stability (see section 5.3.4), which in turn may have masked any subtle increases in stability expected from the experimental protocol. Subsequent experiments by Waddington and more recent experiments that are reported later in this chapter and in Chapter 5, have indicated that developmental stability is a concomitant of genic balance, co-adaptation of gene complexes, and to a lesser extent, overall heterozygosity of the genome.

Very basically, the difference between canalisation and stability is that canalisation represents the control of development under a range of conditions, whereas stability is the resistance of random errors under specified conditions. Canalisation can be regarded as a property of the genome that tends to ensure that a developmental pathway remains on its intended trajectory, whereas developmental stability is the phenotypic outcome of that genetic property. These are important distinctions which have implications for how we should go about measuring developmental stability, which is discussed later in this chapter.

1.3 Developmental stress

We have introduced the concept of developmental stress already. It is the vague spectre that causes a reduction in developmental homeostasis. More usefully, it can be described as a state caused by a factor that increases the 'noise' associated with developmental processes and causes long-term, detrimental change to an organism (definitions of stress are discussed in Chapter 4). Even if an organism possesses the most ideally balanced and structured genome for the environment it occupies, and that environment is completely stable, there will still be some element of randomness in its development. The degree of randomness may be very small, maybe even undetectable to a trained observer, but perfection never exists. This unpredictable generation of developmental errors, or 'noise', is increased by a number of factors that can act from within the genome or from the external environment. In other words, these factors increase the level of developmental stress in the system.

A number of genetic and environmental factors that increase developmental instability have been identified over years of research. Genetic effects such as high levels of inbreeding, extreme homozygosity, hybridisation that splits up co-adapted gene complexes, novel mutations, and periods of intense directional selection have all been shown to break down the balance of a genome and reduce the ability of the organism to buffer its developmental pathways against the production of random errors. An almost limitless number of environmental factors could be cited. Unusual temperatures, deficient food sources, chemical pollution, and even loud sounds can all reduce developmental stability. We review the genetic and environmental factors that give rise to developmental instability in Chapters 5 and 6, respectively. At the moment it is sufficient to say that developmental stability can be broken down in many ways and that developmental pathways are not as rigid as they first may seem. If you look at a 'perfect' structure closely enough, you will find imperfection.

1.4 Frequency indices of developmental stability

Here, we review the pros and cons of measures of developmental stability and critically assess each approach in turn. As there are a number of different methods, we have separated these indices into two convenient categories. First, there are indices that have been derived from frequency estimates within and between populations. These indices we have termed 'frequency indices', which include the frequency of phenodeviants and phenotypic coefficients of variation, and are the subject of this section. Second, there are indices that are based on comparison of measurements taken from repeated elements of the same trait on the same individual. We have named these 'repeated-formation indices' and they are described in section 1.5. All of the indices that we discuss are listed in Table 1.1.

1.4.1 *Frequency of phenodeviants*

A phenodeviant is an abnormal phenotype or an unusual expression of a trait, in other words a trait that displays a growth pattern substantially different to

Table 1.1. Table of published methods for estimating developmental instability. 'Predicted pattern' refers to the ability to predict unstable phenotypes *a priori*. 'Genotypic homogeneity' refers to whether the estimate controls for genetic differences between samples. 'Environmental constancy' refers to whether the estimate controls for environmental differences between samples. 'Partial' indicates that the estimate is constructed from measures that may be made under difference environmental conditions due to the sequential growth of traits.

Estimate	Discriminatory capabilities	Predicted pattern	Sample size required	Genotypic homogeneity	Environmental constancy	Example references
Frequency of phenodeviants	often low	no	larger	no	no	Waddington 1953; Bateman 1959; Rasmuson 1960
Coefficient of variation	high	no	larger	no	no	Mitton 1978; Eanes 1978; Handford 1980
Repeated structures	high	yes	smaller	yes	partial	Shackell and Doyle 1991; Graham *et al.* 1993a; Escós *et al.* 1995
Fluctuating asymmetry	high	yes	smaller	yes	yes	Ludwig 1932; Møller 1990a; Parsons 1992
Frequency of asymmetric traits	sometimes low	yes	smaller	yes	yes	Leary *et al.* 1989; Zakharov 1989; Zakharov and Yablokov 1990
Radial asymmetry	high	yes	smaller	yes	partial	Graham *et al.* 1993a; Freeman *et al.* 1993; Eriksson 1996a
Adaptive asymmetry	high	no	larger	yes	yes	McKenzie and Clarke 1988; Leary and Allendorf 1989; Markow 1992
Fractal dimension	high	varies	smaller	yes	partial	Alados *et al.* 1994; Escós *et al.* 1995

that observed in the mean phenotype. This is not simply a statistical artefact of developmental expression, as phenodeviants are distinct from simple polymorphisms which reflect a range of possible phenotypic expressions; these are extreme deviants and so can be seen as gross indicators of developmental instability. Phenodeviants are normally rare, but they do usually exist at some frequency within populations. Examples of such occurrences are birth deformities seen in humans. Dubinin and Romashov (1932) and Rasmuson (1960) have proposed that the frequency of these abnormal growth patterns in a population can give an index of developmental stability that is comparable with measures such as fluctuating asymmetry (see section 1.5.1). By comparing frequencies of phenodeviants in a contingency table by chi-square analysis, it is possible to test for overall differences in developmental stability among populations. For example, Graham *et al.* (1993a) found a difference in frequency of lower jaw deformities between populations of goldfish, *Carassius auratus*, that occupied polluted and unpolluted lakes. From the lake polluted with methanol, 11% of the goldfish had gross abnormalities, whereas none of the fish from the control lake had abnormal growth patterns.

However, as these abnormalities, by definition, are relatively rare within populations (Jones (1988) defined such an anomaly as a character that appears in less than 4% of the population), they do not represent a particularly sensitive measure of developmental stability (Table 1.1). Development has to be disturbed to a reasonably major degree before these extreme phenotypes arise. By adopting the type of definition that Jones (1988) derived for detecting the phenodeviant traits, it must be presumed that at least 96% of the population are 'developmentally stable' for any given character. This figure will vary greatly among characters and may even be far higher for some very rare growth abnormalities. Therefore, the use of gross abnormalities and frequency of extreme phenodeviants will only give practicable and useful indicators of developmental stability if a large number of traits are measured or if the population is highly stressed; otherwise the great majority of the population will be seen as being equally 'stable', making among-individual comparisons almost impossible. Even in studies among populations, the frequency of phenodeviants may be very small, for example Waddington's (1953) and Bateman's (1959) index of wing-vein mutations in *Drosophila* showed relatively little variation between their data sets. The various mutations they observed occurred in approximately only 1% of normal flies.

1.4.2 *Coefficient of variation*

There are other indicators of developmental instability that could be more sensitive than the frequency of phenodeviants within populations. One such index is the coefficient of variation, or in other words the relative variation in size of a trait both among and within populations.

A large number of authors have used the degree of morphological variation as an index of developmental homeostasis, particularly for detecting differences among populations (e.g. Mitton 1978; Eanes 1978; Handford 1980; Beacham and Withler 1985; King 1985; Mitton and Koehn 1985; McAndrew *et al.* 1986; Govindaraju and Dancik 1987; Yezerinac *et al.* 1992; Hartl *et al.* 1993 cited in Mitton 1993; Ward and Elliott 1993). Under accepted theory, you would predict that if there was strong developmental stability within the organisms' growth processes, then morphological variance would be small. Hence, this has led authors to predict that when developmental processes are highly stabilised, the coefficient of variation is small. In support of this, Rasmuson (1960) found a positive relationship between fluctuating asymmetry and the coefficient of variation in *Drosophila*. However, phenotypic plasticity and developmental instability are not correlated (Scheiner *et al.* 1991; Tampolsky and Scheiner 1994; Tarasjev 1995).

However, there are sometimes problems with applying and interpreting this measure, especially when testing for genetic differences among populations (for example, if trying to elucidate the relationship between heterozygosity and developmental stability). If the genes under investigation determine the expression of the trait that is being assessed in terms of developmental stability, then you may expect morphological variance to rise, rather than fall, with increased heterozygosity (Chakraborty and Ryman 1983; Leary *et al.* 1984, 1985a). In this kind of example, a stability index of morphological variance could only be applied if the genes involved did not directly influence the development of the trait, for example, if the loci were enzyme polymorphisms that coded for proteins that do not directly influence morphological characters (Mitton 1993).

Also, if a trait is subject to strong canalising selection and the peak of the size–fitness function is narrow and steep, there will be little variation in trait sizes both among and within populations. This lack of variation, which may be expected in traits that play a large role in determining and maintaining levels of individual Darwinian fitness, may mean that the coefficient of variation will be an insensitive measure of developmental stability and it will be difficult to detect differences among and within populations. However, characters affected by persistent directional or disruptive selection may be an exception here, as these traits often show large variation among individuals (see section 3.4).

1.4.3 *Problems with frequency indices*

Another major drawback with the population frequency methods detailed above is that the populations need to be thoroughly sampled before any measures are calculated or statements about developmental stability can be made (Table 1.1). It is not possible to start work on a new population of organisms and immediately predict which traits are going to display levels of phenodeviants, let alone what these frequencies are going to be. Likewise for the coefficient of variation. The whole point of these indices is that they vary among and within populations, so *a priori* predictions are not possible. Both the frequency of phenodeviants and coefficient of variation indices are based upon frequencies within populations and frequency differences between sets of populations, therefore the population frequencies have to be calculated accurately for each new population that is studied. This process may be technically difficult to perform in the field, for example if the researcher is studying a particularly elusive population or a population that cannot be disturbed. In the laboratory this may be more feasible, but would be extremely time consuming, as the majority of the entire population would have to be sampled in order to get accurate predictions of population frequency parameters.

It is also fair to criticise the utility of these techniques as measures of developmental stability on the grounds that they have not controlled for genetic differences between individuals or taken the external environmental conditions into account (Table 1.1). Developmental stability is defined in terms of variation of growth patterns resulting from common genotypes in similar environmental conditions. Field studies often do not incorporate studies of genetic variation, and between-population studies will also confound environmental conditions; so the cases in which the applications of these population frequency indices satisfy the criteria of genetic and environmental homogeneity will be rare. In most cases the criteria will be violated, and the indices will not reveal measures purely based on developmental stability. Laboratory-based studies are a different kettle of fish, and if performed on a genetically uniform population (or sets of populations) in controlled environmental conditions, could produce meaningful measures of developmental stability. However, these studies may have limited applications to species in natural populations, so there is a need to develop indices of developmental stability that can control for the influences of genetic and environmental conditions in both the laboratory and field.

1.5 Repeated-formation indices of developmental stability

'Repeated-formation indices' do tend to control for genetic and environmental differences as these estimates of developmental stability are derived from the comparison of measurements from repeated elements of the same trait within the same individual. As the trait is under the influence of a single gene complex, the repeated component elements of that trait will be influenced by a common genome, and so negate the problems of genetic homogeneity that we described earlier. Examples of the types of characters that fulfil these criteria are repeated features, such as the scales of fish (e.g. Shackell and Doyle 1991), and the left and right elements of a bilaterally symmetric traits (fluctuating asymmetry).

1.5.1 *Levels of fluctuating asymmetry*

The most commonly used index of developmental stability is fluctuating asymmetry (Ludwig 1932), which is a direct result of the inability of individuals to undergo identical development on both sides of a bilaterally symmetrical trait. As developmental pathways on either side of a bilateral character are identical (i.e. the trait is the product of a common gene complex), the resulting asymmetry is a product of random, local disturbances in morphogenesis that arise from upsets of internal cellular development or differences in the external environment. As these disturbances are random, the average expression of the trait is symmetry, although notable asymmetries can arise in a small proportion of individuals, giving a normal distribution of signed asymmetry scores that are centred around a mean of zero. In short, the expression of individual asymmetry in a population that displays fluctuating asymmetry provides a measure of how well that individual can buffer its development against internal genetic and external environmental stresses during morphogenesis. Additionally, the breadth of the population distribution of asymmetry scores indicates the stability of the population as a whole; and so the differences between left and right sides of a bilateral trait can provide an excellent index of developmental stability at both individual and population levels, and are often simple and easy to measure.

However, it is worth remarking that, as Swain (1987) has suggested, measuring fluctuating asymmetry in meristic characters (i.e. those that are discrete and countable, such as number of bristles on flies' legs or the number of rays present in the fins of fish) may not be as accurate an indicator of developmental stability as quantifying fluctuating asymmetry in metric (continuous) characters. This may be because meristic characters will only become asymmetric once 'stress' reaches a threshold level; whereas metric characters have the potential to become asymmetric in very subtle ways and so will provide a more sensitive indicator of developmental stability.

1.5.2 *Frequency of asymmetric characters*

If the characters involved are very complicated or are essentially qualitative (e.g. areas of coloured plumage on a bird), it may be difficult to attain scores for the left and right sides of the trait. In these cases, it may only be possible to say whether the trait is symmetric or asymmetric. If only one trait is being studied, then it would be possible to reflect developmental stability as the frequency of asymmetric individuals within the population. However, many evolutionary biologists are interested in among-individual differences in developmental stability, which could only be assessed in this way if many traits are scored as being symmetric or asymmetric on the same individual. The measure of stability could then be calculated as the total number of asymmetric traits on the individual (Leary and Allendorf 1989; Zakharov *et al.* 1989; Zakharov and Yablokov 1990). For example, Zakharov and Yablokov (1990) counted the number of traits displaying asymmetry out of 34 bilateral traits in populations of grey seals, *Halichoerus grypus*, and found that asymmetries increased between 1940 and 1960 in the three Baltic populations that they studied. This approach may also have the additional statistical advantage of yielding data that are suitable for parametric analysis, and can also provide data in the form of a contingency table for chi-square analysis when sample sizes are low (e.g. if only one population is studied). However, the classification of a character as being symmetric or asymmetric is far from straightforward, as threshold asymmetry values need to be determined (Table 1.1). Also, it is important to determine that the traits under investigation display fluctuating asymmetry and not directional asymmetry or antisymmetry, as an asymmetry in the case of the latter two may not indicate developmental instability.

1.5.3 *Fluctuating asymmetry as a population statistic, and an individual property*

As mentioned previously, fluctuating asymmetry can be viewed as a population statistic, and expression of asymmetry within that population can be seen as the property of an individual. This statement may cause a few eyebrows to raise and heads to shake, as fluctuating asymmetry is defined as a population parameter (Ludwig 1932; Van Valen 1962; Palmer and Strobeck 1986), but let us clarify exactly what we mean. The early geneticists and developmental biologists who introduced the concept of asymmetry into studies of developmental stability really intended fluctuating asymmetry to be exclusively a population parameter. The level of asymmetry 'fluctuates' within the population (with an approximate normal distribution), therefore fluctuating asymmetry is a property of the population. However, individuals within a population also have the propensity to display a certain level of asymmetry, which is a reflection of how well their genotype can express the ideal phenotype in the given conditions, which are identical for both sides of a paired trait. Therefore, expression of asymmetry can also be viewed as an individual property and a measure of developmental stability.

The concepts of both individual asymmetry and population-level fluctuating asymmetry are useful to evolutionary biologists. At the population level, it may be important to trace periods of developmental stability and evolutionary change across time, perhaps even geological time (e.g. Møller and Pomiankowski 1993b); or to study population responses to changes in environmental conditions (Leary and Allendorf 1989). Whereas, at the individual level, it may be interesting to employ asymmetry as an indicator of individual condition (Møller 1990a, 1992a; Swaddle and Witter 1994) or to investigate its usefulness as a signalling trait in communication between conspecifics (Møller 1992b, 1993e; Swaddle and Cuthill 1994a, b; Swaddle 1996a). In these two cases you would probably employ different methods of calculating the relevant statistic (see section 1.10). When measuring population-level fluctuating asymmetry for a particular trait it may be best to use the variance in random deviations between left and right sides, i.e. $\text{Var}(R_i-L_i)$ (Palmer and Strobeck 1986; Palmer 1994); whereas at the individual asymmetry level it may be best to use simple deviations from symmetry, i.e. $|(R_i-L_i)|$ (Swaddle *et al.* 1994).

1.5.4 *Developmental instability as a measure of deviations from the mean*

It is possible to view developmental instability as deviations from the mean or average expression of a trait (Freeman *et al.* 1993; Graham *et al.* 1993a). Hence, one of the main differences between the ways of measuring developmental stability hinges around predictions and accurate statements concerning this population mean. With this simple concept in mind, it would be possible to think of other ways of measuring deviation from the norm, and so measures of developmental stability are not necessarily confined to those that we have discussed so far. Indeed, some researchers have applied this notion and derived new measures of developmental stability that may apply even to non-morphological traits.

Graham *et al.* (1993a) have gone so far as to suggest three new methods for measuring developmental stability in different systems that have implications for developmental stability outside of the specific examples that they give (see discussion in Freeman *et al.* 1993). Their first application is in the arrangement of leaves on the stem of a plant, otherwise termed as phyllotaxis. Plants are known to arrange their leaves in regular patterns (sometimes exactly opposite to each other, and sometimes in a Fibonacci sequence) and so deviations from these patterns could indicate disruptions of developmental stability. By studying the distances between rachises of opposing leaflets in the black locust (*Robinia pseudoacacia*), whose leaflets are normally arranged opposite to each other, they demonstrated that developmental stability is weakened at a site closer to an ammonia production facility than a control site some 8 km away. This concept, especially in this particular application, is in practice a measure of developmental instability, as the leaflets are normally arranged opposite to each other. In other plants that show different patterns, the application will greatly depend on the accuracy of the predicted phyllotaxis, so studying completely novel plant systems may be difficult.

The second application was in their study of shell growth in snails (Graham *et al.* 1993a). Snails are known to follow rigid developmental patterns when growing their spiralled shells (Huntley 1970). At each stage of growth, the shell maintains its overall shape and enlarges at a rate represented by a logarithmic function (Coxeter 1961). It is possible to compare the growth of an individual snail with how well it fits the predefined equiangular spiral that is determined by this function. Graham *et al.* (1993a) collected snails, *Cepaea nemoralis*, from three different sites, one of which was believed to be polluted with pesticides. Their findings indicated that snails from the polluted site had larger deviations from the ideal developmental pathway than the other populations, indicating a loss of developmental stability. This was further supported by an increase in gross abnormalities in the polluted population.

Graham *et al.*'s (1993a) final application of the concept of developmental stability reflecting deviations from the norm was a study of fractal dimensions of sutures in the skulls of dorcas gazelles, *Gazella dorca*. Most readers are probably aware of the invariant nature of fractals (e.g. Mandelbrot 1977, 1982; Feder 1988). And as they possess an invariance, they can be seen as a form of symmetry. Fractals have often been applied in nature to describe growth patterns of structures and organisms (e.g. branching of blood vessels and the sutures of the skull); and the loss of fractal structures has recently been associated with disease, illness, and perhaps unstable development (West and Goldberger 1987; Goldberger *et al.* 1990; West 1990b; Weinstein *et al.* 1992; Cross *et al.* 1993; Alados *et al.* 1994). This latter hypothesis (accredited to West 1990a, b) is supported by the results obtained by Graham *et al.* (1993c). They discovered that the fractal dimension of skull sutures in inbred dorcas gazelles was lower than that of less inbred individuals. However, the relationship between fractal dimensionality and developmental stress may not be simple; Emlen, Freeman, and Graham (unpublished but cited in Graham *et al.* 1993c) have produced both empirical and theoretical support that fractal dimensions may increase with stress in certain situations. Their hypothesis predicts that fractal dimensions should decrease with stress in systems where stress slows metabolic pathways. Whereas if the effect of stress is additive to an existing complex pattern (e.g. a complex branching system), then stress may lead to an increase in fractal dimensions. As this hypothesis has not yet been fully developed, it may suffice to say that stress will induce a change in fractal dimension of the structure. If the fractal trait is paired, it is possible that the asymmetry of fractal dimension may also reveal developmental instability, as both sides of the trait should display the same fractal complexity.

If there is a reliable method for detecting or predicting both individual and population-average expression of a trait, then an index of developmental stability can be derived. Such a trait is not necessarily restricted to morphology; Markow and Gottesman (1993) suggest that deviance of behavioural traits can be used when assessing developmental stability. By adopting the same approach, Escós *et al.* (1995) have investigated behavioural phenodeviance in the antipredatory behaviour of grazing Spanish ibex, *Capra pyrenaica*. When under stressful conditions, such as pregnancy and parasitic infection, the fractal dimension of this behaviour decreased.

In general, it is the accuracy and reliability of estimating or predicting the phenotypic average that will greatly determine the usefulness of the measure of developmental stability. For traits displaying bilateral symmetry this is simple; for traits that display a Fibonacci sequence of phyllotaxis, a form of radial symmetry in their shell formation, or a fractal dimension in skull sutures, the predictions are also possible and may be more complicated, but should also return reliable and accurate results (Table 1.1). We will return to this line of reasoning later in section 1.7, when we discuss the possibility of using directional asymmetry and antisymmetry as ancillary indicators of developmental instability.

1.5.5 *Advantages of the repeated-formation indices*

There are a number of advantages to the repeated-formation indices. First, they control for genetic heterogeneity, as all comparisons made to derive the figures are conducted within the same individual. This allows simple estimation of both individual and population levels of developmental stability. Both of these estimates are extremely useful to evolutionary biologists and can be easily compared with each other. Second, by taking measurements from repeated elements of the same trait, environmental conditions will tend to be constant during the development of each element. Environmental constancy is particularly pertinent in the case of fluctuating asymmetry as the two sides of the trait are developed simultaneously and so environmental conditions are as homogeneous as they possibly could be (Table 1.1). Genetic homogeneity and environmental constancy will lead to the detection of morphological differences between the repeated elements that are more likely to be explained by developmental instability rather than a lack of canalisation, and therefore will generally provide more accurate estimates of developmental instability than the frequency indices.

The use of repeated-formation indices also has a clear advantage in that these indices generate *a priori* predictions concerning morphology (Table 1.1). In other words, the optimal phenotype is known before the study has begun. In the example of repeated production of scale morphology on a fish, Shackell and Doyle (1991) were able to predict that each circulus (circular ridge) on the scale should be identical, and so deviations from this expression indicate developmental instability. The same applies to measures of asymmetry (e.g. Mather 1953; Beardmore 1960), as it is a developmental invariant. Before the investigation has started, the researcher can predict that (if the trait is bilaterally symmetrical) deviations from perfect symmetry (i.e. zero asymmetry) will reflect instability during developmental processes, as symmetry is the intended phenotype. We should add at this point, that from a life-history perspective we should expect some minor degree of asymmetry even in the 'optimum' state, as decreasing asymmetry will be traded off against fitness in other traits. The resulting 'optimum' phenotype is then dependent on the 'optimum trade-off' of developmental stability and fitness among traits, but will probably lead to very small asymmetries in most cases. Therefore, especially as measurement error will invariably mask these very small asymmetries (section 1.10.2), this problem may be negligible and lower asymmetry will most often indicate a 'more optimum' phenotype.

We, and many other researchers (e.g. Mather 1953; Beardmore 1960; Valentine and Soulé 1973; Leary *et al.* 1985a, b; Van Valen 1962; Palmer and Strobeck 1986; Clarke and McKenzie 1992; Graham *et al.* 1993a; Mitton 1993) recommend the use of fluctuating asymmetry as the index of developmental stability to adopt. Fluctuating asymmetry provides an analytical tool that can quickly and inexpensively assess levels of developmental stability in both natural field observations and strictly controlled laboratory experiments, the results of which can be compared empirically under the same framework.

1.6 Types of asymmetry

We have concluded that low levels of asymmetry can reveal developmental stability, so it is important to draw clear distinctions between the different types of asymmetry that can arise in nature, and explain how each of these are related to the concept of developmental stability. As we have already discussed the usefulness of fluctuating asymmetry, we shall describe the two forms of adaptive asymmetry: directional asymmetry (section 1.6.1) and antisymmetry (section 1.6.2), and how these forms of asymmetry may also be useful to investigators of developmental stability (section 1.7).

1.6.1 *Directional asymmetry*

Directional asymmetry occurs when there is a propensity for one side of a trait to develop more than the other (Van Valen 1962). In cases of directional asymmetry it is possible to predict which side of the trait will be larger before the character has started growing. Therefore, within a population, there is a handed-bias for one side of a trait to be larger than the other. If we look at a frequency histogram of signed asymmetry (right minus left) values in the population, we will observe a large skew to either the left or the right, the direction of the skew depending on the trait and the species. As it is possible to predict which side will be larger, there must be some significant genetic element to directional asymmetry (reviews in Palmer and Strobeck 1986; McManus 1991), although it is still possible that developmental noise will be present in addition to directional asymmetry and may influence the size of the asymmetry (Markow 1992; Graham *et al.* 1993a, b; Møller 1994a).

Examples of directional asymmetry include the direction of coiling in the shells of snails and the number of lobes in the left and right sides of the human lung. In fact, many mammalian internal organs show directional asymmetry, for example, the heart, brain, testes, arterial arches, and much of the alimentary canal all show a form of handedness. It is strange that our bodies hide so much internal asymmetry with an exterior that is, essentially, symmetric.

Placement of eyes on flatfish is another example of a directional asymmetry. Flatfish live, for most of their lives, with one side of their body resting on the sea floor; so as an adaptation to their peculiar body shape and swimming behaviour, one eye migrates from the under surface to the upper surface during morphogenesis, giving them the advantages of binocular vision (Hubbs and Hubbs 1945).

There are other examples of directional asymmetries naturally occurring in most taxa. Crickets produce their songs by rubbing their wings over each other; what they actually do is stridulate a scraper against a file, which are located on opposite wings. This occurs predominately by the right wing passing over the left, e.g. *Gryllus bimaculatus* and *Acheta veletis*, and hardly ever occurs the other way around (Neville 1976). Most gastropod mollusc shells spiral clockwise (if looking at the shell from above) which is known as a dextral spiral. A smaller number of species possess a sinistrally spiralling shell. These species show a directional rotational asymmetry. There is one gastropod, however, which is striving to stand out from the crowd; *Tamanovalva limax*, resident of the Inland Sea of Japan, has dextrally arranged viscera but a small sinistral spiral at the apex of its shell (Neville 1976).

There are two remarkable examples of directional asymmetry in birds. The wry-billed plover, *Anarhynchus frontalis*, has a bill that is approximately 3 cm long and is always bent to the right at the tip by up to 12° (Neville 1976). This helps them with turning over stones as they search for food along shingle-river beds. Naturally, it may help with flipping stones over to the right, but is a hindrance for any left-handed feeding attempts, so the birds only feed right-handed. The ears of many owl species show directional asymmetry (Norberg 1978). The external ears are positioned in different vertical positions on either side of the head so that left and right ears have different maximum sensitivity to sound coming from different directions and orientations. This allows the owls to accurately pinpoint the horizontal source of sound in complete darkness without having to tilt their heads. By tilting their head they can fix on the vertical location. So their ear asymmetry has provided them with an extremely accurate and reliable way of detecting prey in their naturally dark habitats.

1.6.2 *Antisymmetry*

Antisymmetry occurs when one side of a character is larger than the other, but there is no handed-bias as to which side will be larger (Timofeeff-Ressovsky 1934). Therefore, it is not possible to predict which side of the character will develop to the larger size. Typically, antisymmetry shows a platykurtic (i.e. broad peaked) or bimodal frequency distribution within a population. Examples of this type of asymmetry are relatively rare, the most often cited of which are the larger signalling claws of male fiddler crabs which belong to the genera *Uca* (females have symmetric, small claws). In *Uca musica*, young male crabs start their existence with two large 'male-type', equally sized claws. The laterality of the asymmetry is determined by which of these two claws becomes damaged and drops off first. The chances of this happening on the left or right sides are 50-50, and hence the bimodal frequency distribution of asymmetry scores. If one claw is damaged, it is regrown as a small 'female-type' claw; if two drop off simultaneously, then they both regrow as small chelae; however, if neither is damaged then the male crab has two very large claws. The chances of the latter two situations occurring must be very small, as adult male crabs with symmetric small or large claws are extremely rare. Once the claws are damaged and regrown, the developmental pathway becomes fixed; hence the majority adult phenotype is one small and one large claw, with an equal chance of the larger claw appearing on the left or right side (Neville 1976).

Another example of antisymmetry is the position of the sail-like structure on colonial floating coelenterates. Both the Portuguese man-o'-war, *Physalia physalis*, and the jack sail-by-the-wind, *Vellela vellela*, and many other coelenterates, are blown along the sea surface by the action of a sail that is set at an angle to the axis of the body. The direction that this sail points in varies throughout the population, some pointing to the left and some to the right, generally giving a bimodal distribution of this asymmetry. The difference in position of the sails results in broad dispersal of the population, as the direction that the sail is pointing in will determine the direction in which the organism will move (Neville 1976). In birds, crossbills *Loxia curvirostra* are know to display antisymmetry in their crossed beak. Approximately 58% of birds have a lower mandible that crosses to the left and 42% to the right, giving a bimodal distribution of mandible overlap within a population (Neville 1976).

1.7 Which asymmetries estimate developmental instability?

Traditionally, fluctuating asymmetry is thought of as being the only asymmetry that relates to accidents during morphogenesis and hence developmental instability (e.g. Palmer and Strobeck 1986). The other two types of asymmetry have mainly been interpreted in terms of adaptive and functional asymmetries. However, Leary and Allendorf (1989), McKenzie and Clarke (1988), Markow (1992), Graham *et al.* (1993c) and Møller (1994a) have all suggested that adaptive asymmetries could also be used as indicators of developmental stability, as they are a form of developmental invariance (Table 1.1). Møller (1994a) has demonstrated that testicular directional asymmetry is greatest in individual birds that have the largest sexual ornaments, and are hence in the best phenotypic condition (see Fig. 1.1). This implies that directional selection is not only acting on the size of the sexual ornament, but also on the size of the testicular directional asymmetry. This relationship will also give individuals with the largest directional asymmetry a selective advantage, as they have large sexual ornaments and so will have increased mating success (Møller 1994b). This evidence, along with the work of Graham *et al.* (1993c), has strongly suggested that directional asymmetries can be used as an index of developmental stability. Individuals that possess phenotypes that differ greatly from the norm (independent of whether the norm is symmetry or asymmetry) appear to be less fit and less developmentally stable. Essentially, the premise is that individuals in the long tail of the population frequency diagram are of lower phenotypic, and perhaps genotypic, quality and have unstable patterns of morphogenesis (see Fig. 1.2).

This relationship of deviation from the intended developmental trajectory, fitness, and developmental stability may be more complicated in cases of antisymmetry. Where antisymmetry gives a bimodal frequency distribution of asymmetry scores through the population, it may be fairly straightforward to identify individuals with aberrant growth patterns. In this case, individuals that are symmetric or have very large asymmetries (in both directions) may possess destabilised developmental pathways. However, if antisymmetry gives a platykurtic distribution of asymmetry scores, identifying the developmental 'outliers' may be far more troublesome. In this case are symmetric individuals outliers even though they may fall in the modal frequency group (see Fig. 1.2)? Nevertheless, a number of authors have recommended the use of antisymmetry as a measure of developmental stability (McKenzie and Clarke 1988; Leary and Allendorf 1989) and in some cases it may be applicable. Unbiased estimates of developmental instability could be obtained from characters displaying directional asymmetry or antisymmetry by simply using the residuals of the regression of the size of the left character on the right character (two regression lines in the case of antisymmetry).

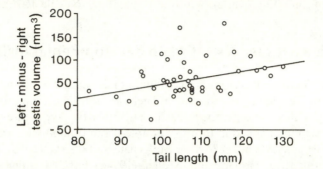

Fig. 1.1 Directional asymmetry in testis volume in relation to tail length of male barn swallows *Hirundo rustica*. There is a significant positive relation as revealed by linear regression, left-minus-right testis volume $(mm^3) = -99.39 + 1.46$ (0.61, s.e.) tail length (mm); $F_{1,44} = 5.76$, $r^2 = 0.12$, $P = 0.021$. Testicular directional asymmetry is greatest in individual birds that have the largest sexual ornaments, and are hence in the best phenotypic condition. Adapted from Møller (1994a).

Palmer and Strobeck (1986, 1992) refute the use of either directional asymmetry or antisymmetry as indicators of developmental stability and stress, as they have a significant but unknown genetic component. At present, there is no indication of there being 'left-handed' and 'right-handed' genes, although there may be genes that switch the direction of handedness (e.g. Brueckner *et al.* 1991). However, there is evidence for mutations that interfere with handedness of development in a rather non-specific way (Morgan 1991). Therefore, the differences between left and right sides of a trait may not result solely from 'developmental noise' or random accidents during morphogenesis, and hence there is a case for these asymmetries not to be used as measures of developmental stability.

However, these views are not necessarily accepted, and 'handedness' may not always need a genetic component (see Neville 1976; Freeman *et al.* 1993). Within coconut trees, *Cocos nucifera*, handedness is not heritable and has been shown to change both within an individual and within a clone (Davis 1962, 1963). The development of chela asymmetry in the snapping shrimp *Alphaeus* has both genetic and environmental influences (Bethe *et al.* 1930). If a female-type (small nipper) claw is removed, a female claw is regrown; similarly, if a male-type (large crusher) claw is removed, the replacement claw is also male-type. However, if both claws are removed with a delay between the two, the female claw can become a male claw if it is given a 40 hour headstart over its homologue. If both claws are removed simultaneously, then the claws maintain their original 'sex'. This implies that there is some interaction and competition between the two sides of the body and claw type is not necessarily predetermined. Another form of environmental influence on claw asymmetry is

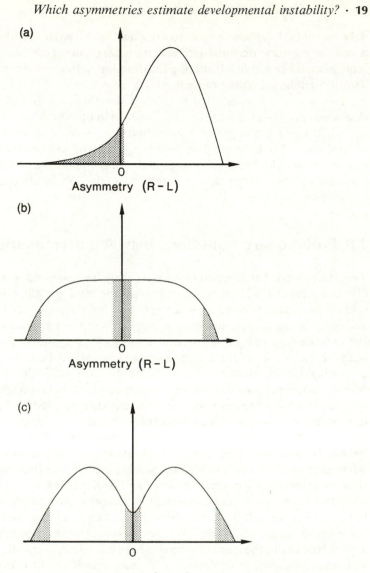

Fig. 1.2 Population frequency diagrams for (a) directional asymmetry; (b) platykurtic antisymmetry; and (c) bimodal antisymmetry, with distribution tails shaded.

observed in the previous example of chela formation of male fiddler crabs, where random damage determines the direction of the asymmetry.

Brown *et al.* (1991) have suggested a three-step model for the production of adaptive asymmetry in mammals that is based on initial molecular asymmetry. Many biological molecules are indeed asymmetric (Pasteur 1860; review in Mason 1991). First, the underlying molecular asymmetry is converted into cellular asymmetry; second, there is random generation of asymmetry, which may be biased by the preceding conversion stage to produce a handed asymmetry; and finally, an interpretation stage in which the organs of the individual are constructed with the information defined in the first two stages. Through empirical testing of their model, Brown *et al.* (1991) have found that there is a sensitive time window during development. If embryos are heat stressed or given a number of chemicals whilst in the time-frame of this sensitive period, the normal directional asymmetry is reversed. Brown *et al.* (1991) suggest that this time window occurs during the conversion step of their model, and would also tend to suggest that directional asymmetries can reveal some information about developmental stability and conditions during morphogenesis.

1.8 Evolutionary transitions between asymmetries

The fossil record has revealed that heterochely, the asymmetry observed in the claws of some crustaceans, is normally only observed in males (Neville 1976). It is also noteworthy that the large degrees of asymmetry observed in fiddler crabs and other similar crustaceans has generally only evolved in those species that are bottom-dwelling and are not free-swimming. This is presumably because the biomechanical cost of these large asymmetries would be too great in a free-swimming species, where balance would be extremely difficult. The robber crab, *Birgus* (Coenobitidae), shows amazing transitions between asymmetry and symmetry as adaptive responses to different stages of its life cycle. It begins its post-larval existence with a symmetrical abdomen. It then occupies a dextral gastropod shell (i.e. one with a left-handed spiral) and adapts to this change in habitat by developing an asymmetric abdomen. Once it has outgrown its adopted protective shell shelter and becomes free-living once more, it reverts back to its original symmetric design (see Neville 1976).

Hermit crabs are a classic example of adaptive asymmetry in response to habitat. They initially arose as a symmetrical creature that occupied symmetrical empty shells. From this they have evolved into crabs that can occupy asymmetric shells, the vast majority of which have dextral spirals, by developing suitably asymmetric body forms. They are so well attuned to their environmental conditions that it has been claimed that there are no sinistral hermit crabs (Neville 1976). In a preference test, Brightwell (1951) found that hermit crabs preferentially choose dextral over sinistral shells.

Hermit crabs are not the only example of asymmetrical adaptation to habitat; there are a number of parasites that have evolved forms of adaptive asymmetry in response to the changing internal environment of their hosts (Neville 1976). For example, a mature female *Bopyrus*, which parasitises the gills of prawn and shrimp, shows asymmetry corresponding to the side of the host that it attaches to. The parasite's preference for left or right gills appears to be related to asymmetric pressures of the gill chamber walls. The copepod *Botryllophilus* parasitises the gill cavity of ascidians and possesses asymmetric limbs that are either adapted for attachment to the gill wall or to free-swimming within the small gill cavity.

Several studies have demonstrated that it may be possible to induce the evolutionary change between fluctuating asymmetry and the two adaptive asymmetries within a relatively small number of generations. These studies have shown that the asymmetry present can change its statistical properties from that of fluctuating asymmetry to directional asymmetry or antisymmetry in response to intense directional selection (e.g. Mather 1953; McKenzie and Clarke 1988; Leary and Allendorf 1989; Graham *et al.* 1993c; see Table 1.2). These findings further support the claim that both directional asymmetry and antisymmetry can reflect developmental instability, as they can be a product of stressed developmental systems.

Table 1.2 Studies that have shown that the asymmetry present can change its statistical properties from that of fluctuating asymmetry (FA) to directional asymmetry (DA) or antisymmetry (AA) in response to directional selection.

Species	Transition	Reference
Drosophila melanogaster	FA to DA	Mather (1953)
Drosophila melanogaster	FA to DA, DA to FA	Graham *et al.* (1993b)
Lucilia cuprina	FA to AA	McKenzie and Clarke (1988)
Lucilia cuprina	FA to AA	McKenzie and Yen (1995)
Salvelinus confluentus	FA to AA	Leary and Allendorf (1989)
Mus musculus	AA to DA	Yokoyama *et al.* (1993)
Theory	FA to DA or AA	Graham *et al.* (1993c)

1.9 Why is adaptive external asymmetry not more common?

The two forms of adaptive asymmetry are relatively rare in comparison with bilateral symmetry and have evolved relatively few times (Ludwig 1932; Van Valen 1962; Palmer 1996). Why is this so? Why are directional asymmetries and antisymmetries so uncommon? The answers to these questions must, at least partly, be related to the adaptive value of asymmetry compared with symmetry. If there is no advantage in being asymmetric, then it will not pay an organism to invest in asymmetry and there will be an increased probability that asymmetric individuals will die or go extinct. Crossbills have their crossed beaks as an adaptation to picking seeds out of fir cones (Benkman and Lindholm 1991). Young crossbills initially have straight bills and they start to cross only after approximately 30 days, building up to the time when they can start to feed themselves (after about 45 days). Here there is a clear advantage in developing an asymmetric bill shape. However, it is possible that the cases where asymmetry is adaptive are uncommon and this may account for the relative scarcity of asymmetric external phenotypes.

One major function of external asymmetry in colour patterns may be in crypsis. Asymmetric colour patterns, like those observed in beetles (e.g. *Lithinus nigrocristatus*), frogs (e.g. *Hyla langsdorfii, Dendrobates tinctorius*), snakes (e.g. *Python regius, Natrix fasciata, Vipera berus*), salamanders (*Salamandra*), geckos (e.g. *Uroplates fimbriatus*), fish (e.g. *Scomber scombrus*), and the eggs of many bird species (e.g. *Vanellus vanellus*) may all act to break up the outline of the animal and so make it less conspicuous (see also discussion in Osorio 1996). Visual detection of prey (or predators) is greatly facilitated by the recognition of a regular, repetitive form, and so colour patterns that deviate from symmetry may decrease detectability (Cott 1940; Osorio 1996). Therefore, there may be a selective advantage for external asymmetry in cases of disruptive colour patterns, as asymmetry may entail an evolutionary advantage over symmetry.

Hence it is possible to envisage scenarios in which asymmetry should, and indeed has, evolved. This still does not tell us why it has not evolved more frequently. Perhaps adaptive asymmetry is particularly difficult to evolve. It could be that there is not sufficient genetic variability for asymmetry to evolve in some systems. Maynard Smith and Sondhi (1960) performed a selection experiment with *Drosophila* and found that it was not possible to increase the levels of directional asymmetry within the population (see also Coyne 1987). This implies that there was no, or very little, additive genetic variance present for directional asymmetry. However, in most cases, there is no immediately obvious reason why we should presume that asymmetry has not had time to evolve, and there are reported cases where significant asymmetries have evolved due to selection processes (e.g. Beardmore 1965; Purnell and Thompson 1973).

Many people have interpreted adaptive asymmetry as a form of division of labour. One side of a trait becomes specially adapted for a function that cannot be conducted by the other. For example, lobsters have one claw that is much heavier and larger than the other, which appears to be related to their relative functions. The larger, heavier claw is adapted for crushing, whereas its much smaller counterpart is used for nipping and picking up food (Neville 1976). In evolutionary terms, it may have been less costly for the lobster to diversify the functions of its paired claws into crushing and pinching rather than opting for an intermediate claw that may not have been very good for either purpose, or evolving a completely separate structure that would have taken on the role of either one of the claws. The lobster has opted for claw diversification (asymmetry) and this appears to be the most logical adaptation; but this strategy must also incur costs, as the loss of a crushing claw would mean that the lobster becomes dysfunctional until a new crushing claw can regrow. In organisms that do not have the ability to regenerate structures this cost would be much greater, perhaps life-threatening. However, if the lobster had a back-up claw (i.e. was bilaterally symmetric) there would be less of a cost associated with the loss of this specialised feature.

Another example of the possible increased cost of directional asymmetry could be seen in the example of the wry-billed plover. If a phenotypic error occurred so that a plover developed a left twist to its bill, it is possible that it may never feed, as they always feed right-handed. This error in development may be very difficult to overcome. Perhaps this is why the adaptive asymmetries are so uncommon; they entail too great a cost for the individual, in this example the costs of co-adaptation between specialised morphology and behaviour. Symmetry, on the other hand, may be an evolutionarily preferred strategy as everything is produced in duplicate. Therefore, if one side of a trait becomes damaged or is lost, there is a functionally identical contingency structure on the opposite side that can immediately fulfil the role of the damaged trait. If this logic is extended, perhaps directional asymmetry and antisymmetry could be viewed as evolutionary dead-ends (cf. Møller 1994a), as the loss of a character from one side of the body cannot be compensated for. However, most paired traits function as a partnership (e.g. birds' wings or horses' legs) and so removal or damage to one will dramatically alter the usefulness and function of the other, and so this 'contingency' argument may not apply in many cases. Notwithstanding this, there is some empirical support for the idea that adaptive asymmetry represents an evolutionary dead-end as fluctuating asymmetry can evolve into directional asymmetry or antisymmetry (e.g. Mather 1953; McKenzie and Clarke 1988; Leary and Allendorf 1989; Graham *et al.* 1993c; but see Maynard Smith and Sondhi 1960; see Table 1.2), but the evolution from adaptive asymmetry to fluctuating asymmetry has rarely been reported.

An alternative view is that perhaps more adaptive asymmetry exists than we actually realise. Wilson (1968) demonstrated that locusts (*Schistocera gregaria*) have asymmetry in their motor nerve output that connects to their flight muscles. Without visual sensory feedback, locusts actually rotate along the roll axis when in flight. The visual feedback they receive from their eyes allows them to compensate by double firing axons on one side of their flight muscles, and so they fly straight. This type of locomotor asymmetry also appears to exist in the mosquito *Aedes aegypti*, dipteran flies *Syrphus viridiceps, Anisopus fenestralis*, and milkweed bugs *Oncopeltus*. Even humans are known to walk, swim, and (most frighteningly) drive in broad circles if they are deprived of sensory feedback to their limbs (Schaeffer 1928). However, many of these asymmetries are hidden and internalised; and the conclusion still remains that external adaptive asymmetries are relatively uncommon and the majority external phenotype is symmetric.

1.10 The measurement and statistical analysis of fluctuating asymmetry

As we mentioned previously, asymmetry can be calculated at the individual and population level. Palmer and Strobeck (1986) assessed the effectiveness of the various indices that have been used to detect population differences in fluctuating asymmetry. They did this by simulating data sets in which they systematically varied mean character size, level of asymmetry, the relationship between trait size and asymmetry, and the levels of directional asymmetry. Through their many simulations, they generally found that indices based on the variance of asymmetry in a population (i.e. $\text{Var}(R–L)$) outperformed those based on the absolute values of unsigned asymmetry ($|R–L|$). However, when studying asymmetry at the individual level, it is not possible to calculate variance of asymmetry (unless large numbers of traits are measured), so absolute values of asymmetry have to be calculated (Swaddle *et al.* 1994). All published indices of population fluctuating asymmetry and individual asymmetry are summarised in Table 1.3.

Table 1.3 Summary of population fluctuating asymmetry (population) and individual asymmetry (individual) indices. Measurements are taken from right (R) and left (L) sides of j traits on an individual i. In the population fluctuating asymmetry indices only one trait is measured on each individual. Sample size is denoted by N. Expanded from Palmer and Strobeck (1986).

Number	Type	Index	Comment
1	population	$\dfrac{S\,(|R_i-L_i|)}{N}$	No scaling by character size
2	population	$\text{var}(R_i-L_i)$	No scaling by character size
3	population	$\dfrac{S\,(|R_i-L_i|)^2}{N}$	No scaling by character size
4	population	$\dfrac{S[(|R_i-L_i|)/0.5(R_i+L_i)]}{N}$	Character size scaled by individual
5	population	$\text{var}[(R_i-L_i)/0.5(R_i+L_i)]$	Character size scaled by individual
6	population	$\text{var}[\log(R_i-L_i)]$	Character size scaled by individual
7	population	$\dfrac{[S(|R_i-L_i|)/N]}{[S(R_i-L_i)/2N]}$	Character size scaled by population
8	population	$\dfrac{\text{var}(R_i-L_i)}{[S(R_i-L_i)/2N]}$	Character size scaled by population
9	individual	$|R_i-L_i|$	No scaling by character size
10	individual	$(|R_i-L_i|)/0.5(R_i+L_i)$	Character size scaled by individual
11	individual	$S\,\text{rank}(|R_j-L_j|)$	No scaling by character size
12	individual	$S\,\text{rank}(|R_j-L_j|)/j$	No scaling by character size
13	individual	$S[(|R_j-L_j|)/0.5(R_j+L_j)]$	Character size scaled by individual
14	individual	$\dfrac{S[(|R_j-L_j|)/0.5(R_j+L_j)]}{j}$	Character size scaled by individual

The studies of Palmer and Strobeck (1986), Palmer (1994) and Swaddle *et al.* (1994) have highlighted the need to take a number of factors into account before fluctuating asymmetry data can be analysed. The criteria that they list are relatively stringent, but their recommendations are important and future studies of asymmetry should, at least, address each of the issues that we discuss in sections 1.10.1–5.

1.10.1 *Statistical properties of fluctuating asymmetry*

Palmer and Strobeck (1986), Palmer (1994) and Swaddle *et al.* (1994) indicate that an important first step in analysis of fluctuating asymmetry data is that the data must be shown to display the statistical properties of fluctuating asymmetry, in other words, display an approximate normal distribution of signed asymmetry ($R–L$) scores around a mean of zero. The most commonly used statistical procedures employed for this purpose have been the Kolmogorov–Smirnov test (Palmer 1994) and the sign test (e.g. Thornhill and Sauer 1992; Balmford *et al.* 1993; Radesäter and Halldórsdóttir 1993; Manning and Chamberlain 1994). However, the sign test considers only the direction of deviations from zero, and neither are particularly effective for detecting deviations from a normal distribution (Siegel and Castellan 1988; Swaddle *et al.* 1994). Swaddle *et al.* (1994) recommend the use of more powerful techniques, namely normal probability plots (e.g. Filliben correlation coefficient (Aitken *et al.* 1989); Lilliefors' test; Ryan–Joiner test; Anderson–Darling test (Ryan *et al.* 1985)), followed by one-sampled *t*-tests to zero that will examine the centrality of the distribution. It may also be useful to examine the level of kurtosis within the population, as antisymmetry often shows a degree of platykurtosis (Palmer and Strobeck 1986), although this deviation from normality should be detected by the normal probability test mentioned above. It is imperative to distinguish fluctuating asymmetry from directional asymmetry and antisymmetry, as conclusions concerning the difference in asymmetry between samples and populations will depend on the type of asymmetry present. Palmer (1994) recommends that researchers should always report skew and kurtosis statistics, as they are easily computed, reveal important information as to the nature of the asymmetry, and are also helpful to other researchers as both statistics are useful descriptors of the data distributions.

Although characters displaying fluctuating asymmetry theoretically have normally distributed right-minus-left character values, this may not always be the case. First, when the units of measurement are large in comparison with the actual asymmetry, there will be a tendency for the population distribution of asymmetry to show leptokurtosis even though the trait may display fluctuating asymmetry. Therefore, using a measurement technique with a high resolution will generally be able to resolve this problem. Second, there appears to be intense natural and sexual selection against asymmetric individuals (see Chapters 8 and 9). Hence, very asymmetric individuals may be relatively rare in field samples, if selection has already acted against the most asymmetric individuals. Platykurtic frequency distributions with few observations in the tails of the distributions may simply reflect that selection has already removed very asymmetric individuals from the sample rather than the trait not displaying the characteristics of fluctuating asymmetry.

In a recent mathematical model, Leung and Forbes (1997a) have indicated that a leptokurtic population distribution of asymmetry scores can arise due to developmental stability, and hence can be thought of as indicating fluctuating asymmetry. The reasons for this appear logical based on their assumption that asymmetry results from the random effect of developmental stability following the equation $FA = Dn/(1 + Ds)$ (where FA denotes fluctuating asymmetry, Dn denotes developmental noise and Ds denotes developmental stability; see Leung and Forbes 1997a for details). As developmental noise contributes a random degree of asymmetry to the phenotype and developmental stability contributes an asymmetry value determined by some genetic component, there will be a tendency for both high and low asymmetry scores to be relatively over-represented in the population. This results in the tails and centre of the normal distribution becoming larger, and hence resulting in leptokurtosis. This model therefore implies that researchers could assume that asymmetry data that follow a normal or leptokurtic distribution represent fluctuating asymmetry. This further highlights the need to examine carefully the properties of asymmetry distributions.

Another approach to this problem of distinguishing fluctuating asymmetry from directional asymmetry and antisymmetry is to tease out directional asymmetry from fluctuating asymmetry within the analysis process. At present, there is no way of statistically correcting for antisymmetry. Directional asymmetry has been accounted for by calculating fluctuating asymmetry as the error variance from a two-way analysis of variance (ANOVA) in which side and individual are the two factors. Essentially, the mean square of the side factor has been interpreted as directional asymmetry, whereas the mean square of the error variance represents the interaction of side and individual, and has hence been used as a population-level estimate of fluctuating asymmetry (Leamy 1984). As stated above, this approach does not test or control for the presence of antisymmetry, and we recommend that the raw data are still tested for the properties of fluctuating asymmetry before any such approach is adopted. It is also of great relevance (and contention) that both directional asymmetry and antisymmetry may reflect aspects of developmental stability in certain situations (section 1.7). This underlines the caution that must be adopted with this kind of statistical methodology. For example, individuals with least asymmetry in a trait that displays directional asymmetry may, in fact, be the most developmentally unstable as they possess the phenotype furthest from the intended pattern of development. Generally, we recommend caution in adopting this kind of *a posteriori* adjustment of asymmetry values. It is more appropriate to test for the statistical properties of the various kinds of asymmetry and then interpret results in terms of the type of asymmetry that the data display.

If the raw data do display the statistical properties of fluctuating asymmetry, the advantage of the variance-partitioning approach is that it calculates fluctuating asymmetry as a population parameter and will be powerful for detecting among-population differences in developmental stability. However, the mean square error-term generated by this method also incorporates any measurement error that may have occurred during the collection of the data, which may artificially inflate the population fluctuating asymmetry value (see Palmer 1994; Swaddle *et al.* 1994).

1.10.2 *Measurement error and fluctuating asymmetry*

As signed fluctuating asymmetry is characterised by a normal distribution around a mean of zero it is likely to be indistinguishable from measurement error, which shows similar properties (Lundström 1960; Greene 1984; Palmer and Strobeck 1986; Palmer 1994; Swaddle *et al.* 1994; Merilä and Björklund 1995; Fields *et al.* 1995). This is exacerbated by the fact that fluctuating asymmetries are normally very small in relation to the trait being measured (typically less than 1% of trait size; Møller and Pomiankowski 1993a). Measurement error has been demonstrated to account for up to 25% of the apparent variation in dental asymmetry in wild mice (Bader 1965) and humans (Greene 1984) and 76% of the variation in wing length asymmetry in nymphalid butterflies (Mason *et al.* 1976).

Therefore, repeated measurements must be taken from the same individual to ascertain the relative influence of measurement error on the asymmetry estimates. Such measurements should be done blindly; preferably conducted one or more days apart to avoid any biases. This point was made explicit by Lundström (1960), Palmer and Strobeck (1986) and Palmer (1994), but appears to have been ignored in some of the recent literature (e.g. Thornhill 1992b; Thornhill and Sauer 1992; Radesäter and Halldórsdóttir 1993). In some cases, measurements have been replicated, but the repeatability of the size of each side of the trait has been erroneously equated to repeatability of the derived asymmetry between the two (e.g. Balmford *et al.* 1993; Manning and Chamberlain 1994). Low measurement error in component traits (i.e. size of left and right sides) does not ensure low measurement error in the derived trait (i.e. asymmetry, unsigned right-minus-left value), especially as the derived trait (asymmetry) is so small (Palmer 1994; Swaddle *et al.* 1994; Fields *et al.* 1995). Swaddle *et al.* (1994) illustrated this point empirically by showing that left and right tarsi of 35 adult European starlings, *Sturnus vulgaris*, can be significantly repeatable (intraclass correlation coefficient, r_I (Zar 1984); right r_I = 0.91, $P < 0.0001$; left r_I = 0.80, $P < 0.0001$) whilst the asymmetry between the two is not repeatable ($F_{34,68}$ = 0.65, P = 0.97; see below for mixed-model ANOVA of asymmetry repeatability).

There are two ways around this problem. First, Palmer and Strobeck (1986) suggest a mixed-model ANOVA approach that is modified from the one proposed by Leamy (1984). The difference between the two models is that Palmer and Strobeck recommend repeating measurements on both sides and so entering an extra factor into the analysis, that will effectively separate measurement error from a combined measure of fluctuating asymmetry and antisymmetry (see Table 1.4). Analysis of the distribution of signed right-minus-left scores will indicate the importance of antisymmetry in the population sampled (Palmer 1994; Swaddle *et al.* 1994). This approach will be very effective and statistically powerful at detecting asymmetry differences among populations (especially if based around the use of variance asymmetry measures rather than right-minus-left values, see Palmer and Strobeck 1986); however, it also falls prey to one of the criticisms aimed at Leamy's (1984) original ANOVA approach, namely the interpretational problems of statistically isolating directional asymmetry from fluctuating asymmetry. As mentioned previously, directional asymmetry may reveal some information concerning developmental stability, and so figures yielded from this approach may be slightly troublesome to interpret. There is also the limitation that this approach can only be applied among populations and not among individuals; this method does not yield individual levels of asymmetry. However, as many researchers study fluctuating asymmetry as a population parameter, Palmer and Strobeck's (1986) mixed-model ANOVA, when combined with exploratory descriptive data analysis, is the most appropriate method currently available.

Table 1.4 Palmer and Strobeck's (1986) mixed-model ANOVA approach that includes factors J (genotypes), S (side) and M (number of repeated measurements). It is presumed that there is only one individual per genotype. Nondirectional asymmetry is a combination of autisymmetry and fluctuating asymmetry. For measurements among samples, measurement error can be partitioned out by subtracting the measurement error (MS_m) from the remainder (MS_{sj}), and dividing by the number of repeated measurements taken (M). Refer to Palmer and Strobeck (1986) for more details concerning this analysis.

Source of variation	Label	df	Mean square	Mean square tests for
Side (S)	MS_s	$(S-1)$	$s^2_m + M(s^2_i + J/(S-1)Sa^2)$	Directional asymmetry
Genotype (J)	MS_j	$(J-1)$	$s^2_m + M(s^2_i + Ss^2_j)$	Shape or size variation
Remainder	MS_{sj}	$(S-1)(J-1)$	$s^2_m + Ms^2_i$	Non-directional asymmetry
Measurements (M)	MS_m	$SJ(M-1)$	s^2_m	Measurement error

Swaddle *et al.* (1994) suggest that the best way to test the accuracy of individual measures of asymmetry is to test the repeatability of individual component (right and left side) and derived trait (asymmetry) scores (see Table 1.5). They too employ a mixed-model ANOVA approach which includes the factors Individual, Side (right or left), and Repeat (each repeated measure for each side of each trait). An important part of this methodology is that repeated measurements of left and right sides are taken in pairs and numbered accordingly when specifying the data for the ANOVA model. The pairing of left and right repeats is important as it is both components that derive an asymmetry estimate and there may be longitudinal biases or errors in measuring either or both sides (cf. Palmer 1994). The ratio of the Individual-by-Side mean square to the combined Individual-by-Side-by-Repeat and Individual-by-Repeat mean square provides an *F*-test of whether between-individual variation in estimated asymmetry is significantly larger than that due to measurement error. In other words, this *F*-test provides a rigorous measure of the repeatability of the derived asymmetry values. If the values are repeatable and show the properties of fluctuating asymmetry, then you have robust measures of individual asymmetry derived as the mean value of the repeated asymmetry measures (Swaddle *et al.* 1994). Table 1.5 provides a worked example of how this *F* ratio is calculated.

Table 1.5 A worked example of how to calculate the repeatability of individual asymmetry estimates. Three repeated measurements were taken from primary nine (outermost) on left and right wings from 10 adult European starlings, *Sturnus vulgaris*. A mixed-model ANOVA with factors *S* (side), *I* (individual) and *R* (repeat measurements) was performed. Repeatability is estimated by dividing the *I*-by-*S* mean square by the combined mean square of *I*-by-*S*-by-*R* and *I*-by-*R*; therefore providing an *F* test of whether asymmetry differences between individuals is greater than that estimated from measurement error (see Swaddle *et al.* 1994). The combined mean square of *I*-by-*S*-by-*R* and *I*-by-*R* is calculated by summing the relevant sum of squares (0.00193 + 0.00256) and dividing by the summed degrees of freedom (18 + 18). Hence, in this example the combined mean square of *I*-by-*S*-by-*R* and *I*-by-*R* = 0.000125. Dividing the mean square for *I*-by-*S* by this combined mean square gives the *F* ratio (0.00186 ÷ 0.000125 = 14.91). Therefore, the asymmetry estimates are significantly repeatable as $F_{9,36} = 14.91$, $P < 0.00001$.

Source of variation	df	Sum of squares	Mean square
I	9	11.00303	1.22256
S	1	0.00033	0.00033
I-by-*S*	9	0.01674	0.00186
I-by-*R*	18	0.000256	0.00014
S-by-*R*	2	0.00010	0.00005
I-by-*S*-by-*R*	18	0.00193	0.00011

It is important to point out that measurement error is not only relevant to assessing asymmetry of metric traits, but also meristic traits. It is obvious how measurement error can occur when assessing metric traits, and how this error will inflate asymmetry estimates; but measurement error can also occur during the quantification of discrete meristic traits. Hubert and Alexander (1995) report a study in which they investigated among- and within-observer measurement error of five meristic characters (pectoral fin rays, pelvic fin rays, gill rakers on the upper and lower first branchial arches, and mandibular pores) in 50 cutthroat trout *Oncorhynchus clarki*. There was large variation in asymmetry estimates between three trained observers, with low concordance among their observations. Additionally, when one observer performed two repeated measures on the same fish, there was also substantial measurement error in trait estimates. This study highlights that even meristic traits can be prone to measurement error and so investigation of error relative to asymmetry should be conducted in all studies. At present, there are no formal recommendations for assessing levels of measurement error in meristic traits where the asymmetries may only be one or two counts. Palmer (1994) has suggested employing a likelihood ratio or *G*-test where the counts in each class of asymmetry are compared with the counts of each class of measurement error (i.e. when there are discrepancies between repeats). This may indicate whether the asymmetry is significantly larger than the variation between repeats, which is similar to the approaches described above for assessing measurement error in metric traits.

But how many repeats should be performed? Both Palmer and Strobeck (1986) and Swaddle *et al.* (1994) provide empirical data to indicate that the accuracy, and hence repeatability, of asymmetry data will generally increase with increasing numbers of repeats. Swaddle *et al.* (1994) present data of tarsal asymmetry from starlings which indicate that when left and right sides are measured twice each, the derived asymmetry was not repeatable (see above); however, when the number of repeats was increased to six, the data become significantly repeatable ($F_{34,340} = 5.44$; $P < 0.001$). This underlines the need for repeated measurements in any study that attempts to quantify asymmetry values. In general, Palmer (1994) recommends that traits are measured at least twice, unless measurement error is less than approximately 25% of the variation between left and right sides. This is a useful rule of thumb, and we also recommend that traits are routinely measured twice. In cases where sample sizes are small (< 20), measurements of left and right sides should be repeated three or more times.

1.10.3 *Relationship between asymmetry and size*

Often it is of interest to compare the size and asymmetry of a given trait; this is important as asymmetry can vary with size in a number of different ways. There are several indices that attempt to 'control' for the effects of trait size on asymmetry (reviews in Palmer and Strobeck 1986; Palmer 1994), the most commonly used is a relative measure of asymmetry (i.e. $[(|R-L|)/(R+L)]$). However, it is important to realise that relative measures of asymmetry control for trait size only if the relationship between asymmetry and trait size is isometric and intercepts the origin (Møller and Höglund 1991; Cuthill *et al.* 1993). In circumstances where there is a clear linear relationship between trait size and asymmetry, Swaddle *et al.* (1994) have suggested that it may be more appropriate to control for trait size by analysis of covariance (cf. Packard and Boardman 1987; but see problems of parametric analysis below). In traits under stabilising selection, the relationship between size and asymmetry is often ∪-shaped (reviews in Soulé 1982; Møller and Pomiankowksi 1993a; Watson and Thornhill 1994) and so relative measures of fluctuating asymmetry will often not be useful or appropriate. The best advice may be to explore the relation between trait asymmetry and size, as recommended by Palmer and Strobeck (1986) and Palmer (1994), and to control for size by dividing asymmetry by size and trying various transformations so that there is no longer any relationship between these two variables. In practice this may be difficult, if there is a ∪-shaped relationship there may be difficulty in finding the most appropriate transformation. There may also be interpretational problems in using relative measures of asymmetry between populations, as populations may differ in overall trait size, which will then influence the level of relative asymmetry. It is also important that when describing the relationship between trait size and asymmetry, parametric techniques should be employed with caution, as fluctuating asymmetry data will often not conform to the assumptions of these tests (section 1.10.4).

Palmer (1994) has raised another interesting issue related to asymmetry measures that control for trait size. He has remarked that body size may reflect some aspect of condition, and so controlling for trait size may partially control for condition and render asymmetry estimates that are condition-independent. Although the validity of this assumption would have to be investigated in each single case, it may be a point that researchers should bear in mind when considering using relative measures of asymmetry and other estimates that control for overall trait size.

In some cases, however, the use of relative measures of asymmetry are relevant and do effectively control for variations in trait size (e.g. Thornhill and Gangestad 1993). This approach may be particularly useful, as composite indices of asymmetry can be calculated across traits on the same individual (e.g. Thornhill and Gangestad 1993; Manning and Ockenden 1994; Leung and Forbes 1997c). These composite indices can either be composed of the mean relative asymmetry values from all traits, or the sum of all relative asymmetry measures. In cases where relative asymmetry measures are not used, composite indices of asymmetry have been composed of summed or averaged ranks of asymmetry (e.g. Graham and Felley 1985). As developmental stability is often not correlated between traits on the same individual, composite indices of asymmetry may reveal the overall developmental stability of the organism more accurately than a measure of asymmetry from a single trait (see also discussion in Palmer 1994; Leung and Forbes 1997c). However, the relative costs and benefits of asymmetry in different traits will vary, especially in terms of their signalling (Chapter 8) and biomechanical (Chapter 7) properties; therefore composite indices of asymmetry may mask specific information from individual traits. Hence, composite indices should be adopted only when it can be predicted, *a priori*, that the relative costs and benefits of asymmetry are fairly homogeneous among the traits being measured.

1.10.4 *Methods for analysing fluctuating asymmetry*

Asymmetry data display two important properties that influence the ways in which they can be analysed statistically. First, absolute (unsigned) asymmetry

($|R-L|$) has a characteristic 'half-normal' distribution (Van Valen 1962; Swaddle *et al.* 1994). Second, in a mathematical model of developmental stability, Whitlock (1996) has demonstrated that means and variances will often be closely associated in estimates of absolute asymmetry. Even though the assumptions of normality and homogeneity of variances for parametric statistics are likely to be violated, *t*-tests, ANOVA, and linear regression have frequently been used to analyse asymmetry data in the recent literature (e.g. Møller 1990a; Thornhill and Sauer 1992; Radesäter and Halldórsdóttir 1993; Wakefield *et al.* 1993; Wilber *et al.* 1993; Manning and Chamberlain 1994; Solberg and Sæther 1994). The residuals from regression analyses may be normally distributed (e.g. Møller 1990a), but more often they will not (e.g. Radesäter and Halldórsdóttir 1993; Thornhill 1992b; Solberg and Sæther 1994). Swaddle *et al.* (1994) suggest that two-parameter Box–Cox transformations of the form ($Y + \lambda_2)^{\lambda_1}$ are often suitable for normalising skewed positive data containing zeroes (Palmer and Strobeck 1986; Aitken *et al.* 1989). Swaddle *et al.* (1994) explored the utility of these transformations using GLIM (Numerical Algorithms Group 1985; Aitken *et al.* 1989) for a variety of feather, tarsus, and randomly generated data. They found that values of λ_1 around 0.3 and λ_2 set to be somewhat smaller than the smallest non-zero asymmetry, work well in transforming absolute asymmetry data to approximate a normal distribution. The limitation of this approach is that small differences in asymmetry near zero have a large influence and, if zero asymmetry is common, then so too does the choice of λ_2 (see Aitken *et al.* 1989). For highly stabilised traits with very small asymmetries this may be a significant problem; however, for sexually selected characters, where average asymmetry is large (Møller and Pomiankowski 1993a), this is less likely to be an issue. Additionally, it is important to point out that under such transformations the error structure of the data is also transformed, which will artificially inflate the relative size of errors at one end of the population distribution compared with the other. Hence, there may be problems with interpreting data from this type of transformation and non-parametric techniques (see Siegel and Castellan 1988) should therefore be routinely employed in analyses of absolute unsigned asymmetry values. Relative measures of asymmetry may also be prone to the same statistical problems as described above, and the normality of residuals from analyses should be inspected. Randomisation procedures will often provide reliable and powerful estimates of probabilities when comparing the asymmetry of different samples (Thomas and Poulin 1997). However, parametric inferential statistics can be highly robust with regard to departures from non-normal skewness (Gangestad and Thornhill 1997b). Use of transformations and non-parametric statistics can result in a lowered power of statistical analyses, and parametric statistics may therefore provide the more robust tools for analysis of absolute asymmetry in some situations (Gangestad and Thornhill 1997b).

Multivariate methods of analysing fluctuating asymmetry have also been developed (for example see Livshits and Smouse 1993a) where measurements are taken from left and right sides of many traits and then entered into a principal components analysis (PCA). The resulting axes of variation are, reputedly, independent and may help with explaining asymmetry levels across a large number of different traits. This method also has advantages in that it may indicate the relative importance of asymmetry in relation to size, although this is not always easy to interpret, especially as isometric size can be spread amongst the first couple of principal components and is not constrained to only the first component, as commonly believed (Burnaby 1966; Lockwood *et al.* 1997). However, this PCA approach is often complicated to perform and results may vary slightly between PCA programs. Additionally, a multivariate approach also falls prey to the measurement error problem and is difficult to interpret or apply at the individual level, although it may be a useful descriptive tool when studying population fluctuating asymmetry differences. If this approach is developed further it may become a useful analytic technique. However, it is difficult to appreciate how a multivariate technique will render asymmetry estimates dramatically different to those obtained from simple univariate methods, as the first principal component (PC1) will most likely represent an axis of trait size (i.e. left = right) and, hence, the second, orthogonal principal component (PC2) should represent the asymmetry (i.e. left–right). Unsigned PC2 scores may well violate the assumptions of normality, as they will display a half-normal frequency distribution across the population, and hence encounter similar problems in statistical analysis as those we have described for unsigned absolute asymmetry.

1.10.5 *Problems of pooled samples*

Biologists have often reported differences in the degree of fluctuating asymmetry among populations of the same species (e.g. Valentine and Soulé 1973; Picton *et al.* 1990; Markowski 1993; Møller 1993b), between years within populations (Zakharov 1981, Møller 1993b) and at the extremities of a population range (Downhower *et al.* 1990; Parsons 1991b). However, some evolutionary biologists are also interested in differences in asymmetry within populations, as fluctuating asymmetry may reveal aspects of individual condition or be a target of mate choice (Møller 1992b; Swaddle and Cuthill 1994a; see Chapters 8 and 9). In cases where samples from populations are pooled, the heterogeneity of the source may invalidate the conclusions drawn from these studies (see Swaddle *et al.* 1994, 1995; Simmons *et al.* 1995). Heterogeneity of source is not a problem that many large-scale studies of fluctuating asymmetry have addressed (e.g. Wayne *et al.* 1986; Møller and Höglund 1991; Møller 1992a, b; Balmford *et al.* 1993; Manning and Chamberlain 1993, 1994; Solberg and Sæther 1994) although this information can be noted and taken into account in the analysis (e.g. Møller 1990a; Tomkins and Simmons 1995).

There may be additional problems associated with large-scale pooled samples of fluctuating asymmetry that apply to both museum-based studies and uncontrolled field sampling (see Swaddle *et al.* 1994). These include the possibility that there may be differential mortality by level of fluctuating asymmetry, trait size, or an interaction of both. Therefore, the measured relationships between asymmetry and size may reflect the action of natural selection rather than developmental constraints or condition-dependent expression. Simmons *et al.* (1995) have suggested that this bias is not important in terms of the use of fluctuating asymmetry in studies of condition and mate choice, as whether a (negative) relationship between trait size and asymmetry exists due to condition-dependent expression or post-developmental viability selection, females will still have a reliable indicator of male quality. However, this would only be true if the population was carefully and comprehensively sampled at the appropriate time of year, i.e. during mate choice (Swaddle *et al.* 1995). Also, discrimination between the various reputed costs that maintain honest signalling is of prime interest for those studying sexual selection, and is currently an area of considerable controversy, particularly to those interested in fluctuating asymmetry. Therefore it is crucially important to be able to understand how the relationship between asymmetry and size arose, which may only be possible by controlled laboratory studies (e.g. Swaddle and Witter 1994) or by cohort analysis in the field (e.g. Zakharov 1981).

Large-scale studies may also incur problems with biased sampling as humans find symmetric objects aesthetically pleasing (Eisenman and Rappaport 1967; Szilagyi and Baird 1977), and so hunters and collectors may (subconsciously) collect unrepresentative samples from a population. A component of the Boone and Crockett score for trophies (e.g. antlers and horns) is degree of symmetry. This may be a particular problem with highly decorative ornaments, but far less of a problem with collections of invertebrates (Simmons *et al.* 1995). It may also be difficult to distinguish wear and damage asymmetry from developmental asymmetry, which may greatly alter conclusions made concerning differences in developmental stability between and within populations (Cuthill *et al.* 1993; Møller 1993a). However, a recent comparative study of fluctuating asymmetry in feather characters based on museum specimens demonstrated a strongly positive correlation between asymmetry estimates from the field and from museum specimens (Cuervo and Møller 1997a). This result suggests that museum collections may hold important asymmetry information suitable for scientific investigations.

Another source of sampling error may occur through over-zealous collection of asymmetry data from several traits on the same individual. If asymmetry is collected from many traits, the probability that asymmetry in one of these traits will be significantly associated with the independent variable (e.g. temperature, inbreeding) is increased. Therefore, we recommend that *a priori* predictions concerning the relation between asymmetry in a particular trait and the independent variable are made. Additionally, where multiple comparisons are made, significance levels should be adjusted accordingly, for example by the Bonferroni adjustment procedure (Rice 1989).

1.10.6 *Suggestions for fluctuating asymmetry analysis*

Palmer and Strobeck (1986) have highlighted the increased power and effectiveness of population fluctuating asymmetry indices that are based on variance of left and right differences, and we recommend that these indices should be employed where possible. However, their ANOVA analysis technique may have interpretational problems if there is significant directional asymmetry within the population, so we recommend that the population asymmetry scores are tested rigorously for normality around a mean of zero. Palmer and Strobeck (1986) detail a number of methods with which among-population differences in fluctuating asymmetry can be tested; but the assumptions of these analyses must be met beforehand (see Swaddle *et al.* 1994). Palmer's (1994) primer has many useful suggestions for analysing fluctuating asymmetry data at the population level. A.R. Palmer has also developed a Web site which contains a spreadsheet that computes a range of fluctuating asymmetry statistics (http://gause.biology.ualberta.ca/palmer.hp/asymmetry.htm).

When investigating individual levels of asymmetry in a trait that displays fluctuating asymmetry, we recommend the use of unsigned absolute asymmetry values ($|R-L|$), as individual variance can often not be calculated. Relative measures of asymmetry (($|R-L|$)/0.5(R + L)) can effectively control for trait size if the relationship between size and asymmetry is isometric and intercepts the origin, and may be particularly useful when constructing composite measures of asymmetry across traits. However, there are a number of criteria to bear in mind when analysing absolute and relative asymmetry data. First, the data should conform to the statistical properties of fluctuating asymmetry. Second, measurement error must be taken into account, for example by testing the repeatability of signed asymmetry scores. Third, these data often have to be analysed by non-parametric techniques, as absolute and relative asymmetry values are likely to violate the assumptions of parametric statistical tests. Once these considerations have been made, meaningful analyses can be performed and robust conclusions can be drawn.

Methodological stringency has implications for the rest of the chapters in this book. The old literature, which serves as a foundation for the work of current students of developmental stability, cannot be blamed for not having adhered to methodological criteria that were previously unknown. However, a fair number of older studies have tested for directional asymmetry or antisymmetry and have estimated measurement errors of asymmetry. We would like to stress that readers should be aware of the fact that studies do differ in methodological quality, and that this should be taken into account when reading the book. Throughout we have tried to emphasise studies of the highest quality.

1.11 Summary

- In this chapter we define developmental stability and review the ways in which developmental instability can be estimated.
- Both the frequency of phenodeviants and the coefficient of variation are limited in their applications as it is not possible to make *a priori* predictions concerning their prevalence and magnitude in populations. Also, these

indices do not control for differences in genotype between samples or environmental conditions during development.

- Indices of developmental instability that are based on the repeated formation of structures on the same individual tend to control for genotypic and environmental differences, and allow *a priori* predictions concerning the form of the most stable phenotypes.
- The differences between the two sides of a bilateral trait (fluctuating asymmetry) render very sensitive measures of developmental stability.
- The utility of fluctuating asymmetry as a population statistic and asymmetry as a property of an individual is discussed.
- We describe the two forms of adaptive asymmetry: directional asymmetry and antisymmetry.
- Adaptive external asymmetry may be relatively uncommon due to the increased costs of this form of morphological and functional specialisation, or because adaptive asymmetry represents an evolutionary dead-end.
- We review the recommendations for the statistical analysis of fluctuating asymmetry data.

2

Ontogeny of asymmetry and phenodeviants

2.1 Introduction

The term developmental stability denotes the reproducible (constant) development of a genotype under given environmental conditions. The nature of ontogenetic processes, and their aberrations, is central to the study of developmental stability and, hence, there is a long history of interest in this subject. The developmental origins of phenodeviants and fluctuating asymmetries were originally considered a 'black box'. For example, Mather (1953) suggested that asymmetries arose as a consequence of differences in the environment or disruption of the development of cells. Later work suggested that random deviations in the activity of molecules or smaller particles were responsible for the formation of developmental instabilities (Reeve 1960; Soulé and Cuzin-Roudy 1982; Lewontin 1983). This explanation may seem akin to the most recent developments suggesting that the accumulation of initially small deviations from perfect symmetry leads to major deviations from perfect symmetry (Emlen *et al.* 1993; Graham *et al.* 1993c). However, all developmental systems may not necessarily be organised in the same way. Different species may have different ways of ensuring developmental stability and so exhibit different vulnerabilities to the loss of bilateral symmetry. For example, growing imaginal discs in holometabolous insects, such as *Drosophila* fruitflies, are probably relatively independent of each other. Genetically based asymmetries in fruitflies, often due to mutations, were discussed extensively by Goldschmidt (1940). In contrast, developing vertebrate embryos show a smaller range of fluctuating asymmetry than developing beetles and flies. Co-development of the two sides in vertebrates can be better ensured through diffusible substances and, quite possibly, neural stimuli (e.g. Gilbert 1991).

There are several ways to analyse the causes of developmental instability. These include the use of genetic approaches, modelling (employing techniques from chaos theory), and experimental developmental studies. The challenge for the future is obviously to integrate these methods. These three approaches are all discussed in detail in this chapter.

The ultimate mechanism leading to the production of unstable phenotypes can be studied at the genetic level of ontogenetic processes. Recent studies in developmental genetics have emphasised the mechanisms giving rise to the stable ontogeny of phenotypes, the ontogeny of asymmetries, and the role of genetic modifiers in restoration of the disruptive effects of novel mutations on genomic co-adaptation (section 2.2).

The second approach to the study of developmental instability is modelling. One of the major developments in biology in recent decades is the application of chaos theory to the analysis of the often complex dynamics of natural systems (e.g. Mandelbrot 1982; Peitgen *et al.* 1992; Hastings and Sugihara 1993). The morphogenetic processes that lead to the formation of organs and whole organisms are incredibly complex, but still remarkably precise. Many complex biological structures demonstrate self-similarity in the sense that a structure can be decomposed into smaller copies of itself, which again can be decomposed into smaller copies of itself, and so on. Complex structures with self-similarity are said to display fractal geometry. Examples include trees that are composed of branches that resemble trees, mammalian lungs that branch into trachea, secondary bronchi, bronchioles, and alveoli, and nervous systems that at a number of levels are composed of branching nerves. Fractals are useful in studies of development because the apparently complex patterns can be described by a single fractal dimension. In other words, the self-similarity of the fractals themselves 'forces the complexity of the objects into the building blocks and describes the inherent regularities through power laws' (Hastings and Sugihara 1993). Even minor deviations from perfect symmetry at an early stage in natural developmental systems can result in the formation of major abnormalities without negative feedback or similar dampening mechanisms. Extensive control mechanisms that are able to restore deviations from perfect symmetry must continuously be at work. Only recently has it been suggested that developmental stability arises from nonlinear dynamics of biosynthetic processes in cells and nonlinear feedback occurring among developing body parts (Emlen *et al.* 1993; Graham *et al.* 1993c) (section 2.3).

The third approach to the study of developmental stability is experimental developmental biology. The prevalence in nature of bilateral symmetry makes the study of developmental processes central to our understanding of the mechanisms that give rise to the stable development of phenotypes. Recent experiments testing the effects of major genes on development and novel experimental approaches have provided a number of interesting results. In section 2.4 we discuss the formation of symmetry and the feedback mechanisms between sides of a plane of bilateral symmetry that result in the stable development of phenotypes.

An inherent assumption of the use of individual asymmetry as a measure of phenotypic quality is that certain genotypes are developmentally more stable than others under given environmental conditions. The phenotype developed by a genotype under different environmental conditions is termed the reaction norm of the genotype. If individual asymmetry reliably reflects the quality of an individual in terms of the ability to control the stable development of a phenotype, then the slope of the reaction norms for individuals with the highest fitness should be less steep than for those with low fitness. The theoretical basis for this argument and some empirical tests are presented (section 2.5).

The final part of this chapter deals with studies of ontogeny of developmental instability in free-living organisms. Such organisms are exposed to a range of different environmental conditions, and the study of developmental processes that lead to stable phenotypes in these organisms has recently become a scientific discipline in its own right. Two types of organisms render themselves particularly suitable for studies of reactions norms of developmental stability because the development of the same structure of an individual can be studied under different environmental conditions. First there are a number of organisms which undergo repeated development of the same structure at different times following subsequent growth episodes. Examples of this category include exoskeletons in crustaceans, feathers in birds, and hair in mammals. A second group of organisms has a number of almost identical copies of a morphological structure, each of which can be considered a replicate that can be used for studying the effects of experimentally variable environmental conditions on morphogenesis. Examples include somites in vertebrates and leaves in plants (section 2.6).

2.2 The genetics of the formation of developmental stability

The genetics of many developmental systems are relatively well understood. The genes involved in formation of patterns are also relatively well known. However, the mutant genes that can disrupt the stable development of phenotypes are not well known, and neither are the kinds of genes that can re-establish a stable phenotype in the presence of mutant genes by means of increased canalisation. Some early studies of the effects of mutations on the expression of the phenotype provided clear support for mutant-induced asymmetry or phenodeviants (review in Goldschmidt 1940). These classical developmental genetic studies provided much knowledge of the generally disruptive effect of mutations on the expression of a developmentally stable phenotype. More recent studies of developmental genetics have provided detailed information on the genes involved in the formation of morphological asymmetry and the genes involved in the re-establishment of symmetry. The presence of mutations may give rise to true fluctuating asymmetry as demonstrated, for example, by a mutant affecting the appearance of eyespots in the butterfly *Bicyclus anynana* (Brakefield and Breuker 1996). A particularly revealing example relating directly to the effect of mutations on developmental instability concerns the Australian sheep blowfly *Lucilia cuprina*.

A study of the disruptive effects of an insecticide resistance allele on the phenotype of the sheep blowfly, and the restoration of a symmetric phenotype by a modifier allele provide a remarkably detailed study of the genetics of developmental instability (Clarke and McKenzie 1987). Two genes (*Rop-1* and *Rmal*) provide resistance to two different pesticides, but the pesticide-resistance alleles originally also gave rise to an elevated level of fluctuating asymmetry in the number of sternopleural bristles. Enzymatic mechanisms of insecticide resistance involve carboxylesterases (McKenzie and Batterham 1994). Interestingly, the two resistance genes (*Rop-1* and *Rmal*) in sheep blowflies encode carboxylesterase enzymes that participate in mediating cell–cell interactions in the differentiation of bristles, which are terminal structures of the peripheral nervous system. The direct and indirect fitness costs of fluctuating asymmetry in sternopleural bristles remain unknown, although asymmetry in bristles may have severe consequences as bristles are part of the sensory system of insects. It is noteworthy that bristle asymmetry is positively correlated with developmental time and negatively related to egg hatchability, both of which indicate a fitness cost associated with asymmetry (McKenzie and O'Farrell 1993). A dominant modifier allele has been selected to control these negative side-effects of the resistance allele (McKenzie *et al.* 1990). The modifier gene, which is an allele of the *Scl* gene, is an over-producer and has been shown to be homologous to the *Notch* gene in *Drosophila* (Davies *et al.* 1996). *Notch* in fruitflies is a major control gene specifically involved in controlling the development of the nervous systems. *Notch* mutants in *Drosophila* generally have elevated levels of asymmetry compared with wild-type individuals. Resistance loci may interact with *Scl* influencing the adhesion properties of cells to modify the phenotypes of bristles (Lauder 1993). Two different genes that cause asymmetry in the sheep blowfly appear to be involved. Wild-type asymmetric phenotypes were exhibited by *faswbBG*, which results from the insertion of different retrotransposons into the same position as the second intron of the *Notch* gene. Flies with the allele *fa* had elevated levels of fluctuating asymmetry, and this effect was also due to a similar insertion. Both these mutations are partial loss-of-function mutations in *Notch* that independently give rise to asymmetry. A difference in the level of the *Notch* product at a specific stage during development may be responsible for differing asymmetry phenotypes of the two alleles. The modifier gene was identified for the sheep blowfly, and modifier and non-modifier strains of flies were crossed to *Scl* + flies. The modifier gene reduced the level of fluctuating asymmetry of *Scl* wing phenotypes in a partially dominant manner, providing an explanation for why the bristle asymmetry, caused by the resistance allele, took a relatively long time to become reduced at the population level.

A large number of papers using modern molecular techniques have recently reported effects of single alleles on the development of an abnormal, asymmetric phenotype in various organisms. Although these effects of single alleles on the expression of the phenotype are mutational, the studies do not allow any firm conclusions about the involvement of single alleles in the production of a developmentally stable phenotype. Here we provide just a couple of examples from the recent literature of such studies of mutational effects on the development of disrupted phenotypes. First, over-expression of the gene *Ets2* in mice gives rise to skeletal asymmetry. Humans suffering from trisomy-21 (Down's syndrome) have severe skeletal abnormalities with increased levels of fluctuating asymmetry (reviewed in Thornhill and Møller 1997). The gene *Ets2* is a proto-oncogene and transcription factor that occurs in a variety of cell types in mammals. It is highly expressed in newly forming cartilage in mouse. Furthermore, it is located on chromosome 21 in humans and over-expressed in Down's syndrome. Over-expression of *Ets2* in transgenic mice resulted in development of neurocranial, viscerocranial, and cervical skeletal abnormalities similar to those of trisomy-16 mice and humans with Down's syndrome (Sumarsono *et al.* 1996). These abnormalities occurred as a result of the peculiar histology of calvaria in the bones. These were disorganised and irregular in thickness with thinner bone surrounded by increased amounts of connective tissue. Similarly, sutures became asymmetrical and disorganised, and bony trabeculae were irregular in shape, size, and mineralisation. The resultant phenotype was highly asymmetric with distorted body proportions. Although a clearly increased dosage of *Ets2* occurs in humans with Down's syndrome and trisomy-16 mice, it is likely that a range of other genes are involved in regulating the expression of the overall symmetric phenotype. A second example concerns the effect of the vascular endothelial growth factor gene (VEGF) on abnormal blood vessel formation and the resultant developmental abnormalities at the tissue level (Carmeliet *et al.* 1996; Ferrara *et al.* 1996).

The genetics of the stable development of patterns have also received attention recently. *Hox* genes play an important role in pattern formation, and they are involved in determining patterns in all animals from sponges to chordates (Akam 1989). A recent experiment based on targeted disruption of *Hox* genes in mice provided clear information on the role of these gene complexes in the stable formation of the phenotype (Davis *et al.* 1995). Mice with targeted disruptions were used to identify the effects of genes in various groups of *Hox* genes. Mice with individual mutations in the genes *hoxa-11* and *hoxd-11* had in the homozygous, double mutant state dramatic phenotypic aberrations with radius and ulna of the forelimb almost entirely eliminated, and the axial skeleton showing homeotic transformations. These effects of *Hox* genes were quantitative because more mutant alleles in the genome resulted in progressively more severe phenotypic aberrations.

A large number of genes could potentially influence the development of a symmetric phenotype either directly or indirectly through pleiotropic effects. Genetic variation at loci affecting normal development of the phenotype is obviously common, as evidenced from the few examples reviewed above. Relatives will therefore resemble each other with respect to development of a symmetric phenotype (see also section 5.2). Future studies of developmental processes will provide information on whether major control genes involved in developmental processes are the main genes responsible for the formation of symmetric phenotypes.

2.3 Developmental stability and chaos theory

Bilateral, radial, and other kinds of symmetry are ubiquitous in nature, despite most Metazoa consisting of enormous numbers of cells that are the product of numerous cell divisions. Before going into detail about developmental processes it may be appropriate to consider the fundamental differences in development among two major kingdoms, animals and plants. Animal development involves growth, differentiation, and morphogenesis. In animals growth involves cell multiplication and expansion. Differentiation occurs through altered physiology of some cells resulting from changes in gene expression, and further cleavage of differentiated cells gives rise to different tissues. Morphogenesis is an early step in development that results in the formation of a particular arrangement of tissues, often caused by direct movement of cells within a developing embryo. Animal development thus is caused by movement and behaviour of cells (Gilbert 1991).

Plants (and many other organisms) are modular in the sense that they consist of repeated structures that are added by indeterminate production of modular units (Watkinson and White 1986). The self-similarity of many plants is obvious from minor parts such as twigs or branches resembling a small tree, as any bonsai enthusiast would know. First, plant cells increase in volume, they then initiate internal cell wall formation, repartitioning the existing volume of the organism. Cell division in plants occurs through the formation of a phragmo-plast rather than by cleavage (Fahn 1974). Cell walls are formed by the phragmoplast while preserving parts of the endoplasmic reticulum and Golgi apparatus between cells. This has important implications for the potential of 'selfish' behaviour by individual cell lineages and the ways in which such behaviour is controlled (Buss 1987). Plants are modular organisms with different parts of an individual potentially differing in genotype, and this could lead to selfish behaviour of cell lineages. However, early segregation of the germ line is precluded in plants due to the absence of cell movement, and the resultant control of differential proliferation of cell lineages (Buss 1987). Furthermore, competition among branches or similar units for the inductive plant hormone signals that limit their relative development may result in true competition among cell lineages without giving rise to problems of control of differential proliferation of cell lineages (Sachs *et al.* 1993; Sachs 1994). In other words, plants may represent an interesting intermediate between true clonal populations and unitary organisms such as animals. Morphogenesis is directed by formation of growth hormone gradients within the developing embryo, or physiological modifications of transport through pre-formed vascular channels. The strength of such gradients is subject to nonlinear dynamics, and nonlinear feedback systems may control developmental processes (Turing 1952; Meinhardt 1982; Held 1992). Self-organisation may generate complex spatial patterns due to a reaction–diffusion process in which one chemical substance enhances its own formation by autocatalysis as well as that of a second chemical, which inhibits the first substance. If the second chemical substance diffuses much faster than the first, this system can act as a creative force in pattern formation. An example of Turing waves giving rise to complicated coat patterns in mammals was put forward by Murray (1990) while Kondo and Asai (1995) reported a potential example from colour patterns of fish.

Some animal structures like lungs, kidneys, nerves, and the vascular system are similar to some structures in plants in terms of certain developmental properties. In particular, they are fractal by showing self-similarity which allows the object to be decomposed into smaller copies of itself. This feature of fractal geometry allows complex structures to display large surface areas in small volumes. The branching pattern of these structures results in a maximum amount of the three-dimensional space being occupied by the structure. A large dimension of the fractal will increase the efficiency of the structure for its physiological function up to a certain limit. Plants also display fractal features (Schroeder 1991). Typical examples are leaf venation, branches, and roots that all tend to maximise physiological efficiency.

The major differences between animal and plant development in terms of modularity, movement, and behaviour of cells have important implications for developmental instability. Generally, animal development is much more sensitive to stresses that affect morphogenesis because disruption of cytokinesis may lead to major disruptions of the phenotype. Particularly, disruptions during early embryogenesis have a major impact on measures of developmental instability, as shown by severe disruption of the developmental programme resulting from stresses early during pregnancy in humans (Gilbert 1991).

The basic thesis raised in this section is that the ontogeny of measures of developmental instability depends on two closely related phenomena. First, the nonlinear dynamics of biosynthetic processes within cells that lead to non-random growth of cells and ultimately, if not checked by feedback control mechanisms, to non-random cell division (Emlen *et al.* 1993). Second, the nonlinear feedback mechanisms occurring among developing body parts (Graham *et al.* 1993c). These processes are the subjects of the following two sub-sections.

2.3.1 *Nonlinear growth dynamics of cells and cell division*

Construction of biological entities will always result in trade-offs between assuring fidelity and optimising speed or efficiency of construction independent of whether the trade-offs are measured in terms of energy, time or materials used. This should be true at the molecular level (e.g. in protein synthesis), but also at higher levels such as development. Growth of morphological structures depends on the growth and proliferation of cells, which again is based on the synthesis of proteins, lipids, and other constituents. Biosynthesis of these constituents relies on smaller precursor building blocks, energy, and enzymes. In particular, the role of enzymes in biosynthesis results in complex, nonlinear dynamics, since a single enzyme may function in many interconnected enzymatic pathways. Chemical processes such as those resulting in biosynthesis are characterised by controlled chaotic processes with a particular fractal dimension. This is for example the case for enzymatic reactions (Olsen and Degn 1977). The preponderance of controlled chaotic processes at the biochemical level of morphogenesis ensures that growth processes do not go awry. Synthesis at a subcellular or cellular level might get out of control and result in deviant morphogenesis. If a small perturbation resulted in a non-random pattern of biosynthesis within a cell, this would result in a non-random distribution of the constituent parts within cells. Magnification of such heterogeneity of biosynthesis over time may lead to deviant patterns of cell proliferation unless checked by intercellular feedback mechanisms, that is if different growth stages are sensitive to deviant environmental conditions that prevail for a long time. It is difficult to imagine what the product of such uncontrolled growth would look like, perhaps it would appear similar to the uncontrolled growth of cancer cells. Random growth of different cell lineages is certainly not the normal pattern of development, and control mechanisms must obviously have evolved to maintain morphology and size within certain limits. If we consider two homologous parts developing on the right and left sides of the body under identical genetic control, fluctuating asymmetry in this paired trait would be controlled by the developmental check mechanisms.

Developmental control can be achieved by at least three different hypothetical mechanisms (Emlen *et al.* 1993): (i) identical genetic and constraint control mechanisms; (ii) negative feedback among growing cells within structures; and (iii) feedback between left and right structures. We do not consider these three explanations to be mutually exclusive because (ii) and (iii) may obviously take place under genetic and constraint control mechanisms. We will briefly consider each of these possibilities, starting with identical genetic and constraint control mechanisms (Emlen *et al.* 1993). Given that there is encoded genetic information in organisms for extremely large numbers of characters, could it be possible that the developmental pathways and the size of the final product would be encoded genetically? This would require an enormous, unrealistic number of pieces of information, much of which would be redundant. Developmental constraints may have evolved to limit deviations in growth trajectories or end products. This may seem unlikely given the wide range of environmental conditions under which most ordinary organisms may happen to develop. Check and feedback mechanisms would certainly be a much more sensitive way of controlling such a growth process, since deviations in size or growth rate from the two parts of a paired structure would be controlled continuously. Therefore, we consider this to be an insufficient explanation to account for controlled morphogenesis when extremely large numbers of cells are involved.

Second, negative feedback among cells within a growing structure depends on the covariation in growth among cells being negative. This would result in a dampening of the cumulative growth pattern over all cells (Emlen *et al.* 1993). Cell–cell feedback is usually assumed to act only at small distances of a few cell layers, and this should be sufficient to suppress biochemical synthesis and cell proliferation within certain bounds. Many examples of such regulatory feedback among growing cells have been reported (e.g. Turkington 1971; Koji *et al.* 1988; Cross *et al.* 1993). This process probably cannot produce the similar development of paired traits, unless the threshold bounds for feedback are extremely narrow. This condition is unlikely to be fulfilled in a relatively large organism containing millions of cells. However, a number of overlapping and interacting feedback mechanisms within a structure could generate stable development. Imagine that the range of feedback is small (e.g. over four cells). The overlapping nature of feedback could lead to regulated development throughout the structure as a whole. This mechanism would create a number of 'feedback units' that regulate growth of cells within their own range, and also provide information to nearby units, as they overlap with each other. This view of development would imply that asymmetries get magnified with increasing stages of morphogenesis.

Third, feedback between right and left structures requires communication between right and left sides via the neuronal or circulatory systems or via hormonal regulation (Emlen *et al.* 1993). Positive feedback, resulting in catch-up growth on the side lagging behind, should promote equal growth rates. The efficiency of such a system would depend on the time lag in communication between sides, the strength of the feedback signal, and the existence of other, systemic feedback systems. Controlled growth that does not result in oscillations could be achieved by maintaining the growth waves between the two sides in phase if feedback time lags between sides were not too large. Suppression of divergence between sides of a particular organ can be achieved by phase-locking the growth process in each of the two sides. However, organs have to act in unison within the body of an individual, and control mechanisms of the growth of different body parts must regulate these within certain limits. Hence, it is likely that the developmental processes of different organs have been entrained within specific ranges determined by their relative roles in the functioning of an individual.

Cascading effects of growth due to the effects of initially small perturbations may become magnified during suboptimal environmental situations and lead to the breakdown of the developmental check mechanisms (Emlen *et al.* 1993). Initially small deviations in growth patterns of identical structures on two sides of the body almost invariably become independent of each other, unless specific control mechanisms maintain the similarity in development. The variance in the growth of the two parts of a paired structure will, due to the contributions of variances in each of a number of growth increments, become greater over time unless controlled by overlapping feedback mechanisms. Random perturbations thus tend to become magnified.

An ordinary growth process relies on the almost continuous synthesis of proteins and other growth precursors. This in itself results in a reduced variation in growth among cells if all cells have access to similar nutrient levels. Adding suboptimal environmental conditions to the system should immediately result in a loss of energy from growth and maintenance to control and repair processes

(Winberg 1936; Mitton and Koehn 1985; Ozernyuk 1989; Wedemeyer *et al.* 1990; Alekseeva *et al.* 1992; Ozernyuk *et al.* 1992). Organisms have evolved a range of antistress responses, such as heat-shock proteins, that stop growth and/ or development until recovery is complete. However, conditions for growth may continuously remain adverse for some individuals. Since the total amount of energy is limiting, and there is a trade-off between different uses, suboptimal environmental conditions should result in a lowered growth rate and raise the variance in growth rate among cells due to nonlinear growth dynamics of the developmental process. The negative feedback control mechanism between cells is likely to be limited only by the efficiency of the cell communication system. Disturbance caused by suboptimal environmental conditions is predicted to lead to decreased resistance to variation in growth rates among cells and therefore increased asymmetry. The phase-lock mechanisms generated by feedback between organs is also likely to break down under suboptimal environmental conditions. Different structures thus may start to develop out of phase with each other, resulting in disruption of allometric relationships. If such deviations in morphology have developed once, they are likely to result in magnified effects on other structures. For example, disruption of the growth of nerves may affect the development of other tissues such as growth of the vascular system, which may further affect the growth of the organs in question.

Patterns of colours may develop in response to standing waves or gradients of morphogen concentrations (molecules whose local concentration determines the local pattern of differentiation) (Meinhardt 1982; Eilbeck 1989; Nagorcka 1989; Castets *et al.* 1990; Agladze *et al.* 1992). The pattern formation may depend on rates of production of chemical activators, sinks, and inhibitors. Minute changes in these parameters, as caused by deviant environmental conditions, may have dramatic effects on the patterns formed. A particularly popular model system for the study of development of patterns is the colour wing patterns of butterflies (reviewed in Nijhout 1991; Brakefield and French 1993). Surgical manipulations can reveal cell interactions that specify the patterns of cell fate on the wing epidermis. Colours on butterfly wings are formed as structural colours or from pigments deposited in the scale cuticle, and particular patterns develop depending upon the position of the cell on the wing surface. Microcautery of the presumptive centre of an eyespot in the early pupal wing can prevent its development, while grafting the focus to a different position results in development of a novel eyespot (Nijhout 1991). The pattern appears to be specified by epidermal cell interactions in the larval imaginal disc which eventually forms the adult butterfly wing. Experiments using grafting experiments have demonstrated phenotypic and genetic differences for developmental mechanisms that give rise to different wing patterns. Morphogen gradients are possibly involved in pattern formation in butterflies, because experiments involving cauterising the presumptive eyespot centre on the pupal wing of the nymphalid butterflies *Bicyclus safitza* and *B. anynana* at different stages of pupation resulted in a major change in eyespot formation (French and Brakefield 1992). Cautery during the early part of pupation resulted in reduction or elimination of the eyespot; while later cautery gave rise to enlarged posterior eyespots and non-focal cautery induced a new ectopic eyespot. A source-diffusion mechanism of a morphogen with time and developmental-stage-dependent components can account for this effect. The focus of the presumptive eyespot centres may be a morphogen source, and cautery may remove the focus and lower the response in surrounding cells. Alternatively, the eyespot centre may generate the gradient by removing morphogen, and cautery can eliminate the focus, but simultaneously cause a transient destruction of the morphogen.

2.3.2 *Nonlinear feedback mechanisms among developing body parts*

While the main aim of the preceding section was to emphasise the consequences of the nonlinear growth dynamics of cell growth and cell division, this section will put this knowledge into a larger framework by emphasising the role of nonlinear feedback mechanisms between developing body parts on the onto-geny of asymmetric phenotypes. This section deals particularly with models and theories to explain bilateral symmetry. A linear growth process with variation in cell division rates would also result in asymmetrical structures or non-constant phenotypes. However, a nonlinear growth process would on average generate larger levels of asymmetry than a process characterised by linear growth dynamics.

Regulatory control systems of development may act by means of morpho-gens, hormones, or neural impulses which will determine the size and shape of morphology during the developmental trajectory. Homologous cells on the two sides of an organism generally have similar cytoplasmic determinants and share a parallel history of cell–cell interactions, and they are therefore similar with respect to their pattern of development (Gilbert 1991). The stable development of a phenotype depends on the control of environmental or genetic perturba-tions from the developmental trajectory, and this is achieved by attractors. Palmer and Strobeck (1992) considered asymmetry to arise as a consequence of a point attractor that automatically gives rise to perfect symmetry in the absence of any kind of perturbation. A less restrictive point of view is that the attractor is a distribution (Zeeman 1989). The developmental trajectory never arrives at the point attractor because of diffusion of morphogens, hormones, and other biochemical molecules. Global stability of a developmental system is therefore unlikely to be achieved, and the absence of asymmetry is thus an unlikely state for any morphological character.

Morphogen oscillations may be involved in determining the number and location of body segments (Turing 1952; Maynard Smith 1960; Goodwin 1971), and Graham *et al.* (1993c) have proposed a similar mechanism for the stable development of bilaterally symmetric phenotypes. Regulatory feedback be-tween sides can result in the synchronisation of morphogen oscillations on right and left sides of a character, and the outcome will be the characteristic development of a phenotype with fluctuating asymmetry. Asynchronous or chaotic oscillations may give rise to phenotypes with non-normal distributions of asymmetry values, which are characteristics of antisymmetry and directional asymmetry. There is no empirical information on the kinds of developmental oscillations involving bilateral parts of an organism. However, oscillations of other biochemicals involved in development have been reported for cyclic AMP concentration in *Dictyostelium* cells and in auxin plant hormones (Devroetes 1989; Wodzicki and Zajaczkowski 1989). Responses to the oscillations of morphogens rather than their absolute concentrations may be common determinants of developmental processes.

The empirical evidence for morphogenesis with nonlinear growth dynamics has been reviewed by Graham *et al.* (1993c). Chaos has been suggested to be important in biochemical and physiological processes (e.g. Olsen and Degn 1977; West 1990a; Glass 1991), and nonlinear dynamics models of development have also been published (Bailly *et al.* 1991; Hademenos *et al.* 1994). Indirect evidence for the importance of nonlinear processes in morphogenesis comes from different sources. The importance of initial conditions for the outcome of morphogenesis (e.g. Govind and Pearce 1986; see also section 2.5) is characteristic of some nonlinear systems. The preponderance of developmental thresholds in morphogenesis (e.g. Gilbert 1991) is characteristic of the chaotic dynamics of nonlinear systems.

Graham *et al.* (1993c) developed a modified Rashevsky–Turing reaction–diffusion model of morphogenesis (Parisi *et al.* 1987) based on the interaction between neighbouring groups of cells on the two sides of the body. Information between the groups can be transferred by means of the circulatory or nervous systems. The model is based on the concentrations of morphogen (or a similar biochemical) on the right and left sides. The rate of increase of an activator depends on its own concentration and that of an inhibitor on that side. However, the rate of increase of the inhibitor depends on its own concentration and that of the activator on that side, plus the concentration of the inhibitor on the other side. Deviations of environmental conditions from the optimum generally results in the dissipation of energy (Alekseeva *et al.* 1992; Ozernyuk *et al.* 1992), and development will generally follow the trajectory with the smallest energy expenditure. Increases in energy dissipation under suboptimal conditions either results from loss of gene regulation or from changes in the kinetics of enzymatic reactions. Under stress, physiological fine-tuning of any system may be disturbed because of changes in membrane potentials. This will have important consequences for the dynamics of the Rashevsky–Turing model of development, as discussed later.

The behaviour of the Rashevsky–Turing developmental model was investigated in a number of simulations by Parisi *et al.* (1987) and Graham *et al.* (1993c). Under optimal environmental conditions, and when starting conditions on the two sides were identical, the concentration of the growth activator oscillated over time, but remained symmetric on the two sides. If starting conditions on the two sides were dissimilar, asymmetry developed with complex dynamics ranging from phase-locked periodicity to chaos (Parisi *et al.* 1987; Graham *et al.* 1993c).

More realistic conditions were achieved when noise was added to the model system. In the absence of any regulatory feedback of the growth activator and inhibitor, the concentration of the activator on the two sides demonstrated a random walk pattern that gave rise to a normal distribution of signed differences in concentration. A random walk process obviously resulted in ever increasing variance in the asymmetry with time and with the extent of the perturbation. Symmetric regulation of growth activator and inhibitor with independence between sides resulted in concentrations of the activator on the two sides staying in phase if starting conditions were similar (Fig. 2.1). An increase in noise resulted in increases in the normally distributed asymmetry of activator concentration, but within specified boundaries (Fig. 2.2a). Moderate amounts of feedback between sides stabilised the oscillations of the growth activator, even in the presence of considerable environmental noise (Fig. 2.3). The average level of asymmetry decreased as a direct result of increased amounts of feedback.

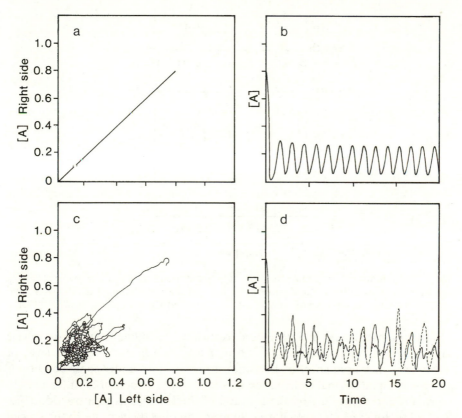

Fig. 2.1 Concentrations of a growth activator on the two sides of a system and temporal oscillations in concentration (a and b) in the absence and (c and d) presence of environmental noise. Part (d) shows time series of concentration on the right side (solid curve) and the left side (dashed curve). Adapted from Graham *et al.* (1993c).

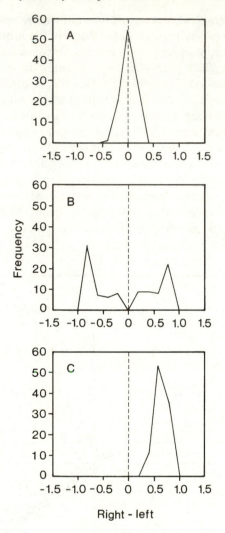

Fig. 2.2 Asymmetries in concentrations of a growth activator on the two sides of a system with (a) no diffusion and with regulation, (b) chaos with moderate feedback, and (c) chaos with biased control between sides. Adapted from Graham *et al.* (1993c).

Increased feedback and increased diffusion constants resulted in phase lagging or chaotic behaviour of the activator concentrations. The level of asymmetry increased with the transition to chaos, and under some conditions switched from a normal distribution to a bimodal distribution of differences in concentration depending on the location of the two basins of attraction (Fig. 2.2b; Graham *et al.* 1993c). This mechanism can directly give rise to the special property of antisymmetry. Directional asymmetry in morphology may arise as a result of slight differences between sides in the control parameters of the Rashevsky–Turing equations. Even a slight bias in parameters between sides may generate a considerable directional bias in asymmetry (Fig. 2.2c; Graham *et al.* 1993c). These results suggest that not only fluctuating asymmetry, but also other kinds of asymmetry may represent measures of developmental instability.

The control of stable development, as generally found in all different kinds of organisms, may seem at odds with the point of view that morphogenesis is chaotic in nature. Graham *et al.* (1993c) have suggested that development represents constrained chaos. In other words, there is order in this disorder. Phenotypes develop as a result of the action of the nonlinear developmental system and the linear genetic system (with each gene always coding for a specific biochemical). The interaction between the two systems will therefore also demonstrate nonlinear dynamics. While the developmental system continuously increases in information content due to the effects of gene regulation, cell–cell interaction, and cell migration, the information of the genetic system during development remains constant.

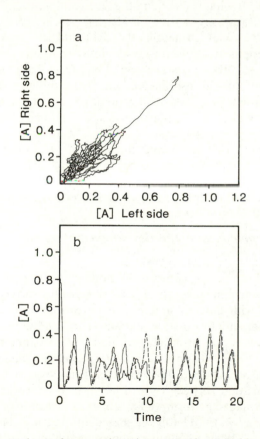

Fig. 2.3 Concentrations of a growth activator on the two sides of a system (a) and temporal oscillations in concentration (b) in the presence of environmental noise and feedback between sides. Part (b) shows time series of concentration on the right side (solid curve) and the left side (dashed curve). Adapted from Graham *et al.* (1993c).

There is an apparent paradox in symmetrical development because good communication between left and right sides can readily explain the high degree of symmetry. Any perturbation affecting levels of morphogen on one side of the body might be readily communicated to the other side with some averaging out of differences, which would act to reduce the potential asymmetry. However, there is often a narrow interval of time in which tissues are competent to experience different signals (whether for growth or differentiation). If the competence interval is shorter than the time required for equilibration between the two sides, one can have a major asymmetry developing, even given the normal communication-adjustment mechanisms. At a functional level, costs of communication and mechanical costs of asymmetry could determine levels of communication and the expression of asymmetry.

In conclusion, development can be induced to undergo transitions from phase-locked periodicity of the two sides of a developing organ, to phase-lagged periodicity, to chaos, by simply changing the levels and nature of feedback and inhibition in the Rashevsky–Turing model of development. The chaotic attractor has two basins of attraction, the right and left sides. With minor disturbance, a developmental trajectory settles into one basin or the other. With increasing disturbance, the trajectory can jump from basin to basin. The changes that lead to phase transitions and chaos are those expected to occur with either perturbations of a coadapted genome or suboptimal environmental conditions. This may also have implications for evolutionary transitions from fluctuating asymmetry to antisymmetry. If we assume that the morphogen influences the behaviour of cell populations, then a transition from phase-locked periodicity of the two sides of a developing trait to chaos in the morphogen produces a corresponding transition from fluctuating asymmetry to antisymmetry in both morphogen concentrations and cell populations.

2.3.3 *Cell-cell communication and developmental stability*

Rather than describing the ontogeny of developmental instability from a developmental biology perspective the development of asymmetries can be understood from a signalling theoretical point of view. Just as reliable signalling among individuals with different interests requires the production of costly signals that are differentially costly with respect to condition (Zahavi 1975), signalling at the level of cells or organs can best be understood from a similar perspective (Pagel 1993; Nahon *et al.* 1995; Krakauer and Pagel 1996). Although cells of a body have almost identical genotypes, cell-cell signalling may allow discrimination between cells differing in phenotypic quality. Only high quality cells are like to perform well in a body, and decisions about cell fate during ontogeny may most reliably be based on signalling of cell quality (Krakauer and Pagel 1996). A similar line of reasoning can be applied to the ontogeny of asymmetry in otherwise symmetrical characters. A high degree of variance in cell quality results in poor communication among cells of the two sides of a body as determined by differences in quality among sides (Møller and

Pagel 1997). In this scenario, canalization can best be understood as selection for increased quality control of cells and hence reduced variance. Several examples of heterogeneity in tissues, and hence developmental instability, are directly related to inferior signals of cells and the incorporation of inferior cells into an organ (review in Møller and Pagel 1997). Cell death often reaches very high levels during ontogeny and results in an overall high level of quality of cells that ensures reliable communication between sides during the developmental processes (Møller and Pagel 1997).

2.3.4 *Individual and population asymmetry parameters*

If developmental stability is a characteristic of genotypes, then different traits should demonstrate consistent deviations from stability or symmetry within individuals. However, there is usually no consistency in the ranking of different individuals within populations for the asymmetry of several characters. A significant positive relationship for individuals within a population would reveal an individual asymmetry parameter, while a positive relationship for the mean developmental instabilities of different populations would reveal a population asymmetry parameter (Soulé 1967).

Consistency in the ranking of symmetry among characters for individuals of a population have been investigated in 56 studies and 27% of these demonstrated a significant positive relationship, while the rest found no relationship (Table W2.1). This proportion tended to be larger in plants (50% of eight studies) than in animals (23% of 48 studies). The concordance of asymmetry was generally low, even in the studies with a statistically significant individual asymmetry parameter. A larger proportion of studies investigating the ranking of symmetry of different characters among populations have found a significant relationship (76% of 17 studies; Table W2.1). Although a relatively large fraction of the studies reported a statistically significant population asymmetry parameter, this pattern was not ubiquitous. Furthermore, the estimates of individual asymmetry parameters may be biased since several studies were based on asymmetries of just a couple of characters. If two characters are phenotypically and genetically closely positively correlated, they may be more likely to develop simultaneously and do so similarly under the influence of the same environmental and genetic conditions, as shown for morphological characters in rice (Sakai and Shimamoto 1965a). Studies reporting a non-significant individual or population asymmetry parameter may have been based on morphological characters that are only weakly phenotypically and genetically correlated.

There might be a number of reasons for the absence of general individual and population asymmetry parameters. First, if two characters developed at different times with certain windows of opportunity, it is likely that they will be subject to different environmental influences and hence develop different asymmetries.

Second, characters may belong to different developmental units and therefore be subject to different developmental problems. Sakai and Shimamoto (1965a) proposed the hypothesis that characters differentiated from a common pre-decessor may, at a later stage, have many of their developmental instabilities in common. A common developmental pathway for two or more characters may be revealed by their positive genetic correlation which may disclose whether two or more characters have a common genetically controlled developmental relationship. This hypothesis was supported by an empirical study of developmental stability of different morphological characters in rice *Oryza sativa*. Characters that were closely correlated genetically and belonged to the same developmental unit demonstrated strong similarities in developmental instability, while that was not the case for characters that were less tightly correlated. A second reason why characters belonging to the same developmental unit have positively correlated asymmetries is the ubiquity of pleiotropy. Most gene products are multiply employed, frequently in seemingly different pathways. If effects of pleiotropy are particularly common among characters belonging to the same developmental unit, it is likely that limiting gene products, which may reduce stable morphogenesis, will simultaneously affect the development of asymmetries of a number of different characters.

Third, different characters may differ in their developmental stability if different developmental processes and different buffering mechanisms control these processes, as suggested by a character-based model of stability that reflects the selection history of the character and the relationship of the character to individual fitness. Developmental processes are likely to be diverse, biochemically, physiologically, and genetically, among different kinds of characters, and it may seem reasonable to assume that stability mechanisms may need to be equally diverse to accommodate such differences in developmental processes. It is inconceivable that a single mechanism could effectively buffer a large number of potentially different developmental processes against a wide variety of environmental or genetic perturbations. A character-specific model would not give rise to correlations among asymmetries in different characters, but when all responses among individuals are averaged within a population a correlated population response among characters is imaginable under certain circumstances (i.e. the same type of characters, in terms of their association with fitness, are examined, or there is genetic variability for stability among populations). The genetic architecture of the genes controlling stability could be predicted to differ among characters, with characters closely linked with fitness being under the influence of directional selection and thus showing predominantly directional dominance and non-allelic interactions between genes responsible for stability (Mather 1973). Stability for characters not closely linked with fitness may be expected to be under stabilising selection and showing predominantly additive variation with any dominance being ambidirectional and thus self-cancelling (Mather 1973). However, it is likely that directional selection will give rise to biased mutation (an excess of mutations with detrimental phenotypic effects as compared to mutations affecting traits predominantly subject to stabilising selection (Mukai 1964)) that will tend to disrupt developmental processes. Depending on the balance of opposing selection pressures such a biased mutation process could maintain additive genetic variance in buffering mechanisms and maintain high levels of asymmetry even in characters closely associated with fitness. The relative importance of these three mechanisms for the lack of correlation between asymmetries in different characters for individuals within a population remains to be assessed.

Finally, and perhaps most likely, fluctuating asymmetry is a measure of developmental instability that is subject to sampling problems that automatically reduces the expected correlation among fluctuating asymmetry of different characters (Whitlock 1996; Leung and Forbes 1997a). The correlation among characters in fluctuating asymmetry is, by definition, simply the covariance between the asymmetries divided by the square-root of the product of the variances of the asymmetries. The covariance term of the underlying developmental instability of the two traits is unbiased by the sampling problem and simply equals the covariance of the fluctuating asymmetries. The variance of the underlying developmental stabilities of a trait is equal to the variance in fluctuating asymmetry divided by the repeatability of the trait. The correlation between the developmental instability of two traits thus equals the correlation between the asymmetries times the square-root of the repeatabilities of the fluctuating asymmetries of the two traits. Therefore, the correlation between the fluctuating asymmetry of two traits will often be a substantial underestimate of the true relation of the developmental instabilities of different characters (Whitlock 1996). However, only the small genetic component of asymmetry should give rise to an individual asymmetry parameter.

These reasons for the absence of an individual asymmetry parameter should be obvious. If some individuals are very unstable developmentally, they would still be completely symmetrical for some of the traits some of the time by mere chance (Palmer 1994). The likelihood of finding positive correlations in asymmetries among individuals in a population will depend on the magnitude of the differences in developmental stability among individuals, and the number of traits measured on each individual. This phenomenon differs considerably from correlations of asymmetries for different traits among populations (the population asymmetry parameter) because the mean asymmetry in a sample of individuals is a much better indicator of average developmental stability of that sample than is the asymmetry of a single individual.

2.4 Developmental biology and the formation of symmetry and handedness

The ontogeny of asymmetry has to some extent been investigated as a problem of developmental biology. In this section we will consider problems of the accuracy of the developmental process to maintain developmental stability, problems of formation of bilateral symmetry and handedness, and evidence for feedback mechanisms between sides of the plane of bilateral symmetry as a control mechanism for the stable development of a symmetric phenotype.

An obvious first question for the formation of a symmetric phenotype is: how accurate should the developmental process of a morphological character be to ensure symmetry (Summerbell and Wolpert 1973)? The level of variation in wing length between left and right sides of *Drosophila melanogaster* as a measure of deviation from perfect symmetry is only 0.40% (Summerbell and Wolpert 1973). Length of elements in wings of fruitflies at age 9 and 11 days have been used to directly estimate the precision of development. Wings demonstrate little fluctuating asymmetry, apparently because of selection for developmental control resulting from the large performance costs of asymmetry (cf. Chapter 7). The development of exoskeletal elements of wings is remarkably precise: 99% of embryos have differences of less than 200 μm, which is a minute difference in size. Errors in the specification of the initial primordium of the wing will be amplified more than later ones. Nonlinearity in growth and control of the growth process will invariably magnify initially small developmental errors. The error in initial specification of element length cannot be greater than 5% in order to maintain a functionally high level of symmetry.

At the molecular level, biological structures such as amino acids are decidedly and consistently biased in a particular direction. Morphological characters are not just bilaterally symmetrical, but also frequently demonstrate handedness in the sense that the two characters are mirror images of each other. A model for the development of handedness in bilaterally symmetrical animals was suggested by Brown and Wolpert (1990). The main assumption was that morphogenesis was based on a morphogenetic gradient caused by a morphogen diffusing symmetrically from the midplane. This provides a credible explanation for the development of bilateral symmetry. This could most readily be accomplished by diffusion-reaction, which may lead to subdivision of an organism into two separate distinct regions with the concentration of the morphogen falling in one half and increasing in another. Their model was based on three different processes. The first step was conversion in which a molecular handedness at the subcellular level was converted into handedness at the cellular level. A specific model for this process was put forward, based on cell polarity and transport of cellular constituents by a handed molecule. The second step involved a mechanism for random generation of asymmetry, which could involve a reaction-diffusion process, so that the concentration of a molecule was higher on one side than the other. The handedness generated by conversion could consistently bias this mechanism to one side. This mechanism may be under genetic control since the *iv* mutation in the mouse *Mus musculus* results in loss of function in biasing, so the direction of asymmetry is random. A recessive, insertion mutation in mice causes development of a body plan that is a mirror image of the normal body plan (Brown and Lander 1993). The third step posits a tissue-specific interpretation process, which responds to the difference between the two sides, and results in the development of different structures on the left and right sides.

Left–right asymmetries are central to most aspects of animal and plant morphology, and the developmental processes generating such asymmetries obviously may provide us with information on the mechanisms generating deviations from perfect symmetry. Two such experiments have been performed. Micromanipulation of embryos at the six-cell stage in *Caenorhabditis elegans* resulted in reversed handedness (Wood 1991). *C. elegans* is characterised by many of its contralaterally analogous cells arising from different cell lineages on the two sides of the embryo (De Pomerai 1991). The manipulation resulted in a mirror image but otherwise normally developing animals, in whom all the left–right asymmetries were reversed. Already in the six-cell embryo the pair of anterior blastomeres on the right is equivalent to the pair on the left, and the extensive differences in fates between linearly homologous derivatives of these cells on the two sides of the animal must be dictated by cell interactions, most likely early in embryogenesis.

The second experiment was performed on the toad *Xenopus laevis*. The location of primordia generating left–right asymmetries in morphology was investigated experimentally on gastrulae of the toad (Yost 1992). Localised perturbation of a small patch of extracellular matrix, by means of microinjection, was correlated with localised randomisation of left–right asymmetries in morphology. Left–right axial information is contained in the extracellular matrix early during development, and this information can be manipulated experimentally. Experiments on different parts of the gastrula clearly showed that left–right information is independently transmitted to cardiac and visceral primordia. Large developmental asymmetries in morphology are generated very early during morphogenesis, and this is also the time when the effects of nonlinear growth processes will give rise to the largest magnitude of phenotypic asymmetries.

Larger Metazoa consist of very large numbers of cells that are the end product of numerous cell divisions. If such complex biological systems are the outcome of the nonlinear growth dynamics described in section 2.3.1, then even small deviations from perfect symmetry during early ontogeny should regularly be magnified into abnormal phenotypes. The fact that this is usually not the case suggests that efficient control mechanisms are at work. The most common control mechanism is feedback which relies on the transfer of information between sides of a system. Bidirectional feedback is the most likely mechanism although unidirectional feedback which relies on the dominance of one side is also a possibility. A number of empirical studies have provided evidence for the existence of feedback during the formation of morphological characters. For example, a study of the formation of claw phenotypes in the American lobster *Homarus americanus* investigated the mechanisms leading to the asymmetric development of the large crusher claw and the smaller cutter claw (Govind and Pearce 1986, 1992). Experiments revealed that the claw receiving most use during early development was transformed into the crusher claw, while the other became the cutter. This process was mediated through regulatory feedback via the central nervous system. A subsequent study on *Cancer productus* revealed that both the quality of food and restrained use of a claw independently contributed to the formation of claw phenotypes (Smith and Palmer 1994).

Similar evidence of feedback control mechanisms in development exists for bilaterally symmetric morphological characters. Experimental removal of all the cells of the frog *Xenopus laevis* that would eventually develop into the right dorsal side of the embryo during early ontogeny, was countered to a remarkable extent by two kinds of compensation (Yuge and Yamana 1989). Cell migration from the part of the embryo that would develop into the left side and compensatory increase in the size of notochord cells resulted in the formation of embryos with perfectly normal internal and external phenotypes.

Phase locking of developmental systems only requires a very weak linkage between the two developing sides to ensure bilateral symmetry. Connection of parts of an organism by the circulatory or nervous system in animals and the water transport system in plants is sufficient to ensure efficient feedback control mechanisms of development (Bergé *et al.* 1984; Freeman *et al.* 1993; Graham *et al.* 1993c). In animals, regulatory feedback mechanisms acting through the neural or circulatory systems thus appear to be extremely efficient determinants of developmentally stable phenotypes.

2.5 Fitness, phenotypic quality, and reaction norms of asymmetry

Development of a symmetric phenotype has in the previous paragraphs been treated as a problem of stable development of a phenotype in a particular environment. However, environments encountered by, for example, parents and their offspring are not necessarily identical, and genotypes that are able to produce stable phenotypes under a range of frequently encountered environmental conditions are likely to multiply at a higher rate than those with a stronger susceptibility to environmental conditions. The use of individual asymmetry as measure of phenotypic quality is based on the assumption that certain genotypes are better able than others to produce a stable phenotype under a range of different environmental conditions. Unless this is the case, developmental stability will not evolve, and, for example, choosy individuals with a mate preference for symmetric partners will be unable to acquire indirect fitness benefits in terms of the ability of offspring to undergo a stable development. The phenotypes developed by a genotype under different environmental conditions are termed the reaction norm of the genotype. If asymmetry is supposed to reliably reflect the quality of a genotype under a variety of environmental conditions, then developmentally superior genotypes should demonstrate less variability in asymmetry along an environmental gradient, and individuals with such genotypes should also enjoy the highest fitness. In other words, the slope of the reaction norms of asymmetry for individuals with the highest fitness should be shallower than for those with low fitness. This assumption of developmental stability reliably reflecting the quality of individuals with specific genotypes can only be tested by allowing individuals with the same genotype to undergo ontogeny of a trait under a range of environmental conditions. This estimate of the reaction norm can subsequently be related to fitness.

Two experiments have tested this critical assumption of developmental stability reliably reflecting the quality of a genotype in terms of ability of environmental buffering capacity. The long outermost tail feathers of male barn swallows *Hirundo rustica* are subject to a directional mate preference. European barn swallows moult these tail feathers once a year in the African winter quarters. However, birds can be induced to regrow their feathers by simply plucking the feathers any time during the annual cycle. Environmental conditions for feather growth are relatively poor during the breeding season as compared to the normal moulting season in winter, because reproduction severely constrains the availability of energy for other activities such as moult, and birds have generally evolved temporally well-separated breeding and moult schedules (Shykoff and Møller 1997). The size and asymmetry of the outermost tail feathers grown in the winter quarters by a number of males were compared with those of feathers grown at the breeding grounds. Both tail feather length and asymmetry were strongly positively correlated in the two different environments, although the feathers developed at the breeding grounds were shorter and more asymmetric. The reaction norms of tail asymmetry were steeper for males with short tails, and these males also had lower fitness measured as reproductive success in the year preceding the feather plucking experiment (Fig. 2.4). These results imply that individuals with greater buffering ability against adverse environmental conditions as demonstrated by their more shallow reaction norm indeed had higher fitness.

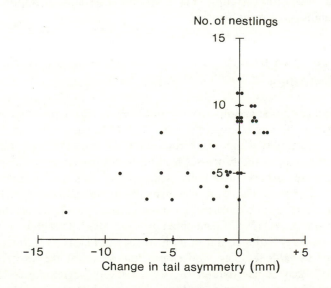

Fig. 2.4 Seasonal reproductive success of male barn swallows *Hirundo rustica* during the year preceding tail feather removal in relation to tail feather asymmetry developed after removal of the outermost tail feathers during the breeding season. Adapted from Shykoff and Møller (1997).

A second experiment was based on the effects of manipulation of the mite contents of the nests of barn swallows on the size and asymmetry of the outermost tail feathers of male and female barn swallows (Shykoff and Møller 1997). A haematophagous mite has severe negative effects on the fitness of barn swallows by delaying breeding and reducing the quantity and quality of offspring produced (Møller 1990b), and it affects the length and the asymmetry of the outermost tail feathers of adults (Møller 1992c, 1994b). Mite contents of nests were randomly either reduced by fumigation, kept at natural levels, or increased by addition of mites to the nest contents during the egg-laying period. The phenotypes of the outermost tail feathers of adult males attending these nests were compared between the year when mite numbers were manipulated with those of the previous year. Again, there was a strongly positive correlation in tail length and asymmetry between environments with tails being shorter and more asymmetric in a year following presence of mites, and longer and more symmetric in the control year (Shykoff and Møller 1997). Importantly, the reaction norm of tail asymmetry was much steeper for males with short tails in the year preceding the experiment. Males with large tail asymmetries following mite treatment also had lower reproductive success in the preceding year than males with less tail asymmetry (Shykoff and Møller 1997).

In conclusion, the two experiments on buffering ability of genotypes and fitness both supported the critical assumption that genotypes better able to buffer adverse environmental conditions also had higher fitness. Obviously, experimental tests of this assumption in other model systems are required before any firm conclusions can be reached.

2.6 Ontogeny of individual asymmetry and phenodeviants: empirical studies

Several theoretical investigations have recently addressed questions concerning the development of individual asymmetry and other individual measures of developmental instability. Empirical studies have also made major contributions to our understanding of the processes leading to the formation of a symmetric morphology. Three groups of organisms are particularly well suited for such studies of development. These are organisms with several copies of the same structure that develop simultaneously or in succession. Examples of this type of organism include several taxa of segmented invertebrates with segments of almost identical morphology and modular organisms such as plants. A second group of organisms well-suited for studies of developmental processes are parthenogenetic organisms with genocopies that develop in spatially or temporally different environments. Examples of this kind of organism include aphids, *Daphnia*, and gynogenetic fish. A third group of organisms particularly well-suited for studies of developmental processes generates the same structure repeatedly during successive years. Examples include exoskeletons of crustaceans that undergo regular moults, feathers developed by birds during subsequent moults, and antlers of deer. Such organisms have been exploited in the following studies of the processes leading to the formation of asymmetric phenotypes.

Empirical studies can readily address questions about sidedness of asymmetries in morphology, whether asymmetries developed during subsequent developmental events are statistically repeatable, and whether the ontogeny of asymmetry may arise as a result of accumulation of a number of minor asymmetries for individual growth increments that result in a cumulative large morphological asymmetry. Soulé (1982) suggested that the number of component parts of a character may affect its level of asymmetry. Asymmetries generated during a large number of independent growth increments should even out and result in the formation of a more symmetric character as the number of increments increased. It remains unknown whether individual growth increments really can be considered statistically independent. Another problem concerns whether there is an effect of the duration of the developmental period on asymmetry. A relatively short developmental period for a given character size should more often result in the formation of large asymmetries, simply because all else being equal, energy more often would be a limiting factor during short bursts of growth and there would be less opportunity for negative feedback and compensatory growth.

Price *et al.* (1991) reviewed the literature on developmental processes of feathers. Feather formation is initiated as a conical pinching of the epidermis. Feather follicles are formed from a mesodermal core and an epidermal ring, and feather growth is the result of cell division in a ring of epidermal tissue at the base of the pinching known as the collar pushing out the previously formed cells (Lucas and Stettenheim 1972). These cells differentiate and either become keratinised to form the barbs and central rachis or die and thereby form the spaces between the barbs. Feathers grow by first forming the tip and later cells result in the formation of subsequent barbs. A study of the development of feather tips in the yellow-browed warbler *Phylloscopus inornatus* showed that population variation in the earliest formed parts of the feather was larger than in later formed parts (Price *et al.* 1991). The developmental noise part of the variance showed a similar pattern. The early and late formed parts of the size of feather characters were positively correlated, while that was not the case for asymmetry of the same characters. In other words, development of the early and late parts of feathers in different individuals was consistent, while that was not the case for measures of developmental instability arising during the early and late parts of development of feathers (Price *et al.* 1991).

Individuals of the crab *Hemigrapsus nudus* were measured and followed through three moults in order to test for temporal consistency in developmental instability (Chippindale and Palmer 1993). Most morphological traits measured had asymmetries that remained constant in sign of asymmetry during subsequent moults (Fig. 2.5), while some asymmetries increased in magnitude with increasing size and hence growth. Although this study suggests that individuals differ in their ability to reproduce a stable phenotype, the design of the study does not allow us to draw any conclusions about the relative importance of genetic and environmental factors.

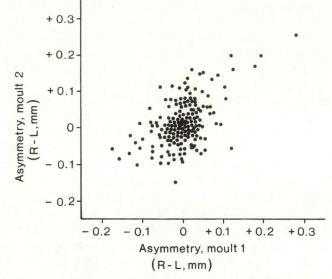

Fig. 2.5 Asymmetries in limb segments between first and second moults in individual brachyuran shore crabs *Hemigrapsus nudus*. Values are signed differences between sides pooled across all limb segments. Adapted from Chippindale and Palmer (1993).

The development of fluctuating asymmetry in a morphological character was investigated in a growth model by Aparicio (1995). Character size was assumed to depend nonlinearly on investment levels, and both sides of the bilaterally symmetrical character were assumed to grow with a similar pattern. Allocation of resources within the growing individual may be unbalanced due to small deviations in the pattern of nonlinear growth between sides and this may directly cause development of asymmetrical characters, because nonlinearity *per se* makes differences between the two sides larger than when growth is linearly dependent on investment. The degree of fluctuating asymmetry will then increase with growth rate at a specific investment level. These predictions were tested on growth of morphological characters in manipulated broods of the European kestrel *Falco tinnunculus*. This was investigated at two different levels. First, the growth pattern and fluctuating asymmetry of individual nestlings was investigated during their development. Nestling growth rates peaked relative to parental investment levels during the middle of the nestling period, and this coincided with the time when fluctuating asymmetry reached an absolute maximum. When growth rates levelled off later during the nestling period, the degree of fluctuating asymmetry decreased continuously. A similar result was found for feather growth in starlings *Sturnus vulgaris* (Swaddle and Witter 1994). Second, individual nestlings growing fast at a specific parental investment level, determined from the experimentally manipulated brood size, developed larger asymmetries in their tarsi than did slower growing nestlings (Fig. 2.6). These results clearly suggest that faster growth rates for a given level of parental investment are associated with larger deviations from perfect symmetry.

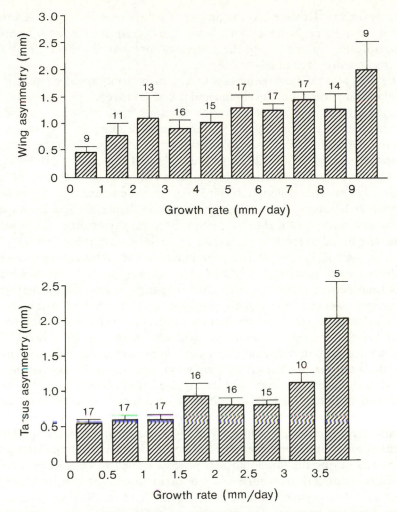

Fig. 2.6 Fluctuating asymmetry in the length of the tarsi of individual nestlings of the European kestrel *Falco tinnunculus* in relation to instant growth rate. Values are means per nest (s.e.). Adapted from Aparicio (1995).

The functional significance of symmetry may change during the development of a trait. In other words, the relative costs and benefits of symmetry may differ between growth stages. Therefore, the ontogeny of asymmetry is not only affected by internal growth-regulation mechanisms, but this may also be due to different consequences of asymmetry over developmental time. For example, development of asymmetry in the growth of primary feathers in starlings changes over time (Swaddle and Witter 1994, 1997b). Initially, asymmetries are extremely large, but get much smaller as the feathers reach their final length. A functional explanation for this phenomenon is that very small feathers do not contribute much to the wing profile, and so there is little cost associated with their asymmetry. However, as the primary feathers become larger, the costs of feather asymmetry increase. The functional significance of the asymmetry changes during ontogeny, and the pattern of 'optimal' asymmetry may therefore also change over time.

The ontogeny of fluctuating asymmetry in morphological characters may reveal information about developmental processes (Møller 1996b). Asymmetry in feather characters of the barn swallow develops repeatedly during subsequent annual moults, and consistency in the process leading to asymmetries can therefore be studied. First, the side developing the larger character value was found to be biased within individuals under natural and experimentally induced growth after removal of feathers during the breeding season. Removed feathers are usually replaced, although the bird does not moult naturally at that time of the year. Individuals also showed consistency in their ranking with respect to asymmetry across years (Møller 1994c). Second, asymmetric morphological traits were found to result from asymmetric daily growth increments (Fig. 2.7), and the size of the increments developing under different environmental conditions were positively correlated. This implies that initially small deviations from perfect symmetry were often magnified during subsequent growth increments leading to a cumulative increase in developmental instability. This result is consistent with the view that the ontogenetic process of developmental instability results from the nonlinear dynamics of the growth process. Third, asymmetries of morphological characters developing under different environmental conditions as reflected by feathers grown during the normal moulting period in the winter quarters and during the breeding season were positively correlated; although the level of asymmetry was larger under the energetically more adverse environmental conditions of the breeding season. Fourth, asymmetries in tail length and daily growth increment size were unrelated to the duration of the developmental period of the feather, although later growth increments during the developmental process tended to be smaller and more symmetric than early increments. These observations are consistent with the hypothesis that asymmetry in feathers of the barn swallow arises as a consequence of the accumulation of stochastic errors during early development, particularly under poor environmental conditions.

Another study on the ontogeny of asymmetry in tree swallows *Tachycineta bicolor* provides some additional information (Teather 1996). A laboratory study of the ontogeny of fluctuating asymmetry in the primary feathers of the European Starling revealed that (i) signed asymmetries were not consistently biased toward either left or right side among feathers of the same individual; (ii) growth increments were also not sided; and (iii) both absolute and relative asymmetry decreased as the feathers developed (Swaddle and Witter 1997b). These data are most consistent with a developmental regulatory system that involves feedback between left and right sides and episodes of compensatory growth to correct large asymmetries.

Consistency in the pattern of the expression of developmental instability can readily be studied in species with repeated structures of the same kind, such as

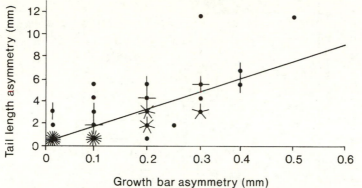

Fig. 2.7 Fluctuating asymmetry in the length of the outermost tail feathers of male barn swallows *Hirundo rustica* in relation to fluctuating asymmetry in the average length of daily growth increments. Observations are shown by cellulation. Adapted from Møller (1996b).

leaves and flowers of plants. The ontogeny of leaf asymmetry was studied in developing leaves of broad-leaved elm *Ulmus glabra* (Møller 1997c). The maximum width of the two halves of six individual leaves with different positions (apical position, second to the apical position, and third to the apical position) on two different branches of each tree were measured daily until reaching full size. The variation in the developmental process was studied with respect to position, branch, and tree. Leaf positions, branches, and individual trees differed in their ontogeny of asymmetry, although by far the largest majority of the variation was found among trees. Individual leaves were consistent in the growth dynamics of their fluctuating asymmetries, with initial, small deviations from perfect symmetry in a particular direction being consistently magnified in this direction during the entire growth process (Fig. 2.8; Møller 1997c). This result is clearly consistent with the predictions from the nonlinear growth models and feedback dynamics proposed in section 2.3. Finally, the amount of environmental stress influenced the ontogeny and the magnitude of asymmetry. Saplings subject to a saline water treatment grew more slowly and developed higher asymmetry than saplings subject to a pure water control treatment. Hence, environmental stress tended to modify ontogeny, ultimately giving rise to increased asymmetry. These results are similar to the growth dynamics of abnormal leaves of grape *Vitis vinifera*, which also demonstrated consistency in development of originally minor deviations from perfect symmetry (Wolf *et al.* 1986).

In conclusion, a number of recent empirical studies have found evidence of developmental processes that are consistent with the predictions of the nonlinear growth and feedback dynamics of chaos theory. The next step, obviously, is to make quantitative predictions of developmental processes and to test these predictions experimentally under varying levels of environmental adversity.

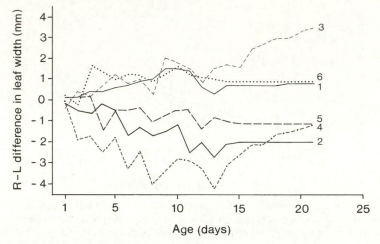

Fig. 2.8 Signed asymmetry in maximum width of the two halves of a leaf of wych elm *Ulmus glabra* during the growth period. Each line and symbol represents a single leaf. Adapted from Møller (1997c).

2.7 Summary

- The ability to ensure stable development of morphological characters is presumed to have a genetic basis. Long-term research on insecticide resistance in sheep blowflies has revealed that resistance alleles disrupt genomic co-adaptation and give rise to increased levels of developmental instability, while a modifier allele restores developmental stability.
- Developmental stability may be the result of the nonlinear dynamics of biosynthetic processes in cells and the nonlinear feedback occurring among developing body parts, as suggested by chaos models of morphogenesis.
- Developmental models of symmetry and handedness and their feedback mechanisms have provided insights into the mechanisms that give rise to developmental stability.
- Developmental studies of the two sides of a number of different morphological characters of plants and animals in nature are consistent with the chaos theory models of ontogeny of developmental instability.

3

Developmental stability and mode of selection

3.1 Introduction

Micro-evolutionary processes involve different genotypes producing phenotypes upon which selection can act. If phenotypes are non-randomly associated with fitness components, this will result in an evolutionary response to selection, provided the characters in question have an additive genetic basis, and that there are no strong negative genetic correlations preventing or delaying a response. In this chapter, we introduce the concept that the prevailing patterns of selection may affect both the phenotypes and the genotypes producing particular phenotypes. This may take place by means of genetic modifiers that control the phenotypic expression of a particular genotype, but also modify the amount of additive genetic variance being expressed phenotypically.

Canalisation reflects the capability for stable development of a genotype under a range of different environmental conditions. The process of canalisation is under genetic control, and selection is thus able to modify the canalisation of a particular trait depending on the prevailing pattern of selection. We briefly review the literature on canalisation and how it is related to the evolutionary process of phenotypic change (section 3.2).

Patterns of prevailing modes of selection such as directional, stabilising, and disruptive selection affect the phenotypic variance of traits, including fluctuating asymmetry. This is brought about by the effects of genetic modifiers on the expression of the genotype. Schmalhausen (1949) and Dobzhansky (1951) emphasised how stabilising selection could maintain additive genetic variance. They argued that if selection affecting a particular trait changed from stabilising to directional in mode, this would have major consequences for the exposure of additive genetic variance hidden by genetic modifiers. Subsequently, Thoday (1958; Tebb and Thoday 1954) has shown that as populations of fruitflies became adapted to particular new conditions in a novel environment, the asymmetry in sternopleural bristles decreased significantly as an indication of the process of adaptation. The theoretical basis for this argument is presented in section 3.3. We briefly review the literature on phenotypic and additive genetic variance and developmental instability in relation to the prevailing mode of selection based on directional, stabilising, and disruptive selection experiments (section 3.4). The effects of prevailing patterns of selection on phenotypic and additive genetic variances are discussed in section 3.5 for secondary sexual characters and other traits with a net directional evolutionary change.

3.2 Canalisation and the micro-evolutionary process

The way in which many biologists are taught genetics in undergraduate courses is based on the model that genotypes produce phenotypes on which selection acts. This view of the biological world is correct in the sense that we now know that different genotypes for a particular trait produce phenotypes. However, it has been known for a long time that the same genotype for a particular trait may develop a range of phenotypes. Standard examples include different kinds of leaves of trees produced in the shade and the sun and asexually reproducing *Daphnia* with more or less pronounced spines depending on the presence of predators. Many more examples of such phenotypic variation are listed by Schmalhausen (1949). Small genotypic changes can have small or large phenotypic effects depending on the relationship between a specific genotype and its phenotypic expression (Fraser 1962; Moreno 1994).

The standard relationship among genotype, phenotype, and selection is presented in Fig. 3.1: the phenotype on which selection acts being an expression of the genotype. The prevailing pattern of selection may act not only on the phenotype, but also on the genotype. Imagine the situation where a genotype invariably gives rise to a specific phenotype. If the phenotypic trait is body size, this trait is likely to involve information from many loci. Individual genotypes with additive allelic effects may produce a large or a small phenotype. The expression of a genotype may not invariably be beneficial for an individual, if, for example, large size is only beneficial under favourable environmental conditions. A gene that makes the expression of the genotype conditional on environmental conditions would be selectively advantageous. Such a genetic modifier that controlled the expression of the genotype would on average give rise to a phenotype that is better suited to environmental conditions. This is phenotypic plasticity; the development of a range of phenotypes under different environmental conditions.

(a) genotype ⟶ phenotype ⟶ selection

(b) genotype ⟵ phenotype ⟵ selection

Fig. 3.1 The relationship between genotype, phenotype, and selection.

Developmental canalisation, which is defined as the stabilised flow of a developmental trajectory of a genotype under different environmental conditions, was suggested independently by Schmalhausen (1940) and Waddington (1940). It provided a mechanism that resulted in the production of a specific phenotype through buffering mechanisms along a specific developmental pathway (a creode in the terminology of Waddington) despite genotypic and environmental differences. Canalisation may break down when the ontogeny switches from one developmental pathway to another as some threshold value of the genotypic or environmental effects is attained, as described in Chapter 1. Selection for canalisation ensures similar development of a phenotype due to invariability of the developmental pathway under a range of environmental conditions. Canalising selection tends to remove those alleles which render the developing organism sensitive to the potentially disturbing effects of adverse environmental conditions, and it will generate genotypes which produce the optimal phenotype even under suboptimal or adverse environmental conditions.

The existence of such selection for canalisation was subsequently investigated experimentally by Waddington (1960), who provided evidence for its existence as a real biological phenomenon. The development of fruitflies is highly canalised, so that exposure to most kinds of environmental stresses produce little phenotypic divergence from normal phenotypes. Waddington (1960) kept *Drosophila melanogaster* that were temperature sensitive to the expression of the *Bar* mutation (which is a tandem duplication of a homeobox gene) at two temperatures, because the phenotype of such mutants is often much more variable and sensitive to environmental conditions than is that of the wild-type. Temperature sensitivity of the *Bar* mutant is more extreme at 25°C than 18°C. Selection was introduced to reduce temperature sensitivity of the *Bar* mutant (which has a markedly reduced number of facets in the eye), to a larger extent in the homozygous females than in the hemizygous males. The number of facets is known to be greater at lower temperatures. Selection at alternating temperatures between generations was not very effective in reducing temperature sensitivity of phenotypes in terms of the number of eye facets of the compound eyes. With family selection in which the progeny was divided into two lots at the two different temperatures, a reduction in temperature sensitivity to selection was rapidly achieved (Fig. 3.2). In other words, the mutant phenotype was more closely the same at both temperatures because the flies became more temperature sensitive at the lower temperature and less sensitive at the higher one. Although this experiment was termed 'selection for canalisation' by Waddington, this claim was at conflict with his own definition in the same paper of canalisation as being the genetic mechanisms that ensure fidelity of wild-type development. However, the results of the experiments are in accordance with our definition of canalisation as a property of the genotype that tends to insure developmental fidelity under a range of environmental conditions. Although he succeeded in making his *Bar* eye stocks less temperature sensitive, the degree of asymmetry between the eyes was not reduced by the selection experiment (Waddington 1960). While the 18 °C *Bar* eyes are more normal than the

25 °C ones, they are still far from wild-type in phenotype; the former had about 150 facets per eye, while the latter had only about 60 facets, which compares to a normal facet number of about 700–750. After selection, Waddington obtained flies with about 100 facets per eye at both temperatures and this is still substantially mutant. Since all of classical developmental genetics shows that as gene function for genes affecting morphology is reduced, bilateral asymmetry is increased, it is not surprising that Waddington, with his selection procedure, did not obtain a substantial reduction in asymmetry. Experiments with three other kinds of mutants gave qualitatively similar results (Waddington 1960).

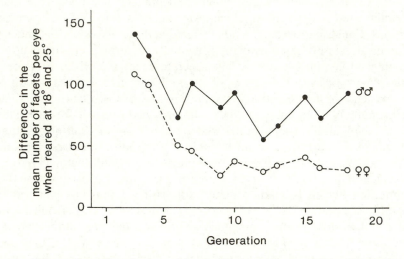

Fig. 3.2 Effects of family selection of *Drosophila melanogaster* of the *Bar* stock at two different temperatures on temperature sensitivity of the development of the number of eye facets. Adapted from Waddington (1960).

Subsequent experiments on canalisation have generally resulted in rapid responses to selection and have thus provided evidence for the existence of additive genetic variation for canalisation (Table 3.1). The number of genes involved in canalisation of two characters in mice is small, as suggested by experiments performed by Kindred (1967). Models show that stabilising selection increases the level of canalisation, assuming that the population is close to the optimum (Vogl 1996), while disruptive and directional selection reduce canalisation (Kieser 1987; Gavrilets and Hastings 1994). Experiments investigating the effects of stabilising and disruptive selection on temperature sensitivity of development of the fourth wing vein in *Drosophila* have supported these theoretical predictions with stabilising selection resulting in increased canalisation and disruptive selection giving rise to a reduction in canalisation (Scharloo *et al.* 1972).

The exact mechanisms giving rise to canalisation are not well understood. Rendel (1967) proposed a quantitative genetics model for canalisation in which an increasing amount of a developmental gene product, such as a morphogen, could elicit phenotypic responses once a certain threshold had been reached.

Table 3.1 Experiments on canalizing selection.

Species	Character	Experimental design	Result	Reference
Drosophila melanogaster	facet eyes	temperature-sensitivity	canalisation altered	Waddington (1960)
Drosophila melanogaster	eye size	temperature-sensitivity	canalisation altered	Waddington and Robertson (1966)
Drosophila melanogaster	bristles	altered numbers in the scute phenotype	canalisation altered	Rendel and Sheldon (1959)
Drosophila melanogaster	bristles	temperature-sensitivity	canalisation altered	Kindred (1965)
Drosophila melanogaster	4th vein	temperature-sensitivity	directional selection decreased and stabilising selection increased canalization	Scharloo *et al.* (1972)
Drosophila melanogaster	bristles	temperature-sensitivity	resistance to canalisation depends on level of canalisation	García-Vázquez and Rubio (1988)
Drosophila melanogaster	bristles	increased number of bristles	canalisation of two symmetrical extra bristles	Piñeiro (1992)
Mus musculus	vibrissae	increased vibrissae number	canalisation depends on genotype	Fraser and Kindred (1960) Kindred (1963, 1967)
Mus musculus	toe number	increased and decreased numbers	canalisation depends on a few genes	Kindred (1967)

Modifications of regulatory thresholds could potentially also affect dominance relationships among alleles, which would result in masking of genetic variation in phenotypes. Phenotypes may be expressed relatively invariantly across a wide range of environments, although considerable amounts of low-penetrance genetic variation may still be present (e.g. Schmalhausen 1949). Canalisation is associated with these common environmental conditions, while unusual environmental conditions may disrupt canalisation and fully express previously unexposed genetic variation phenotypically. If any novel variant expressed under these extreme conditions is superior in the new environment, this may result in genetic assimilation of the phenotype. The expression of hidden genetic variation under novel environmental conditions may initiate genetic revolutions or speciation. The greatest degree of canalisation occurs under optimal environmental conditions, and more extreme phenotypes will be expressed when conditions deviate from the optima. The canalisation model proposed by Rendel (1967) may be too simplistic, because more than a single epigenetic response may be involved, and different morphological traits may have independent thresholds for their full phenotypic expression (Scharloo 1991).

3.3 Phenotypic and genetic variance under different modes of selection

According to standard quantitative genetics theory directional selection on quantitative traits will lead to a decrease in genetic variance and rapid exhaustion of the variance due to fixation of alleles, if the number of loci and alleles is finite (Bulmer 1985). However, this relationship between the prevailing mode of selection and the phenotypic and genetic variance components may not always apply, as demonstrated by a number of models. Kieser (1987) based his qualitative model on the effects of the mode of selection on the evolution of the degree of canalisation and hence suggested that there is a direct evolutionary relationship between canalisation on one hand and phenotypic and genetic variances on the other. Pomiankowski and Møller (1995) addressed the question by suggesting that the prevailing selection pattern affected the evolution of phenotypic and genetic variances through condition-dependent expression of alleles. In other words, a directional selection pressure was supposed to simultaneously select against genetic modifiers that reduce the effects of single gene changes. The model by Gavrilets and Hastings (1994) was based on the assumption that for an additive genetic polygenic trait the contribution of the loci to the trait depends on the micro-environment in a linear fashion. The model was analysed for the effects of short-term stabilising and directional selection on the genotypic and environmental components of phenotypic variance.

The qualitative model of Kieser (1987) assumes that directional selection will result in a destabilisation of the developmental pathway caused by adverse environmental conditions and in selection against genetic modifiers that control the expression of a particular character. Both destabilisation of the developmental pathway and the unmasking of hidden additive genetic variation will result in a rapid increase in phenotypic variance on which natural selection can act. Provided there is a persistent pattern of directional selection on phenotypes arising from an elevated level of phenotypic and genetic variance in the population, this will give rise to a sudden increase in the rate of evolutionary change. Stabilising selection will contribute to phenotypic stasis by increasing concealment of additive genetic variance, deepened canalisation, and evolution of developmental constraints (Kieser 1987). These two mechanisms are the basis for a proposed minimax model of phenotypic evolution: the part of an organism with the minimum potential for phenotypic change (hence the maximum extent of canalisation) will govern the direction of the maximum phenotypic change of other parts of the phenotype. Canalisation masks the underlying genetic variability and transforms the morphogenic programme into a stable developmental pathway. If individuals of a species are exposed to novel environments such as those at the periphery of the distributional range, this will result in exposure beyond the buffering ability of the canalisation system. Such exposure of individuals to extreme environmental conditions may produce aberrant, non-canalised phenocopies on which natural selection can act, and this may result in a micro-evolutionary response if there is a heritable basis for the trait. This scenario apparently fits with evolutionary novelties having been shown to arise rapidly during times of adverse environmental conditions (Gould and Eldredge 1977; Mayr 1982), and rapid evolutionary change having been linked to increased phenotypic variance during the period of transition (Hoffmann 1982; Williamson 1981; see also section 3.4). A similar line of arguments was proposed by Møller (1993d) for the effects of adverse environmental conditions in peripheral populations on developmental stability, sexual selection, and speciation. It is important to note that the qualitative model of Kieser for evolutionary change is at conflict with that proposed by Parsons (1993a, 1994b), who emphasised that only moderately adverse environmental conditions would allow evolutionary change (see section 4.3).

The models of Pomiankowski and Møller (1995) and Gavrilets and Hastings (1994) both provide a mechanism for the evolutionary processes envisaged by Kieser. The basic assumption of the model by Pomiankowski and Møller (1995) is that the phenotypic expression and the additive genetic variance of a character depend on the presence of genetic modifiers. Such modifiers are assumed to control the amount of additive genetic variation and hence the phenotypic variation (Fig. 3.3). Stabilising selection is presumed to conserve the additive genetic variation, as suggested by Schmalhausen (1940, 1949), by selecting for genetic modifiers preventing the full phenotypic expression of a genotype. Alleles masked by genetic modifiers under stabilising selection will not give rise to phenotypic expression and will therefore be maintained. Directional selection is presumed to have the opposite effect. This scenario is consistent with directional and stabilising selection experiments (see section 3.4) and with empirical evidence on phenotypic and genetic variation of characters subject to sexual selection (see section 3.5.1).

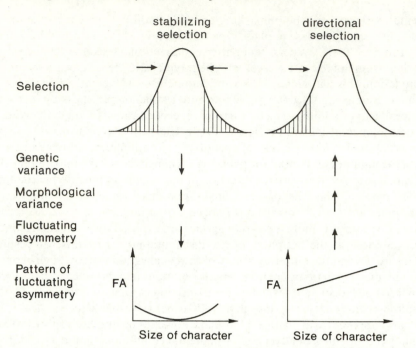

Fig. 3.3 The model of phenotypic and additive genetic variation of characters subject to stabilising and directional selection. Adapted from Møller and Pomiankowski (1993b).

The model by Gavrilets and Hastings (1994) is based on the assumption that the sensitivity of phenotypic expression to fluctuations in the micro-environment has a genetic basis. The contribution of the loci to the expression of the phenotype may depend on the environment in a nonlinear fashion. An understanding of the effects of selection on phenotypic expression depends on the ways in which genotypes are transformed into phenotypes. The shape of the function that maps genotypes to phenotypes may be directly affected by selection processes, as first suggested by Fraser (1962). The model analyses the effects of short-term stabilising and directional selection on the genotypic and environmental components of phenotypic variance. The model predicts that stabilising selection should increase developmental canalisation while directional selection should have the opposite effect. The effects of short-term stabilising and directional selection on the genetic and the environmental components of phenotypic variance may have consequences for their heritability. Stabilising selection will reduce both components, but since a component with a larger value experiences a larger reduction, this can result in an increase in the heritability under stabilising selection, as shown by some studies cited by Gavrilets and Hastings. The additive genetic coefficient of variation (which reflects the additive genetic variance standardized for character size) is considered by some to be a better measure of the evolvability of a trait than the heritability (Houle 1992). Although no predictions were made by Gavrilets and

Hastings, the additive genetic coefficient of variation is likely to become reduced as a consequence of stabilising selection (Pomiankowski and Møller 1995). In accordance with the latter view, stabilising selection generally tends to reduce the heritability (see section 3.4).

Prevailing modes of selection may have important implications for the relationship between genotype and phenotype (Fraser 1962; Moreno 1994). Fraser, in a neglected but very important paper entitled 'The survival of the mediocre', envisaged two types of genetic systems: those with a simple, linear relationship and those with a complex relationship of the pattern of epistasis. In computer simulations, a marked reduction in phenotypic variability through stabilising selection gave rise to modification of the patterns of epistasis from a simple linear function under directional selection to a complex sigmoid function under stabilising selection. This change in epistasis caused many of the possible genotypes to produce the same phenotype. Stabilising selection will tend to modify patterns of genetic interaction towards a sigmoid genotype–phenotype relationship, while directional selection will produce a linear relationship.

In accordance with this suggestion, experiments on small effect genetic variation in *Drosophila melanogaster* showed a potential for strong epistasis (Moreno 1994). Strong interactions are common among mutants of the development of the central nervous system, wing veins, and sternopleural bristles (the latter is a trait known to be under stabilising selection). The results showed that a similar increase in bristle number, which requires a large change in activity at one locus (*emc-*), can be produced by combining functionally related mutations that have small phenotypic effects on their own. A second experiment, comparing the genetic variance resulting from the same set of wild third chromosomes in wild-type and mutant backgrounds, revealed substantial amounts of unexpressed genetic variance. The mutation (a weak allele of the locus *Abruptex*) was introduced at an unlinked locus that mildly affected the mean of the number of sternopleural bristles. The expectation was that the penetrance of mutations at functionally related loci that act as small effect variation because of a nonlinear genotype–phenotype relationship will be increased, resulting in an increase in the genetic variance generated by the wild-type chromosomes. There was indeed a two-fold increase in the additive genetic variance produced by the chromosomes in the mutant background, suggesting that wild polygenic mutations in quantitative characters hide a significant amount of cryptic variability in the form of low-penetrance mutations (Moreno 1994), a point that was already made by Schmalhausen (1949). Such cryptic interaction effects may become important when genotypic frequencies are disturbed by selection, inbreeding, or population bottlenecks. Stabilising selection, because it favours an epistatic concealment of additive variation, will produce a genetic system with long-range adaptive value. Canalisation will reduce variability and allow radiation to all the specific niches of this environment. Directional selection will produce aberrant, non-canalised phenocopies under adverse environmental conditions. There will be a wide range of phenotypic variability with a genetic basis, which will allow for a rapid shift to novel environments. A new canalisation system may subsequently evolve.

In conclusion, the four models of phenotypic and genetic variation in relation to the prevailing mode of selection all suggest that the balance between directional and stabilising selection can have profound effects on the phenotypic and additive genetic variation of quantitative characters. The models by Pomiankowski and Møller (1995) and Gavrilets and Hastings (1994) both stress the importance of the environmental sensitivity of characters. The models by Fraser (1962) and Moreno (1994) suggest how the prevailing mode of selection may have important consequences for the way in which the phenotype is expressed from the genotype.

3.4 Phenotypic and genetic variance: lessons from selection experiments

Several experiments have either directly or indirectly assessed the effects of different modes of selection on phenotypic and additive genetic variation and developmental stability. A review and synthesis of this literature is provided in Table W3.1 on the Web page. Results from selection experiments in the laboratory have to be interpreted cautiously because effective population sizes are often so small that inbreeding and genetic drift may have confounded the results. Furthermore, if the effects of individual genes on a metric trait investigated are small, many experiments may have been terminated before the effects of selection could have been measured given the sampling errors. The following summary of the results of selection experiments should be interpreted with these caveats in mind.

Directional selection in almost all studies has resulted in an increase in all variance components including the phenotypic variation, the additive genetic variation, the phenotypic variance generated by environmental effects, and the developmental instability (within-individual) variation (Table W3.1). This result is consistent with the suggestion that directional selection, generally, simultaneously selects against genetic modifiers that control the expression of the genotype, as suggested in section 3.3.

Relatively few experiments have imposed stabilising selection on characters. Generally, stabilising selection has the opposite effect of directional selection, with a reduction in all variance components (Table W3.1). However, this result should be interpreted with caution because many of the studies have focused on sternopleural bristles in *Drosophila*. Again, this result is consistent with the idea that stabilising selection has a conserving effect on the genetic variation and therefore also on the phenotypic expression of the genotype.

The number of disruptive selection experiments is even smaller than the number of stabilising selection experiments. However, the general pattern shown by the few studies published is similar to the outcome of directional selection experiments; a general increase in all variance components (Table W3.1). Disruptive selection does not differ qualitatively from directional selection in two directions. Therefore, this result is as expected from a theoretical point of view.

Finally, it is important to notice that the majority of the studies have been performed under relatively benign laboratory conditions. This fact will have important consequences for the phenotypic and genetic variances which will generally differ from values obtained under field conditions (see sections 4.3.4 and 4.3.5). The heritability will generally be larger under laboratory conditions, and the response to selection will therefore be larger for a given intensity of selection.

In conclusion, there is consistency in the effects of different patterns of selection on phenotypic and genetic components of variation. While directional and disruptive selection generally result in an increase in the variance components, stabilising selection has the opposite effect with a reduction in variance components.

3.5 Phenotypic and genetic variance: lessons from sexual selection and other directional selection phenomena

The enormous micro-evolutionary potential of organisms has never been realised more clearly than by people engaged in breeding experiments. Darwin (1868) in his book on domestication of animals emphasised the considerable phenotypic variability achieved through artificial selection by humans. A quick glance at the dogs at an exhibition readily reveals the enormous divergence attained under different regimes of directional selection imposed by man. The within-species phenotypic variability created under a range of environmental conditions in nature is at the best small in comparison. The inevitable conclusion from this comparison of phenotypic variability under natural and domestic conditions, and from similar results in dozens of other domesticated animals and plants, is that (i) there is an unrealised potential for micro-evolutionary change within almost any species, and (ii) organisms in the wild generally are subject to intense canalising and stabilising selection. A closer inspection of free-living organisms reveals a large number of similar examples of how intense directional selection uncovers almost unimaginable amounts of phenotypic variation. We will review examples from sexual selection, life-history evolution, palaeontological evidence for directional evolution, and herbivore adaptations to plants in the following sections. These examples concern cases of evolution caused by intense directional selection pressures, although a second feature of these examples is the trade-off between the advantages of the most extreme trait values and other components of fitness. Such trade-offs may give rise to the maintenance of considerable amounts of additive genetic variance.

3.5.1 *Sexual selection*

Sexual selection arises from 'the advantages that certain individuals have over others of the same species and sex, in exclusive relation to reproduction' (Darwin 1871). Sexual selection has resulted in the evolution of extravagant secondary sexual characters such as the train of the blue peacock *Pavo cristatus* and the horns of many beetle species. Closely related species generally differ much more in secondary sexual characters than in other morphological traits, providing evidence for a net directional evolutionary change in secondary sexual characters during recent evolutionary history. Intense directional selection results from the most extravagantly ornamented males having a mating advantage. This selection should simultaneously reduce the amount of additive genetic variation by driving alleles to fixation. However, intense directional sexual selection may unmask hidden genetic variation conserved by stabilising selection and increase the mutational input for additive genetic variation, and additive genetic variation thus may be generated faster than it is depleted by selection (Pomiankowski and Møller 1995). Therefore, secondary sexual characters and other traits subject to intense directional selection may be predicted to have *more*, rather than *less*, additive genetic variation than characters not subject to directional selection. Additive genetic variation can be measured in terms of heritabilities, which are useful for predicting the response to selection (Falconer 1981). Since the heritability is the ratio between the additive genetic variance and the total phenotypic variance, the heritability may not reflect changes in genetic variation if accompanied by changes in the phenotypic variance generated by environmental effects. The additive genetic coefficient of variation is a relative measure of genetic variability standardized for character size (Charlesworth 1984). A recent review of the literature on the quantitative genetics of secondary sexual characters revealed large amounts of additive genetic variation (Pomiankowski and Møller 1995). This was the case independent of whether studies were based on cross-fostering or selection experiments, or if parent–offspring regression or sib analyses were used for estimating the genetic variation. The additive genetic coefficient of variation in secondary sexual characters was considerably larger than for comparable traits not subject to a history of directional selection, or for traits known to be subject to stabilising selection (Fig. 3.4).

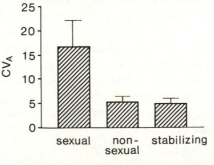

Fig. 3.4 Additive genetic coefficients of variation in secondary sexual characters, in comparable traits not currently subject to directional selection, and in traits known to currently be subject to stabilising selection. Values are means (+ s.e.). Adapted from Pomiankowski and Møller (1995).

The phenotypic variability of secondary sexual characters has been hypothesised to be higher than that of ordinary morphological traits (Alatalo *et al.* 1988). Recent comparative studies of the phenotypic variation of secondary sexual characters demonstrated higher phenotypic coefficients of variation than those of homologous traits in females of the same species, or in males of closely related species without an exaggerated secondary sexual character (Fig. 3.5; Møller and Höglund 1991; Cuervo and Møller 1997a). Extravagant feather ornaments in birds have evolved independently a large number of times, and an analysis of phenotypic variability in secondary sexual characters in relation to the mating system revealed a clear pattern (Fig. 3.5). The intensity of current sexual selection can be assumed to increase with increasing skew in male mating success from monogamy over facultative and obligate polygyny to lek mating systems. There was an initial increase in phenotypic variance, when morphological traits became the target of sexual selection, and the phenotypic variance was high in monogamous bird species with relatively weak sexual selection (Fig. 3.5; Cuervo and Møller 1997a). There was a clear decrease in phenotypic variation in secondary sexual characters in species with a lek mating system in which the intensity of sexual selection is supposedly intense (Fig. 3.5; Cuervo and Møller 1997a). This result can be interpreted in the following way: in species with an extreme mating skew, selection may have tipped the mutation–selection balance in favour of depletion of additive genetic variance resulting in a reduction in the additive genetic and phenotypic variance of the trait. The distribution of male mating success is presumably less skewed in socially monogamous species, which may result in the maintenance of a considerable amount of additive genetic and phenotypic variance (Cuervo and Møller 1997a).

Finally, the relationship between developmental instability and sexual selection was investigated for the same independent evolutionary events of extravagant secondary sexual characters. Relative fluctuating asymmetry in bilaterally symmetrical characters was considerably higher for secondary sexual characters than for homologous characters in females of the same species, or for homologous characters in males of closely related species without a sexual exaggeration of the morphological trait (Fig. 3.6; Cuervo and Møller 1997a). This result suggests that directional selection gives rise to an increased sensitivity of the developmental processes to the effects of adverse environmental conditions and the disruption of co-adapted genomic complexes that both generate developmental instability. The relationship between the intensity of sexual selection as estimated from the mating system and the relative asymmetry increased from non-ornamented species to ornamented, monogamous, and slightly polygynous species exhibiting higher fluctuating asymmetry than lekking species (Cuervo and Møller 1997a). Again, this result is consistent with sexual selection in lekking species having depleted the additive genetic variance, thereby rendering the traits less susceptible to the influences of environmental and genetic stresses.

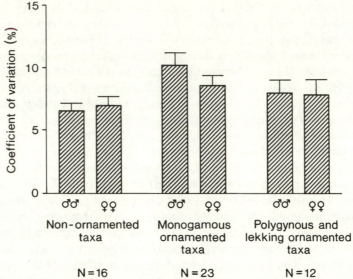

Fig. 3.5 Phenotypic coefficients of variation for male secondary sexual characters in monogamous and polygynous birds (morphological traits composed of feathers), in homologous traits in females of the same species, and in homologous, non-sexual traits in males and females of closely related species. Values are means (+ s.e.) and numbers are sample sizes (number of independent evolutionary events). Adapted from Møller and Höglund (1991) and Cuervo and Møller (1997a).

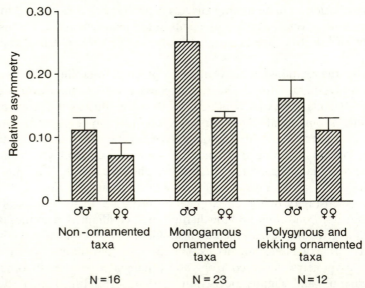

Fig. 3.6 Relative fluctuating asymmetry for secondary sexual characters in male monogamous and polygynous birds (morphological traits composed of feathers), in homologous traits in females of the same species, and in homologous, non-sexual traits in males and females of closely related species. Values are means (+ s.e.) and numbers are sample sizes (number of independent evolutionary events). Adapted from Cuervo and Møller (1997a).

3.5.2 *Life-history traits*

Life-history traits, like secondary sexual traits, are subject to persistent, intense directional selection. For example, individuals that reproduce early, lay large clutches, or produce many clutches are at a selective advantage (reviews in Roff 1992; Stearns 1992). The fact that large differences in life-history traits among closely related taxa can be found supports the claim that the directional selection pressures have actually given rise to evolutionary responses. Such persistent directional selection may be predicted to result in depletion of the additive genetic variation for life-history traits. However, a recent review has revealed that the amount of variation is larger, not smaller, than in other characters (Fig. 3.7; Houle 1992). This is exactly the situation that would give rise to increased additive genetic variation due to strong directional selection acting against genetic modifiers that control the expression of genes for life-history traits (cf. Pomiankowski and Møller 1995). Measures of developmental instability and canalization of life-history traits have revealed considerable degrees of instability, with greater canalization of traits more closely associated with fitness (Stearns and Kawecki 1994; Stearns *et al.* 1995), although this result depends on the degree of stablising selection (Wagner *et al.* 1997). Developmental mechanisms appear to buffer the phenotype against both genetic and environmental disturbance.

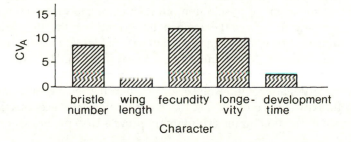

Fig. 3.7 Additive genetic coefficients of variation for life-history traits and a range of other traits supposed to be subject to less intense directional selection in *Drosophila*. Adapted from Houle (1992).

Price and Schluter (1991) provide another explanation for why life-history traits have high additive genetic variation but low heritability despite being subject to intense directional selection. Life-history traits have relatively low heritability and they are subject to seemingly consistent selection pressure and direction, but show little response to selection. Causal relationships between metric traits and life-history traits are expected to give rise to a very high environmental variation in life-history traits because they will be subject to all the environmental variation in the metric traits that affect them and additional environmental variation (i.e. they will express condition-dependence). This will result in a reduction of the heritability, but a low heritability of life-history traits does not appear to be based on a low additive genetic variance.

3.5.3 *Palaeontological evidence of directional evolution*

Evolutionary trends in body size or specific morphological characters of a particular lineage are common (Simpson 1944), and the frequent tendency for the average body size within lineages to increase over time is known as Cope's rule. Gradual increase in body size within a mammalian lineage during evolutionary history provides clear evidence for persistent directional selection, and if directional selection gives rise to increased phenotypic and genetic variance and developmental instability it should be possible to detect such increases. Several examples of increasing body size within a lineage are associated with an increased phenotypic coefficient of variation during the long periods of transition (reviewed in Møller and Pomiankowski 1993b; Hoffmann 1983). For example, Guthrie (1965) showed that the teeth of both fossil and recent specimens of those species of *Microtus* rodents that had increased in size since the Pleistocene were more variable than the teeth of those species that had remained unchanged in size. Unfortunately, there is no information in the palaeontological literature on developmental stability during such evolutionary transitions, although such information should be readily available from fossil remains.

The controversy over whether evolutionary history is best described as a result of gradual anagenesis or rapid change interspersed with long periods of stasis (the model of punctuated equilibria) is still very much alive (Gould and Eldredge 1993). Given that the mode of selection affects phenotypic variation, it may be possible to empirically assess the relative importance of the two models of evolution from studies of the temporal patterns of developmental instability and phenotypic variance (Møller and Pomiankowski 1993b). However, this will only be the case if new mutants have disrupted developmental homeostasis during a period of directional evolution much longer than the few generations in the example of the Australian blowfly. The rich palaeontological collections thus may provide more information on the evolutionary process than previously expected.

3.5.4 *Herbivore adaptations to plants*

Aphids are often specialist herbivores on one or a few plant species, and there is considerable evidence for adaptations among aphids to overcome the anti-herbivore defences of plants (e.g. Shaposhnikov 1987a). A particularly revealing example of how the phenotypic variation of an aphid changed in response to directional selection is demonstrated by an ingenious series of long-term experiments performed by Shaposhnikov (1965, 1966, 1987b). The aphid *Dysaphis anthrisci majkopica* was reared for 50 generations on its host plant *Anthriscus* and on a less suitable plant *Chaerophyllum bulbosum*. The rearing of the aphid on the less suitable host can be considered a directional selection experiment for adaptations to cope with a novel environment. The experiment resulted in an almost immediate increase in the phenotypic coefficient of variation in morphology and apparently strong viability selection arising from the novel association (Fig. 3.8). After only 9–10 generations, there was an

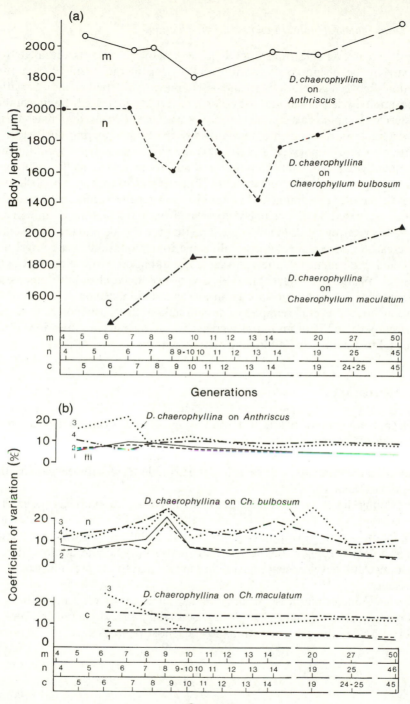

Fig. 3.8 (a) Body size and (b) phenotypic coefficient of variation in the aphid *Dysaphis anthrisci majkopica* reared for 50 generations on its host plant *Anthriscus* and two less suitable plant hosts *Chaerophyllum bulbosum* and *Ch. maculatum*. Adapted from Shaposhnikov (1965).

abrupt and irreversible transition to a new adaptive state as revealed by a decrease in the phenotypic variation and a drop in the intensity of natural selection. Between generations 10 and 46 the phenotypic coefficient of variation in morphology decreased and body size and fertility increased. Shaposhnikov transferred the aphid to a second new host *Chaerophyllum maculatum* which for the second time resulted in a new increase in the phenotypic coefficient of variation in morphology (Fig. 3.8). This increased phenotypic variation was subsequently reduced, while body size and fertility increased, as the aphid once more managed to adapt to the novel host plant. Unfortunately, Shaposhnikov estimated neither developmental instability nor genetic variances. Given that phenotypic variation is positively correlated with fluctuating asymmetry in *Drosophila* (Rasmuson 1960), it is possible that the aphids initially experienced an increase in developmental instability due to exploitation of the novel host plant, and that genetic modifiers subsequently stabilised the development of the disrupted phenotypes. This kind of divergence may also have important consequences for sympatric speciation as suggested in section 4.5.

In conclusion, several examples of directional selection, as reported in section 3.5, are associated with an initial increase in the phenotypic variation, developmental instability, and additive genetic variation.

3.6 Summary

- The prevailing mode of selection will have profound effects on the expression of additive genetic and phenotypic variation.
- Canalisation is under genetic control, as demonstrated by a number of selection experiments, and selection thus is able to modify the canalisation of a particular trait.
- Prevailing directional, stabilising, or disruptive selection affects the level of phenotypic variance of traits including the level of developmental instability.
- Selection experiments imposing directional, stabilising, or disruptive selection on traits support the notion that the phenotypic and additive genetic variance and developmental instability are related to the prevailing mode of selection.
- Similar conclusions arise from a review of a range of traits subject to directional evolutionary change (e.g. secondary sexual characters, life-history traits, herbivore adaptations to their host plants).

4

Adverse environmental conditions and evolution

4.1 Introduction

Stress is a state well known to most readers of this book. There are numerous definitions (most people have their own), but here we adopt the definition of stress as a state caused by a factor imposing a potentially lasting injurious change to a biological system. This implies that a range of phenomena such as starvation, irradiation, and hypoxia can be considered to be stressors. Stress is a common state affecting free-living organisms, and it thereby imposes important selection pressures on individual organisms, but also affects the development of phenotypes upon which selection acts.

Stress generally results from deviations from normal, local conditions to which an organism is adapted. In other words, stress can be considered to arise from suboptimal environmental conditions. This concept of optimal conditions for biological processes such as thermoregulation, biochemical processes, and growth and reproduction has been advocated by Peter Parsons as a framework on which ecological and evolutionary problems can be studied. In section 4.2, we briefly review the literature on this optimality approach and its biological consequences. We also address the consequences of the variation in the ability of organisms to cope with suboptimal environmental conditions, and the potential costs of such tolerance.

A host of potentially evolutionarily important processes can be affected by suboptimal conditions experienced by individuals. These include phenomena associated with the genetic variation that may give rise to evolutionary change: increased rates of recombination, mutation, and transposition, and increased additive genetic variance. The phenotypic consequences of suboptimal conditions are the other side of the coin, potentially resulting in evolutionary change: an elevated phenotypic variance on which selection can act. The outcome will often be more intense selection under suboptimal environmental conditions (section 4.3).

Given that the effects of suboptimal conditions are ubiquitous in nature, and that genetic and phenotypic conditions are altered by such conditions, which are the conditions particularly permissive for evolutionary change? The distributional limits of species are often closely correlated with physical parameters such as temperature and rainfall, and an energetics approach to evolution may thus be appropriate. Organisms are not simply the victims of environmental conditions because the conditions under which organisms live are affected by habitat features and behaviour. Organisms have to allocate a very large fraction of their metabolism to cope with extreme environments, and little excess energy will be available for the stable development of phenotypes and for reproduction. Availability of excess energy for development of stable phenotypes and reproduction will, particularly, be found under moderately suboptimal habitats, and these may be the conditions most conducive to an evolutionary response to selection. These lines of reasoning obviously also have important implications for conservation biology and the ecological and evolutionary consequences of global change (section 4.4).

Speciation and cladogenesis has been a central theme in evolutionary biology since Darwin's 'Origin' was published in 1859. This interest in the study of speciation has increased in recent years as shown by a number of reviews (e.g. Otte and Endler 1988; Coyne 1992; Rice and Hostert 1994). The speciation process is characterised by relatively rapid evolutionary change associated with reproductive isolation. Such rapid evolutionary change should have a number of consequences including a decrease in developmental stability (section 4.5).

The last couple of decades have seen a revival of an old controversy evolving on the relative role of gradual and rapid evolutionary change (reviewed in Gould and Eldredge 1993). This controversy in fact has a much longer history dating back to the last century when T. H. Huxley stated that Charles Darwin 'loaded' his theory of natural selection with an unnecessary difficulty by his emphasis of gradualism (Huxley 1860). This controversy re-emerged when the geneticist R. B. Goldschmidt (1940, 1955) proposed that major evolutionary innovations might arise as a consequence of extreme environmental conditions. This idea met with much scepticism from most parts of the scientific community. However, it was clear from epigenetic reasoning that this process might in fact account for major evolutionary innovations. Interestingly, recent studies of transposable elements primarily in *Drosophila*, but also in other organisms, have provided a mechanism by which such major changes may arise (section 4.6).

4.2 Consequences of adverse environmental conditions

4.2.1 *Energetic consequences of adverse environmental conditions*

Energy and other resources necessary for growth, maintenance, survival, and reproduction are usually limited. Such suboptimal conditions generate stress, if stress is viewed as arising from any factor causing a potentially injurious change to a biological system (Parsons 1991b). Actually, the adenylate energy charge, which is a ratio based on the concentrations of ATP, ADP, and AMP, has been suggested as a direct measure of stress (Ivanovics and Wiebe 1981). Measures of stress at the proximate level include increased metabolic rates, elevated concentrations of specific hormones such as corticosterone in vertebrates (Selye 1974), and higher concentrations of various metabolites such as ADP and AMP. Elevated metabolic rates may result in allocation of less energy to other important activities like maintenance of developmental homeostasis. In a similar vein, elevated concentrations of corticosterone may cause lasting damage to tissue and eventually death.

Deviations from local conditions under which an organism is usually living generally result in stress. This phenomenon of optimal performance at the usual environmental conditions and poorer performance at deviant environmental conditions is termed hormesis. One of the most extraordinary, but also most controversial, cases of hormesis concerns the negative effects of low-dose radiation on animals. Hormesis effects are apparent from exposure to low doses of radiation (Luckey 1980). Optimal effects of radiation in terms of performance at intermediate, normal doses can only be explained by positive effects of radiation that balance the negative effects of mutations. There are several ways in which this might by possible (Sagan 1989). First, the concentration of free-radical scavengers increases after low-dose radiation. This enhanced production might increase defences against free radicals arising from other environmental mutagens. Another possibility is that radiation-exposed DNA may be more rapidly repaired after subsequent exposure to mutagens. Immune cell production may also be enhanced by low-dose radiation. Kuzin (1993) established direct and indirect effects of low-dosage radiation on membrane receptors. Radiation activates membrane-bound enzymes which control many vitally important processes, and such positive effects of exposure to low-dose radiation may balance any negative effects.

The relationship between environmental conditions and optimal performance during ontogeny is well-studied at the metabolic level. The first study indicating an optimum performance for development dates back to Crescitelli (1935) who found that total oxygen consumption of pupae of the wax moth *Galleria mellonella* increased dramatically at both high and low temperatures. Minimum metabolism can therefore be used as a criterion for determining the optimal temperature for development (Winberg 1936). Subsequent experiments on a range of insect and fish species have found that minimum metabolism for a given developmental stage corresponds to the most favourable temperature conditions for development of a symmetric phenotype in poikilothermal animals. For example, the level of asymmetry and the frequency of phenodeviants in the lizard *Lacerta agilis* reach a minimum at intermediate temperatures which are close to the optimal temperature of embryo development (Fig. 4.1; Alekseeva *et al.* 1992). Developmental processes progress most precisely at this optimal temperature giving rise to the smallest number of developmental errors, while deviations from the optimum in either direction disrupt the precision of canalisation (Ozernyuk 1989; Ozernyuk *et al.* 1992). Temperature has been used as an example here because energy metabolism is such a sensitive indicator of the state of a developing organism with respect to factors of the external environment. Similar arguments can readily be made for other aspects of the physiology of organisms, with deviations from the modal state being suboptimal and hence resulting in reduced abilities to control developmental pathways (e.g. Zakharov 1989).

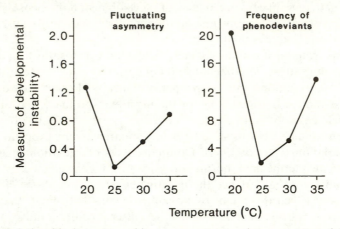

Fig. 4.1 Relationship between ambient temperature and asymmetry and frequency of phenodeviants in the lizard *Lacerta agilis*. Adapted from Alekseeva *et al.* (1992).

Natural populations are usually exposed to suboptimal conditions, and the consequences include elevated metabolic costs (Parsons 1993c). Preferred habitats should be located in environments corresponding to the minimum total energy expenditure. This is supported by field studies of *Drosophila* in relation to temperature and humidity (Parsons 1993d) and studies of performance in lizards (Garland and Carter 1994). Preferred habitats by individual organisms are a compromise between the intensity of stress experienced in the habitat, the magnitude of environmental fluctuations resulting in poor performance, and the amount of energy available (Parsons 1994b). Organisms should evolve broader tolerance features defined by such physical characteristics of the environment.

Energetic arguments are central for understanding the conditions giving rise to deviant and asymmetric phenotypes under inferior environmental conditions. Evolutionary change involves a directional response to selection and a change in the frequency of alleles. As shown in Chapter 3, prevailing directional selection generates the conditions for high developmental instability. How is suboptimal environmental conditions and thus availability of energy related to the potential for evolutionary change? There is a continuum of situations from near-optimal habitats to habitats with extreme environmental conditions close to what is tolerable, and these are directly associated with energy expenditure and thus inversely with the amount of energy available for development, growth, maintenance, and reproduction. Conditions allowing evolutionary change are of two kinds. First, moderately suboptimal habitats will stimulate gradual evolutionary changes permitting a continuous tracking of environmental conditions (Parsons 1991b, 1993e, 1994a). Second, major evolutionary innovations may mainly occur in response to severe environmental conditions that will be sudden and energetically expensive and will rely on unmasking of genetic variability and loss of canalisation (Fraser 1962; Levin 1970a, b; Hoffmann 1983; Kieser 1987, 1993). Peter Parsons has argued convincingly that only moderately suboptimal environmental conditions will allow permissible amounts of energy available for the stable development of phenotypes, reproduction, and hence evolutionary change. If availability of energy is the limiting factor for stable developmental pathways, then when energy is limiting under extreme environmental conditions, or in widely fluctuating environments, little evolutionary change is expected. Organisms will generally have a certain range of environmental conditions that permit optimal performance with respect to growth, survival, and reproduction, while less and particularly more extreme environmental conditions at the most allow existence due to the energetic costs of stress tolerance (Fig. 4.2; Parsons 1994a). Maximum evolutionary rates may occur in moderately suboptimal regions of the distribution of a species where the phenotypic variability would be higher than in benign regions, and the metabolic costs of tolerance of suboptimal conditions are not unduly restrictive. Phenotypic variability, upon which selection may act, is expected to be the highest under moderately stressful environmental conditions. An example of organisms exposed to extremely stressful environmental conditions are relict species and living fossils, and they are characterised by extreme evolutionary conservatism. Such relict species (for example, horseshoe crabs *Limulus* that are able to endure water that varies tremendously in salinity, temperature, and oxygen level, and the brachiopod *Lingula* that inhabits similarly stressful marine environments) spend a considerable amount of their energy budget on stress tolerance and have undergone very little evolutionary change during hundreds of millions of years (Ward 1992). A great capacity for 'physiological plasticity' that permits stress tolerance thus provides an alternative to the need to evolve, at least for some species. If energy availability is limiting under extremely adverse environmental conditions due to the costs of stress tolerance, the evolutionary change that may lead to speciation and adaptive radiation would happen in stable, moderately stressful environments.

An increase in the adversity of environmental conditions could underlie extinctions of environmentally intolerant endemic species (Parsons 1993e). Synergistic interactions among environmental conditions should increase the likelihood of extinctions, where simultaneous exposure to a range of adverse environmental conditions considerably increases the costs of tolerance. Evidence for a role of stress resistance in the ability of organisms to cope with adverse environmental conditions during a mass extinction event comes from some marine organisms that survived the Cretaceous–Tertiary mass extinction because of mechanisms such as tolerant life-cycle stages with low metabolic rates (Kitchell 1990; Briggs 1991).

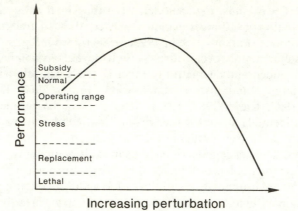

Fig. 4.2 Effects of environmental conditions on performance in a hypothetical organism. Adapted from Hoffmann and Parsons (1991).

In conclusion, organisms appear to perform optimally under those environmental conditions most usually encountered within the normal distributional range, while deviations from such conditions generally result in increased metabolic costs and an inability to follow a stable developmental pathway. Therefore, moderately suboptimal environmental conditions are assumed to be permissive for gradual evolutionary change.

4.2.2 *Tolerance of adverse environments and its costs*

If adverse environmental conditions are ubiquitous, organisms should have developed ways in which to avoid or minimise the costs of tolerating such conditions. Such tolerance could have a behavioural or physiological basis that could be partly genetic. The most readily available way of reducing the impact of the environment is by behavioural means. If particular environmental conditions are adverse, avoidance of these conditions by, for example, habitat choice or thermal preference will reduce the impact of stress. Behavioural modification can alter the effects of environmental conditions on individuals (Wcislo 1989). Complex feedback mechanisms exist between changes in behaviour and morphology, both at the ontogenetic and the phylogenetic levels.

Physiological modifications as a result of the release of stress-related hormones like corticosterone are another way of reducing the impact of adverse environmental conditions because it diminishes the consequences of being stressed (Selye 1974; Harvey *et al.* 1984). Endocrine responses will result in physiological modifications that allow the individual to survive the adverse conditions at a minimum cost. For example, adverse weather conditions give rise to elevated levels of circulating corticosterone in birds, and this apparently promotes feeding behaviour during conditions that may result in starvation (Astheimer *et al.* 1992). A rapid return to normal body weight is thus ensured after food has been located. If an individual is prevented from avoiding adverse conditions and hence suffers from continuous exposure to high circulating levels of corticosterone or similar hormones, this may result in serious damage at the physiological level, or eventually death (Selye 1974; Harvey *et al.* 1984).

Individuals may vary for genetically based behavioural and physiological responses to adverse environmental conditions. Selection experiments on *Drosophila* suggest that the ability of larvae to cope with elevated levels of alcohol in the food substrate has an additive genetic basis (Hoffmann and Parsons 1989). Interestingly, the ability to tolerate a suboptimal food substrate resulted in a correlated ability to cope with desiccation. Although this experiment is based on a particular model system with its own peculiarities, the results may suggest that environmental tolerance of high alcohol levels and desiccation, respectively, reflects a general ability to cope with adverse conditions rather than a specific ability to cope with a particular kind of environmental condition. Subsequent selection experiments for increased resistance to desiccation demonstrated high realised heritabilities of tolerance (Hoffmann and Parsons 1993). In *Drosophila melanogaster* the heritability was greater at higher selection intensities. This finding implies a more rapid response to selection under more intense selection because the two components defining the response (intensity of selection and heritability) both increased under more stressful environmental conditions. Selected lines were also more resistant to starvation and toxic ethanol than the control lines, again providing support for the notion of a general stress resistance syndrome. The experiments also provided some evidence for the mechanisms generating resistance. Selected lines of flies lost water at a slower rate and had reduced activity levels, but did not differ in wet or dry body mass or water content.

If tolerance provides protection against adverse environmental conditions, and if there is plenty of additive genetic variance for tolerance, why are not all individuals better able to cope with an adverse environment? There are a number of potential reasons, but two explanations are particularly likely. First, given that organisms are better adapted to cope with local environmental conditions, gene flow among environments may prevent perfect adaptation to current conditions. The second reason is that there is an antagonistic pleiotropic cost of tolerance. Selection experiments on *Drosophila* have shown that tolerance of adverse environmental conditions is acquired at a fitness cost (Parsons 1990c; Krebs and Loeschcke 1994a; Hoffmann 1995). Acclimation experiments in bacteria and *Drosophila* have shown that short- and long-term acclimation to adverse environments is possible, but at a fitness cost since reproductive rates are reduced after acclimation (Krebs and Loeschcke 1994b; Leroi *et al.* 1994). The benefits of tolerance are therefore balanced by the antagonistic pleiotropic costs of tolerance.

There are some striking parallels between the costs of tolerance of adverse environmental conditions and the costs of pesticide and parasite resistance (McKenzie and Batterham 1994; Polak 1997a). All three phenomena provide individuals with a considerable fitness advantage in the presence of the selective agent, but individuals pay a cost for resistance in the absence of the selective agent. There is evidence of increased individual morphological asymmetry caused by resistance alleles for both pesticide and parasite resistance. Such antagonistic pleiotropy can maintain additive genetic variation for resistance.

4.3 Evolutionary consequences of adverse environments

Optimal environmental conditions are by definition the least adverse for organisms, and tolerant forms suffer the least negative effects from environmental deviations from the optimum. Adverse environmental conditions can be very powerful factors in evolution apart from the evolution of tolerance. Adverse environments have been suggested to affect the level of recombination, mutation, and transposition, the additive genetic and phenotypic variation, the intensity of selection, and hence the speed of micro-evolutionary change. Of course, this does not imply that organisms exposed to more benign environments will not have considerable additive genetic variance. These topics are briefly reviewed in sections 4.3.1–7.

4.3.1 *Recombination and adverse environmental conditions*

Recombination arises as a consequence of crossing over during meiosis, and recombination rates can be estimated from the frequency of chiasmata. Rates of recombination show considerable variation among species, and this may be of evolutionary importance because high rates of recombination may relate to the ability to cope with particular environmental conditions or fight antagonists such as parasites (review in Michod and Levin 1988).

Interestingly, rates of recombination change readily in response to a range of adverse environmental conditions (Table 4.1; Parsons 1988). For example, recombination changes in a ∪-shaped fashion with temperature above and below normally encountered developmental temperatures in *Drosophila melanogaster* (Plough 1917, 1921). Suboptimal nutrition increases recombination in fruitflies (Neel 1941), and behavioural stress caused by exposure to open field situations without cover increases recombination in mice (Belyaev and Borodin 1982). In *Drosophila* the recombination rate is higher in a novel environment than in a standard laboratory environment (Lawrence 1963). Genetic variability generated by recombination in marginal habitats may therefore increase (Levin 1970a; Parsons 1988). The increase in recombination rate caused by deviations from normal environmental conditions can be viewed as a response that will facilitate adaptation to environmental change. Alternatively, increased recombination rates can be considered non-adaptive responses to novel environmental conditions that may render DNA molecules unstable.

Disruption of co-adapted genomic complexes also has an important effect on recombination rates, and the effects of several different disrupting factors tend to enhance the rate of recombination beyond the predicted cumulative effects

(Parsons 1988). Structural heterozygosity due to inversions in one part of the genome tends to increase recombination in the remainder of the genome (Suzuki 1963; Valentin 1972; Lucchesi 1976). Substantial recombination should be inducible under combinations of karyotypes and environments deviating from prevailing circumstances.

Adverse environmental conditions will thus tend to increase the frequency of novel gene combinations, some of which may prove superior under such environmental conditions.

Table 4.1 Rates of recombination in relation to environmental conditions. A '+' signifies an increase in recombination along the environmental gradient, and '−' signifies a decrease. Only the sign of the change is indicated. Adapted from Parsons (1988) and later references.

Species	Environmental factor	Effect	Reference
FUNGI			
Neurospora crassa	temperature	+	Rifaat (1959)
Neurospora crassa	temperature	+	Towe and Stadler (1964)
Neurospora crassa	temperature	+	McNelly-Ingle *et al.* (1966)
Neurospora crassa	infection by parasitic DNA	+	Selker (1990)
PLANTS			
Allium ursinum	temperature	+	Loidl (1989)
Sordaria brevicolis	temperature	+	Lamb (1969)
Coprinus lagopus	temperature	+	Lu (1969, 1974)
Shizophyllum commune	temperature	+/−	Stamberg and Simchen (1970)
ANIMALS			
Various micro-organisms	temperature	+	Kushev (1971)
Caenorhabditis elegans	temperature	+	Rose and Baillie (1979)
Drosophila melanogaster	temperature	+	Plough (1917, 1921)
Drosophila melanogaster	temperature	+	Stern (1926)
Drosophila melanogaster	temperature	+	Graubard (1932)
Drosophila melanogaster	temperature	+	Hayman and Parsons (1960)
Drosophila melanogaster	temperature	+/−	Chandley (1968)
Drosophila melanogaster	temperature	+	Grell (1978a,b)
Drosophila melanogaster	starvation	+	Neel (1941)
Drosophila melanogaster	ethanol	+	Kilias *et al.* (1979)
Mus musculus	behavioural stress	+	Belyaev and Borodin (1982)

4.3.2 *Mutation and adverse environmental conditions*

Novel genetic material arises from mutations in the germ line, and although the majority of mutations are deleterious, point mutations may rapidly result in novel adaptations that allow organisms to cope with a hostile environment. Examples of such rapid change include the evolution of pesticide resistance and resistance to parasites (e.g. McKenzie and Batterham 1994; Wakelin and Blackwell 1988).

The effects of adverse environmental conditions and mutations often have some striking similarities. Experimental exposure of a range of organisms to abnormal conditions during critical periods of development often results in the production of phenotypes with deviant morphology. Interestingly, such phenodeviants, termed phenocopies, often resemble to a remarkable degree the phenotypes produced by particular mutations (reviewed in Ancel 1950; Goldschmidt 1955). This provides evidence for the idea that mutations affect biochemical or physiological pathways, and that these same pathways also can be affected in a similar way by a range of adverse environmental conditions (Goldschmidt 1920; Zeleny 1933). Subsequent research has firmly established this mechanism.

Adverse environments experienced by organisms are often associated with considerable changes in the hormonal balance mediated through the central nervous system (Selye 1974). Intracellular homeostasis may also be affected by other physiological changes associated with environmental adversity. Such disturbances have by themselves been hypothesised to increase the mutation rate (Kerkis 1940; Lobashov 1976). Since adverse conditions eventually lead to severe disturbance of the physiological balance, it is reasonable to assume that the frequency of mutations will increase under such conditions. Deviations from physiologically normal levels of corticosteroids result in dramatic increases in the frequency of chromosome aberrations in both somatic and germ cells (Kerkis 1975; Serova and Kerkis 1974). Even minor increases in the adversity of environmental conditions, for example, as obtained by placing mice in an open field experiment, result in a significant increase in the frequency of chromosomal aberrations in bone marrow cells (Seredenin *et al.* 1980). Hence, there appears to be a direct link between hormonal states induced by suboptimal environmental conditions and mutation rates of germ cells. A particularly well-studied example of the effects of adverse environmental conditions on mutagenesis concerns the SOS system which stimulates mutation rates in *Escherichia coli* (Taddei *et al.* 1996)).

Mutation rates increase under a range of environmental conditions that deviate from the most commonly encountered environments. A list of categories of mutagens that can be considered adverse environmental conditions is provided in Table 4.2. Mutagens are of natural origin such as extreme climatic conditions, teratogens found in food, chemical substances in the environment, radiation, and artificial agents such as pesticides (Loveless 1966; Auerbach 1976). An extensive review of experiments on micro-organisms, fungi, plants, and fruitflies revealed a positive relationship between mutation rate and temperature (Lindgren 1972). Environmental factors resulting in elevated mutation rates are most often found in extreme environments such as those near the distributional limit, in marginal habitats, and in aberrant years. These are the conditions promoting an increased mutational input to the genome.

Table 4.2 Mutation rates in relation to environmental conditions. A '+' means an increase in mutation along the environmental gradient. Only the sign of the change is indicated.

Taxa	Mutagenic agent	Effect	Reference
Bacteria, plants, animals	temperature	+	Drake (1970), Lindgren (1972)
Virus, bacteria, fungi, plants, animals	ultraviolet and visible light	+	Auerbach (1976)
Virus, bacteria, fungi, plants, animals	ionising radiation	+	Auerbach (1976), Sankaranarayanan (1982)
Virus, bacteria, fungi, plants, animals	alkylating agents	+	Loveless (1966), Auerbach (1976)
Virus, bacteria, fungi, plants, animals	other chemicals	+	Auerbach (1976)
Bacteria, plants	nutritional condition	+	Durrant (1971), Cullis (1984)
Bacteria	starvation	+	Cairns *et al.* (1988), Hall (1988, 1990, 1991)
Mammals	corticosteriods	+	Kerkis (1975), Serova and Kerkis (1974)
Fungi	infection with parasitic DNA	+	Selker (1990)
Mammals	open-field exposure	+	Seredenin *et al.* (1980)

A range of factors are known to increase mutation rates as shown in Table 4.2. Adverse environmental conditions increase the mutational input, and this can be considered either an adaptive consequence of adverse environmental conditions or merely a non-adaptive by-product of damage caused by adverse conditions. The adaptive explanation has been termed directed mutations as caused by specific environmental conditions. Such mutations appear to generate genetic variability that allows the organism to cope with the adverse condition in question (e.g. Cairns *et al.* 1988). Possible mechanisms giving rise to apparently directed mutations are reviewed by Jablonka and Lamb (1995). This theory of directed mutation is highly controversial, and neither the frequency of mutations nor the molecular models developed to explain such apparently directed mutations are able to stand critical scrutiny (Lenski and Mittler 1993; Sniegowski and Lenski 1995). Effects of specific suboptimal conditions sometimes increase mutation rates, and mutagenesis in *Escherichia coli* in a structured environment, but not in an unstructured environment (agitated liquid cultures), suggests that the mutation frequency of chromosomal genes depends both on cAMP and the SOS response (Taddei *et al*. 1996). These findings suggest that the mutation rate of the whole chromosome is under the control of the bacterial genome which triggers a specific mutagenic response under stressful environmental conditions (Taddei *et al*. 1996). Such an increased mutation rate is adaptive when there is a misfit between the genotype and the environment, and selection for adaptive mutations therefore selects indirectly for the molecular mechanisms that generated them (Taddei *et al*. 1996). While there is no evidence of evolution being facilitated in a particular direction thereby allowing organisms to cope with the causes of the environment being experienced as suboptimal, there is experimental evidence that molecular mechanisms of adaptive mutations under adverse environmental conditions respond specifically, albeit 'blindly', to environmental challenge.

The effects of mutations are generally accepted to be neutral or detrimental in most cases. Deleterious mutations often severely disrupt the expression of the phenotype, as suggested by Schmalhausen (1949), who emphasised an important role of mutations in expressing deviant phenotypes. Such deviant phenotypes often resemble developmentally unstable individuals, raising the possibility that a certain amount of developmental instability encountered in nature may have a direct origin in mutations.

4.3.3 *Transposition and adverse environmental conditions*

Transposable elements are so-called jumping genes that may move from one part of the genome to another (McClintock 1984). These elements were first recorded from maize *Zea mays* and demonstrated the ability to move around the genome and affect the expression of the genes into which they inserted. Transposable elements are therefore able to produce linkage groups and changes in chromosome complements. During development some transposable elements make regular, heritable, and reversible transitions between active and passive stages. These changes in state are caused by changes in the level of methylation of the transposable elements (Fedoroff 1989). Obviously, the mere existence of transposable elements leads to generation of new genetic variation. Interestingly, a couple of adverse environmental conditions have been shown to increase the transcriptional activity of a particular locus and its rate of transposition (Table 4.3). This is for example the case for a mobile element in *Drosophila melanogaster* following a heat shock, which is known to change gene activity dramatically (Ratner *et al*. 1992). Transposition frequency in the germ line of heat-shock treated males was increased by up to two orders of magnitude. The rate of transposition was affected by the transcriptional activity of potential sites of insertion and by the chromatin structure of the transposable element itself (Fedoroff 1989). Adverse environmental conditions may thus increase the rate of transposition to loci that are particularly active during such environmental states.

Table 4.3 Transportation rate in relation to environmental conditions. A '+' means an increase in transposition along the environmental gradient. Only the sign of the change is indicated.

Species	Environmental gradient	Effect	Reference
Drosophila melanogaster	temperature	+	Ratner *et al.* (1992)
Bacteria	starvation	+	J. A. Shapiro (1984, 1992), Hall (1990)
Zea mays	demethylation	+	Fedoroff (1989)

4.3.4 *Additive genetic variation and adverse environmental conditions*

The additive genetic variance is one of the components determining the heritability of a trait (in the broad sense), the other being the phenotypic variance:

$$h^2 = V_G / V_P,$$

where V_G is the total genetic variance and V_P is the phenotypic variance (Falconer 1981). Heritability in the narrow sense, which is used in the present context, is defined as the additive genetic variance (V_A) divided by the phenotypic variance:

$$h^2 = V_A / V_P.$$

Heritabilities reflect the proportion of phenotypic variance which can respond to selection, such that

$$R = h^2 \times S,$$

where R is the response to selection, h^2 is the narrow-sense heritability, and S is the intensity of selection measured in units of standard deviations (Falconer 1981). The effects of adverse environmental conditions on the additive genetic variance are treated in this section; whereas the influence of adverse conditions on phenotypic variance and the intensity of selection are discussed in sections 4.3.5 and 4.3.6, respectively.

Numerous studies have directly assessed the relationship between environmental conditions, relative to the optimum, and the expression of genetic variance of tolerance factors which permit tolerance of particular environmental conditions. Extremely adverse conditions may bring about realisation of potentials for coping with adversity that are not manifest under more benign environmental conditions. For example, elevated social stress caused by increased density during pregnancy in mice resulted in elevated estimates of additive genetic variances and heritabilities of reproductive parameters (Belyaev and Borodin 1982). Characters that increased compared with controls included pre-implantation loss and litter size and the size of organs involved in stress tolerance such as thymus and adrenals. Similar experiments on rats found elevated genetic variances for the secretory activity of adrenals and for arterial blood pressure under adverse conditions as compared with control situations (Markel and Borodin 1980). Analogous results for corticosteroid concentrations in cattle were reported by Eisner and Reznichenko (1977). In other words, adverse environmental conditions increase the genetic variance and the heritability of traits that facilitate tolerance of such environments and, given the influence of a persistent adverse environment in a particular direction, this will allow for fast responses to selection as well as allow continuous tolerance during the evolutionary transition.

Estimates of heritabilities are often based on the regression of offspring character values on those of one or both parental values, on correlations among sibs, or on the outcome of selection experiments (Falconer 1981). Estimates may be biased in a number of different ways including bias due to maternal or common environment effects. Such effects can be partially controlled in common environments such as the laboratory, but it is important to note that the phenotypic variance also may be affected by the environmental conditions, as shown in section 4.3.4, and the estimate of heritability thus may be subject to another kind of bias under standardised laboratory conditions. It is also important to remember that the heritability can only be used for predicting the response to selection in the environment in which it has been estimated (Falconer 1981). In other words, if the heritability has been estimated in a laboratory setting, this value cannot be extrapolated to the field condition for predicting response to selection.

Two different points of view have been raised concerning the relationship between adversity of environmental conditions and heritability. Parsons (1974, 1983) has advocated the point of view that adverse conditions increase the magnitude of heritability, while others have expressed the opposite view (Johnson and Frey 1967; Blum 1988). Several studies have estimated additive genetic variation and heritability under a range of environmental conditions, and the results are summarised in W4.1 on the Web page. For example, a laboratory study of the rice weevil *Sitophilus oryzae* showed an increased additive genetic variance and heritability in the duration of the developmental period when introduced into a novel, toxic environment as compared with a control environment without the presence of the toxic substance (Holloway *et al.* 1990). Similar results have been obtained in a number of other studies. The overall pattern was that the genetic variance and/or the heritability decreased in poor environments in 37% of 52 studies, increased in 60% of the studies, and remained constant in 4%. These patterns differed somehow among groups of organisms with 94% of 16 studies of invertebrates showing an increase, 54% of 24 studies of vertebrates showing an increase, and only 25% of 12 studies of plants showing an increase.

Environments in some of the studies are not easily classified with respect to the optimum, and it is therefore difficult to determine whether the outcome of an experiment is consistent with an increase or a decrease in additive genetic variance and heritability under more adverse environmental conditions. We have generally assumed that a more variable environment is more adverse than a constant environment because physiological adjustment to changing environmental conditions should on average be more costly. This is not necessarily the case if organisms have evolved specific adaptations to fluctuating environments, and if a constant environment therefore is experienced to be suboptimal by the organism.

There is considerable variation in the relationship between environmental conditions, additive genetic variance, and heritability in the studies reviewed in Table W4.1, and a number of reasons for the lack of consistency can be stated. First, heritabilities estimated over a range of environments have often been compared without the optimum environmental conditions being defined. Second, studies based on the comparison of additive genetic variance and heritability from different varieties do not allow distinction between additive and dominance variance, and heritabilities from such studies thus will be biased. Finally, many experiments only investigated environmental conditions at two levels of an environmental factor, and this will obviously lead to little ability to detect changes in variance components at environmental extremes. Further studies in which the effects of well-defined environmental conditions on the amount of additive genetic variance are obviously needed.

Rapid changes in the additive genetic variation under adverse environmental conditions may be due to directional selection changing the modifying properties of genes (Møller and Pomiankowski 1993a, b). Every gene may be able to act as a modifier or have its phenotypic action modified, depending on allelic conditions. Most phenotypic traits develop from the actions of more than one gene, and the great majority of genes encode products that act in more than one developmental or physiological context. For any two or more gene pathways, any gene in that pathway can, in principle, act as a modifier of hypomorphic mutations of other genes in that pathway. In effect, the set of genes in a genotype can be posited to be virtually co-extensive with the set of potential modifiers. A simple model for the modifier action of genes is based on how the gene products form heterochromatin (Locke *et al.* 1988) and thereby inactivate other genes. Genes with modifying action may code for components of a multimeric complex that spreads with the addition of new units. The number of copies of each gene affects the concentration of each component, and hence the spread of heterochromatin into neighbouring regions of the chromosome. While stabilising selection will result in inactivation of genes by means of heterochromatin, directional selection will have the opposite effect of selecting against parts of the chromosome that have become inactivated by heterochromatin. Evidence for this model is provided by Tartof *et al.* (1989) and Tartof and Bremer (1990).

An alternative explanation for increased heritabilities, and the relative strength of the additive genetic component of the phenotypic variance under extreme environmental conditions, is that such conditions promote the average value of the environmental contribution to the phenotype (Ward 1994). If the phenotypic value is the outcome of metabolic flux through a complex biochemical network, then the function relating phenotype to genotype is always a nonlinear function of the additive genetic component (see also Moreno (1994) for examples of nonlinear relationships between phenotype and genotype). Such nonlinearity will invariably result in the parent–offspring regression coefficient being a function of the genetic and environmental parameters.

4.3.5 *Phenotypic variance and adverse environmental conditions*

The phenotypic variance is the second component affecting the magnitude of the heritability (see section 4.3.4). Furthermore, an increase in the phenotypic variance may increase the probability of a shift from one adaptive peak to another, at a rate higher than that predicted from Wright's model of shifting balance (Whitlock 1995). A summary of the effect of environmental conditions on phenotypic variability is provided in Table 4.4. The general pattern emerging from the studies available is that the amount of phenotypic variation increases in response to deviations from optimal environmental conditions. For example, this is the case for the variation in the number of vertebrae among a range of fish species exposed to different developmental temperatures (reviewed in Tåning 1952). Another example concerns body size of barn swallow *Hirundo rustica* offspring raised in nest environments differing in their abundance of experimentally manipulated levels of ectoparasitism (Møller 1990b). Nestlings raised in poor environments with lots of haematophagous mites were phenotypically more variable than those raised in parasite-free nests. A total of 73% of the 30 studies reported an increase in phenotypic variance under more adverse environmental conditions. This pattern appeared to differ between plants and animals with 30% of 10 studies of plants showing an increase and 95% of 20 studies of animals demonstrating an increase.

Therefore, it is generally thought that variation in the expression of phenotypes increases as environmental conditions deteriorate and environments deviate from the optimum.

4.3.6 *Selection under adverse environmental conditions*

Selection arises as an effect of a discrepancy between environmental conditions experienced by an individual and the conditions optimal for maintenance, growth, mating, reproduction, and survival. The intensity of selection is generally believed to increase as the deviation from the optimal conditions increases (Endler 1986). Therefore, the intensity of selection is predicted to increase under deviant environmental conditions with rapidly increasing intensities as the deviance increases. This relationship is so obvious that examples may seem superfluous. Examples of increased selection as deviations in environmental conditions increase include larval survival in relation to

Table 4.4 Phenotypic variance in relation to environmental conditions. A '+' means an increase in phenotypic variance along the environmental gradient, a '0' means no change, and a '−' a decrease. Only the sign of the change is indicated.

Species	Environmental gradient	Effect	Reference
PLANTS			
Hordeum vulgare	climate	−	Allen *et al.* (1978)
Avena sativa	climate	−	Allen *et al.* (1978)
Avena sativa	nutrition, date	−	Johnson and Frey (1967)
Linathus androsaceus	time of season	+	Huether (1969)
Arabidopsis thaliana	temperature	+	Langridge and Griffing (1959)
Triticum aestivum	climate	−	Allen *et al.* (1978)
Phaseolus vulgaris	climate	−	Allen *et al.* (1978)
Linum catharticum	climate	−	Allen *et al.* (1978)
Brassica campestris	drought	−	Richards (1978)
Brassica napi	drought	+	Richards (1978)
ANIMALS			
Drosophila melanogaster	varying environment	+	Mackay (1980)
Drosophila melanogaster	varying environment	+	Gibson and Bradley (1974)
Drosophila melanogaster	temperature	0	Kindred (1965)
Drosophila melanogaster	temperature	+	Waddington and Robertson (1966)
Drosophila simulans	temperature	+	Murphy *et al.* (1983)
Tribolium castaneum	humidity	+	McNary and Bell (1962)
Tribolium castaneum	nutrition	+	Hardin and Bell (1967)
Dysdercus bimaculatus	desiccation	+	Derr (1980)
Various species of fish	temperature	+	Hubbs (1922)
Various species of fish	temperature, salinity, oxygen, carbon dioxide	+	Tåning (1952)
Fundulus heteroclitus	temperature	+	Gabriel (1945)
Rivulus marmoratus	temperature	+	Lindsey and Harrington (1972)
Pseudacris ornata	temperature	+	Harvey and Semlitsch (1988)
Natrix fasciata	temperature	+	Osgood (1978)
Gallus domesticus	light regime	+	Osborne (1954)
Branta leucopsis	varying environment	−	Larsson (1993)
Tachycineta bicolor	food availability	+	Wiggins (1989)

Table 4.4—*contd*

Species	Environmental gradient	Effect	Reference
ANIMALS—*contd*			
Hirundo rustica	ectoparasites	+	Gustafsson *et al* (1996)
Hirundo rustica	ectoparasites	+	Møller (1990b)
Parus major	brood size	+	van Noordwijk (1984)
Parus major	food availability	+	van Noordwijk *et al.* (1988)
Parus major	brood size	+	Gebhardt-Heinrich and van Noordwijk (1991)
Parus caeruleus	brood size	+	Merilä (1996)
Parus montanus	years of different quality	+	Thesing and Ekman (1994)
Ficedula albicollis	brood size	+	Lindén *et al* (1992)
Ficedula albicollis	brood size	+	Merilä (1996)
Geospiza fortis	years of different quality	+	Price (1985)
Mus musculus	food	+	Falconer and Latyszewski (1952)
Mus musculus	food	+	Falconer (1960)

alcohol content of the food substrate in fruitflies *Drosophila melanogaster* (McKenzie and McKechnie 1978; McKenzie and Parsons 1974), embryo survival in relation to incubation temperature in the loach *Misgurnus fossilis* (Ozernyuk *et al.* 1992), and nestling survival in relation to food availability and hence body condition in two passerine bird species, the great tit *Parus major* and the collared flycatcher *Ficedula albicollis* (Lindén *et al.* 1992).

Given that the intensity of selection increases with increasing deviations from the environmental optimum, and that the amount of exposed genetic and phenotypic variation increases under adverse environmental conditions, the response to selection is likely to increase. Evolution in adverse environments such as the margins of the distributional range, marginal habitats, and in abnormal years is therefore likely to proceed at a higher rate than under more benign environmental conditions.

4.3.7 *Concluding remarks on genetic variance and selection*

The genome has been considered a flexible structure with properties that depend on the cellular environment (McClintock 1984). Cells experience different kinds of adverse environmental conditions, and such conditions will activate particular loci associated with tolerance responses. Interestingly, altered gene activity has been shown to be associated with elevated mutation rates and

DNA repair (Hanawalt 1987; Bohr and Wassermann 1988), increased levels of transposition (Jaenisch 1988; Bownes 1990), and possibly increased levels of recombination (Thomas and Rothstein 1991). The adverse environmental condition *per se*, may thus give rise to elevated gene activity and may therefore play an important role in the generation of genetic variation. Increased additive genetic variation will, together with more intense selection, give rise to an elevated response to selection.

Rapid responses of the genome to adverse environmental conditions may be mediated through environmental effects on chromosome methylation. Demethylation has been shown to affect rates of mutation (Ho *et al.* 1989), transposition (Fedoroff 1989), and recombination (Hsieh and Lieber 1992). Rapid and apparently 'directed' intracellular changes to persistent selection pressures may thus give rise to rapid evolutionary responses.

4.4 Adverse environmental conditions and the distribution of species

Distributional limits of species are important in that they reveal the conditions that prevent further range expansion. Furthermore, peripheral populations play an important role in several theories of peripatric speciation. The adverse conditions in such marginal ranges may affect the evolutionary potential in a number of different ways, as suggested by section 4.3. In this section, we will consider the effects of environmental condition on the distribution of species and how conditions in marginal areas potentially affect evolution.

Species margins are often related to climatic extremes, where metabolic costs may be sufficient to preclude major range expansions. A number of examples illustrate this point. The winter range of North American passerine birds is closely correlated with the isotherms of the coldest month (Root 1988a, b) and the distribution of vampire bats shows a similar correlation with temperature (McNab 1973). The correlation between early spring arrival by migratory birds and early high spring temperatures has been known since the beginning of the present century (Alerstam 1990). Recent climatic change has resulted in a clear range expansion of thermophilic insects such a several species of butterflies in the British Isles (Dennis 1993). Elevated spring and summer temperatures during the 1980s and 1990s resulted in an increased frequency of sightings of rare, thermophilic butterflies outside the normal distributional range of these species. The timing of flowering by plants in temperate climates provides a final example. Fitter *et al.* (1995) studied the appearance of the first flower of 243 species of plants in Oxfordshire, England, during a period of 36 years. A total of 219 of these species showed a clear relationship between temperature in the months preceding flowering and the timing of flowering, and any climatic trends thus are likely to affect the timing of reproductive events in temperate plant species. Climatic conditions appear to limit the distributional range of species and the timing of major events in the life cycle of the species. Such limiting effects possibly have physiological as well as genetic bases.

If the adversity of environmental conditions in marginal habitats is limiting the distribution of species and their timing of reproduction, this may have important implications for evolution in peripheral isolates. Adverse conditions sometimes increase levels of recombination and mutation, increase the amount of additive genetic and phenotypic variation, and result in intensified directional selection, as shown in section 4.3. In peripheral isolates, phenotypic variability may increase through inbreeding effects including the exposure of recessive alleles, the breakdown of developmental canalisation processes, or through increased rates of recombination. Selection for extreme variants is often accompanied by correlated genetic responses which rest upon epistatic interactions of linked loci, or upon pleiotropic effects of modifiers accumulated during stabilising selection (Schmalhausen 1949; Levin 1970a). Novel phenotypic variants may be evoked by disruption of the developmental process by extreme environmental conditions (Levin 1970a). The extent to which evolutionary change is predicted to arise as a result of the elevated levels of genetic and phenotypic variation in marginal parts of the distributional range is a topic of controversy. Parsons (1991b) has emphasised that maximum evolutionary rates are most likely to occur in moderately suboptimal regions of the distribution of a species, where phenotypic and genetic variability will be higher than in benign regions, and the metabolic costs of adverse environments for maintenance and repair are not unduly restrictive. Parsons (1993e) has predicted that little evolutionary change will take place at species borders due to the extremely high costs of accommodating adverse environmental conditions. Even though gradual evolutionary change may be promoted under moderately suboptimal conditions, the role of micro-evolution in major evolutionary innovations and changes remains a topic of great controversy (see, for example, Gould and Eldredge (1993) and Rhodes (1983) for these opposite views). Two evolutionary phenomena may be facilitated by extremely adverse environmental conditions, the destabilisation of the developmental process, and the resultant novel major deviations in phenotypes: speciation and major evolutionary innovations. These phenomena are the subjects of sections 4.5 and 4.6.

4.5 Adverse environmental conditions and the speciation process

Speciation is associated with major evolutionary divergence between populations, and such divergence must have a directional component. Secondary sexual characters differ the most among species, and these are the traits most likely to be involved in pre- and post-mating isolation (Darwin 1871). There must therefore have been rapid, recent evolutionary changes in these characters. Such characters are likely to demonstrate elevated levels of developmental instability (Møller and Pomiankowski 1993a; Chapter 2), and it is therefore likely that components of the speciation process such as the evolution of pre- and post-mating isolation mechanisms are associated with increased susceptibility to the causes of developmental instability. While the speciation process *per se* may increase developmental instability, the increased susceptibility to disruption of co-adapted genetic complexes and adverse environmental conditions may, by itself, play a major role in speciation events (Møller 1993d).

A number of different models of speciation have been proposed, including various kinds of allopatric and sympatric speciation (Endler 1977; Otte and Endler 1988; Lambert and Spencer 1995). The relative importance of various speciation processes is an area of controversy (review in Otte and Endler 1988; Rice and Hostert 1993). A recent review of laboratory tests of mechanisms involved in divergence between populations and initial steps in the speciation process suggested that allopatric speciation may not be predominant, but diminished gene flow due to geographical isolation in combination with strong divergent selection are likely to be of great importance (Rice and Hostert 1993). Sexual selection may have been an important factor in speciation, and this conclusion is supported by comparative evidence of associations between sexual selection and cladogenesis (Barraclough *et al.* 1995). Sexual selection and directional evolutionary change may both promote increased phenotypic variation (including greater developmental instability) (Møller and Pomiankowski 1993a; Chapters 2 and 7), and developmental stability may thus play a key role in the speciation process.

In this context we will discuss three different modes of speciation: (i) allopatric speciation by which the range of a species is broken up into two or more fairly large populations by a geographical barrier; (ii) allopatric speciation through founder effects (peripatric speciation); and (iii) sympatric speciation. Allopatric speciation occurs when the range of a species is split into two or more large populations by a geographical barrier such as a river or a glacier. Each of the separate populations may become subject to different selective forces, mutations, and population perturbations, and therefore show slow adaptive divergence. Reproductive isolation may thus evolve partially or fully as a consequence of long-term divergent selection in different environments.

Models of founder effect speciation (peripatric speciation) have recently come under attack for theoretical reasons (Barton 1989), but the importance of colonisation in speciation is indicated by the large amount of evidence suggesting that isolated island species often are highly divergent in phenotype from their mainland relatives (Mayr 1942, 1963). In peripheral isolates, phenotypic variation may increase due to exposure of recessive alleles and increased rates of recombination, mutation, and transposition (Levin 1970a; sections 4.3.1–3). Adverse environmental conditions may undermine canalisation and produce extreme phenotypes. In other words, novel phenotypic variants may be evoked by disruption of the developmental process through adverse environmental conditions (Møller 1993d).

Models of sympatric speciation have mainly emphasised divergence based on polyploidy and selection for divergence of niches, when habitat and mate selection are associated (White 1978). The prime examples of the latter kind of process concern the association of insects with particular host plants. If herbivorous insects are associated with specific food plants, partial reproductive isolation could arise from adaptation to a new host plant. A particularly illuminating example concerns a long-term experiment on aphid adaptations to novel host plants (Shaposhnikov 1965, 1966, 1987b; see also section 3.5.4). After the aphid *Dysaphis anthrisci majkopica* had been reared on an uncommon host plant for a number of generations, the initially high phenotypic variance decreased, the initially low viability increased, and natural selection became less intense as the aphid clone became adapted to the new host. Reproductive isolation evolved between the aphid clones adapted to the new and the traditional host plants. Even more interestingly, the transfer experiment to the uncommon host plant *Chaerophyllum bulbosum* resulted in morphological convergence to the aphid *Dysaphis chaerophyllina*, the aphid species that usually lives on the host plant to which *D. anthrisci majkopica* had become adapted. Surprisingly, the newly adapted strain of *D. anthrisci majkopica* produced fertile offspring in crosses with *D. chaerophyllina*. Under normal conditions, the two species do not produce offspring despite sympatric distributions on the same host plants with overlapping reproductive seasons (Shaposhnikov 1965, 1966, 1987b). After a few generations of selection in the parthenogenetic stage on the new host plant, one species was transformed into a new form that resembled the other species in morphology and was able to interbreed with it, although many offspring were of aberrant morphology, i.e. they were developmentally unstable. The results cannot be explained by contamination of the *D. anthrisci majkopica* strain with *D. chaerophyllina* (Shaposhnikov 1987b), as suggested by Blackman (1979). The most likely explanation for these observations is that the morphological and reproductive features of the two species are inherited primarily through epigenetic inheritance systems mediated by the host plant (Jablonka and Lamb 1995). Similar kinds of divergence following adaptation to novel host plants may be expected in other insect species with specialised host selection.

The events leading to speciation are obviously difficult to study in detail under natural conditions (with a few notable exceptions such as the one described in detail above), but laboratory experiments aimed at testing specific predictions may provide some clues to the most likely mechanisms. These studies have recently been reviewed by Rice and Hostert (1993). It is characteristic of the laboratory studies that selection experiments for increased divergence between populations and hence incipient reproductive isolation have often resulted in an increase of additive genetic and phenotypic variation (see Table W3.1). This result suggests that the phenotypes of individuals from populations under divergence exhibit an increased level of variation, and that the genetic architecture of such diverging populations may also be altered. Provided that increased levels of phenotypic variation are positively associated with increased levels of developmental instability, as suggested by a comparative study of birds (Cuervo and Møller 1997b), diverging populations are also expected to demonstrate elevated levels of fluctuating asymmetry, if disruptive selection results in the loss of genetic modifiers that dampen the effects of single gene replacements. Evidence from disruptive selection experiments supports this line of reasoning, since lines subject to disruptive selection had higher asymmetry than control lines (see Table W3.1). This interpretation is also supported by comparative data on fluctuating asymmetry in secondary sexual feather characters of bird species from central and marginal populations. The asymmetry for secondary sexual characters was considerably higher in marginal than in central populations of the same species, while that was not the case for ordinary morphological characters (Møller 1995e). This difference between characters is expected if secondary sexual characters are affected by persistent, disruptive directional selection, while ordinary morphological traits mainly are subject to stabilising selection. Although differences in developmental stability of marginal and central populations may be due to differences in genetic constitution, population density, and levels of parasitism and predation, these differences may influence selection and micro-evolution in central populations and peripheral isolates. If asymmetry reliably reflects the ability of individuals to cope with current environmental conditions, symmetric individuals from a peripheral isolate will be those best suited to those conditions because they have had time to adapt, while individuals from the central part of the distributional range will be less able to cope with the adverse conditions of the distributional fringe.

Hybridisation often results in the production of non-viable and infertile offspring with abnormal phenotypes. Hybrids usually demonstrate more asymmetry, but in a few cases less asymmetry, than the parental species (see section 5.3.3). This could result from different degrees of divergence between species, and thus have important implications for the role of developmental stability in speciation processes. If two taxa are distantly related, and therefore have high genetic divergence, increased developmental instability in hybrids may result from the disruption of a co-adapted genome. Disturbances in the schedules of gene expression in developing hybrids have been used to draw inferences about the degree to which the gene regulatory divergence is correlated with structural gene divergence (as estimated from genetic distances) (Parker *et al.* 1985). In one laboratory experiment, sperm from ten species of six genera of fish from the family Centrarchidae were used to fertilise eggs of the Florida largemouth bass *Micropterus salmonides floridanus*. As the genetic distance between paternal species and the Florida bass increased, there was a general decline in developmental success as demonstrated by reductions in percentage of hatching and progressively earlier and more extensive morphological abnormalities reflecting a disturbed developmental programme. However, a few species appeared to have uncoupled rates and modes of evolution of structural and developmentally regulatory genes (Parker *et al.* 1985). The speciation process thus may sometimes involve changes in developmental regulatory genes, thereby reducing or preventing the frequency of hybrids between recently diverged populations.

4.6 Adverse environmental conditions and major evolutionary innovations

Major evolutionary innovations are relatively rare, and they have been considered by some biologists to belong to a different category than the gradual, minor evolutionary changes seen in most lineages (Maynard Smith and Szathmáry 1995). If this line of reasoning is correct, then micro-evolutionary processes are going to differ qualitatively from those leading to major evolutionary innovations. Central to this dichotomy are two contrasting modes of evolution, gradual evolution versus punctuated equilibria. The punctuated equilibria model presupposes that relatively abrupt evolutionary changes (only involving tens or hundreds of thousands of generations depending upon the generation time of the organism(s) and the stratigraphic conditions) are interspersed with long periods of stasis characterised by no or only little evolutionary change (reviewed in Gould and Eldredge 1993). In contrast, the model of gradual evolutionary change presumes that differences among taxa can be accounted for by gradual, micro-evolutionary changes during evolutionary history (e.g. Rhodes 1983). The two models of evolution are not necessarily mutually exclusive, and we would emphasise the possibility that rates of evolutionary change may, among other factors, depend on the nature of environmental change.

The two modes of evolutionary change can be unified if the phenotypic response to adverse environmental conditions depends on the adversity of such conditions. Parsons (1991b, 1993b, 1993e, 1994b) has emphasised that evolutionary change due to suboptimal conditions is more likely when there is an excess of energy available, and thus when environmental conditions are moderately adverse. Although this may apply to characters that are subject to gradual, micro-evolutionary change, more extreme environmental conditions may only infrequently give rise to major phenotypic changes that can be assimilated (Waddington 1961). Two different scenarios are therefore possible:

(1) Long-term changes in response to adverse environmental conditions that are gradual and energetically relatively inexpensive in terms of maintenance and repair.

(2) Short-term changes in response to severe environmental conditions that will be sudden and energetically expensive to the individual in terms of maintenance and repair and will rely on environmentally induced unmasking of genetic variability and loss of canalisation (Fraser 1962; Levin 1970a, b; Hoffmann 1983; Kieser 1987, 1993).

Extremely adverse, novel environments may expose individuals beyond the buffering capacity of the canalisation system. This could result in production of aberrant, non-canalised phenocopies on which natural selection could act.

Evolutionary novelties have been shown to arise relatively rapidly (on a palaeontological time scale) during adverse environmental conditions (Gould and Eldredge 1977; Mayr 1982), in agreement with this idea.

The most extreme version of the point of view that adverse environmental conditions sometimes may result in development of radically different evolutionary novelties is the 'hopeful monsters' of Goldschmidt (1940), who suggested that extreme environmental conditions may sometimes result in mutations with major, beneficial effects on phenotypes. A potential mechanism giving rise to such major changes in phenotypes may be phenotypic and genotypic alterations due to the effects of transposable elements. Transposable elements are chromosome pieces that can change position within the genome (McDonald 1995). Changes in the genetic background of a mobile genetic element such as a transposable element can trigger bursts of transposition (Gvozdev 1986), and conditions within the cell as well as in the external environment may trigger rapid genomic reorganisation via transposons (McClintock 1984). The genetic background may also change as a result of elevated levels of recombination or mutation caused by adverse environmental conditions (see sections 4.3.1–2). Some of these transpositions may influence genes directly involved in the developmental process, while others may produce somatic mutations that increase bilateral asymmetry by creating a mosaic of cell lines (Graham 1992). Insertions of transposable elements occur in transpositional bursts leading to very high mutation rates several orders of magnitude above the background level of mutation. The molecular basis of these bursts is a regulated process that represents a response to inbreeding, disruption of co-adapted genomic complexes, and adverse environmental conditions (McDonald 1995). These may resemble the situations under which major evolutionary innovations take place. Mutations induced by transposable element insertions may be expressed in the phenotype and could therefore be subject to selection. Selection for suppresser alleles usually results in the masking of the effects of transposable elements at the phenotypic level, and transposable elements and suppresser alleles may give rise to regulatory evolution (McDonald 1995). Periods of severe inbreeding, as associated with the founding of genetically isolated populations, could result in rapid loss of suppresser alleles and sudden release of new transposable element-mediated regulatory phenotypes.

Future studies of transposable elements may reveal to which extent phenotypes of organisms may change in response to severe environmental stress. The results will have theoretical implications for our understanding of phenotypic evolution, but may potentially also be of importance for applied aspects related to animal and plant breeding and conservation.

4.7 Summary

- Adverse environmental conditions are ubiquitous factors affecting all life stages of most organisms.

- Energy available for maintenance, growth, survival, and reproduction is more plentiful under optimal environmental conditions, and reductions in availability of energy for biological processes under more adverse conditions result in a higher frequency of asymmetric phenotypes.
- Organisms can avoid adverse environmental conditions by means of habitat selection, behavioural adjustment, and tolerance. The fitness costs of tolerance prevent individuals from attaining high levels of resistance in the absence of the selective environmental agent.
- Adverse environmental conditions affect the evolutionary process at a range of different levels. Rates of recombination, mutation, and transposition increase under adverse environmental conditions, and so does the additive quantitative genetic variation in a number of experimental studies. The phenotypic variation also generally increases under adverse conditions. The intensity of selection generally increases steeply as conditions deviate from the optimum. Adverse environmental conditions thus tend to facilitate adaptation to adverse environmental conditions.
- Moderate suboptimal conditions result in moderate phenotypic changes that facilitate gradual evolutionary change, while extreme environmental conditions tend to reveal major phenotypic deviants that may give rise to major phenotypic changes as seen in connection with speciation and the evolution of major innovations.

5

Causes of developmental instability.
I. Genetic factors

5.1 Introduction

The factors that cause fluctuating asymmetry can be either genetic or environmental in origin. As fluctuating asymmetry represents the ability of an individual to resist genetic and environmental factors during growth and development, then it could be viewed as a general health certificate of an individual. In this chapter we will review the heritability of developmental stability (as measured by fluctuating asymmetry, frequency of phenodeviants, and coefficient of variation) as well as detail the various genetic factors that can cause a breakdown in developmental stability.

In section 5.2 we review the evidence that indicates the heritability of developmental stability. It appears that fluctuating asymmetry does have a heritable component, as revealed by a meta-analysis of all published reports.

In section 5.3 we describe the various genetic factors that can give rise to a loss in developmental stability. Loss of genetic variation due to inbreeding may increase levels of fluctuating asymmetry (section 5.3.1). Additionally, protein heterozygosity may also be related to developmental stability, but in a fairly subtle way. It would appear that heterozygous individuals have a number of fitness advantages over homozygotes (section 5.3.2.1) and that heterozygotes may be more symmetric (section 5.3.2.2). However, this evidence is not compelling and there are many conflicting reports. In a meta-analysis of all published reports that link heterozygosity with fluctuating asymmetry, it appears that heterozygosity is weakly associated with increased developmental stability among populations, whereas there is no general relation within populations. Additionally, there may be taxonomic differences in this relationship: poikilotherms may show a negative relation between heterozygosity and fluctuating asymmetry, whereas there is no relation in homeotherms.

In section 5.3.3 we review the evidence that indicates that hybridisation, which splits up co-adapted gene complexes, can reduce developmental stability. In cases where hybridisation has led to an increase in developmental stability, hybrids may be more heterozygous than parents without disrupting a co-adapted genome. For example, if two species have diverged only very recently, hybridisation may not seriously disrupt co-adaptation. Finally, we discuss the situation where developmental processes can be destabilised through the incorporation of mutant genes into the genome (section 5.3.4). The influence of directional selection on developmental stability has already been discussed in Chapter 3.

Many of these genetic factors are not isolated from each other. For example, hybridisation can both separate co-adapted gene complexes and increase the levels of heterozygosity, within the same genome. These two factors may act in opposite directions, one decreasing and the other increasing developmental stability. This scenario may be further complicated by the introduction of novel genes (mutants), and also effective directional selection (see Chapter 3) brought about by the mixing of inbred strains. If genetic changes are tracked in naturalistic populations, it may be very difficult to predict how developmental stability will change with any one alteration to, or action by, a population. Therefore, the results obtained from the carefully controlled genetic experiments that we have reported above must be applied with extreme caution when interpreting and predicting the relationship between genetic factors and developmental stability in uncontrolled conditions. This is particularly important as uncontrolled conditions will probably occur in many of the situations where evolutionary biologists will be interested in monitoring levels of developmental stability and hence fluctuating asymmetry. However, these studies are extremely useful in that they illustrate the intimate ways in which developmental stability can reflect aspects of the genetic make-up. This may allow the researcher, through simple measurements of fluctuating asymmetry, to gain crucial insights into the genetic background and evolutionary history of study populations.

5.2 Heritability of developmental instability

The heritability of fluctuating asymmetry (and other measures of developmental instability) has been estimated by comparing levels of trait-specific fluctuating asymmetry between parents and offspring, or by changing levels of fluctuating asymmetry in selection experiments (for reviews see Palmer and Strobeck 1986; Parsons 1990a). There is a large amount of disagreement as to whether developmental stability is heritable or not. Some studies report a low, but significant, heritable component to developmental stability (e.g. Hagen 1973; Thornhill and Sauer 1992; Møller 1994c), whereas others report no significant heritability (e.g. Thoday 1958; Tuinstra *et al.* 1990). However, small heritabilities are usually not significant even with large sample sizes because of large standard errors. In an attempt to clarify this issue, Møller and Thornhill (1997a) conducted a meta-analysis of published and unpublished estimates of fluctuating asymmetry from 34 studies of 17 species (see Table W5.1 on the Web site). The raw data indicate the presence of heritability more frequently than expected by chance as nine (41%) of 22 heritability estimates were statistically significant, while only 1.1 studies (5%) were expected to be significant by chance.

Møller and Thornhill (1997a) obtained heritabilities as estimates of the additive genetic component of fluctuating asymmetry and other measures of developmental instability. In cases where repeated studies were performed on the same species (e.g. dermatoglyphics in *Homo sapiens* and sternopleural chaeta in *Drosophila melanogaster*), or where there were repeated samples taken within the same study, mean effect sizes of heritability were calculated. This meant that each species was represented by only one estimate of the effect size of heritability. As meta-analyses have often been criticised for combining studies with inconsistent methodologies and for utilising weak statistical methods (for review see Arnqvist and Wooster 1995), Møller and Thornhill added a number of variables to their analysis that helped to control for potentially confounding factors. These included variables for (i) type of study (i.e. selection experiment, parent–offspring regression, or sib analysis); (ii) whether the original authors tested for the statistical properties of fluctuating asymmetry, as other forms of asymmetry may not reflect developmental stability; (iii) internal validity (scored as high if rearing conditions of each generation were standardised or sample sizes were large, i.e. greater than 50); and (iv) external validity (scored as high if the study population was stable and not subjected to a selection regime as part of the experiment). External validity was not included in their final analysis as most studies were scored as 'high'. The 'file drawer problem' (see Arnqvist and Wooster 1995) of publication bias may have influenced Møller and Thornhill's meta-analysis. However, in the case of published estimates of heritability of fluctuating asymmetry, this may not be the case. The vast majority of these studies do not report heritability as their primary research goal or most important result, upon which the publication of the manuscript may (or may not) hinge. Hence, there is less likely to be a publication bias toward 'positive' or 'negative' results, which is reflected in the range of heritability estimates published in the literature. Møller and Thornhill also addressed this directly in their manuscript and demonstrated that more than 1000 unpublished results that show no heritability of fluctuating asymmetry would have to exist before the conclusions of their meta-analysis would alter significantly.

The results of their meta-analysis revealed that there is a relatively small (0.19), but highly statistically significant ($P < 0.0001$), additive genetic component to fluctuating asymmetry of morphological characters across the species studied. In studies where internal validity was high, heritability was low as expected, but there was still an overall significant effect. There was also an indication that heritability estimates in parent–offspring regression studies were artificially inflated due to maternal or common environment effects (see Whitlock and Fowler 1997; Swaddle 1997b), and that membership in a particular taxonomic group may influence (or be correlated with) the heritability of fluctuating asymmetry. The analysis reported by Møller and Thornhill (1997a) was, of course, retrospective in terms of these factors, and so it would be more valid to design specific experiments to investigate these phenomena directly.

Several researchers have taken exception to Møller and Thornhill's approach to quantifying the heritability of developmental stability through meta-analysis (Leamy 1997; Markow and Clarke 1997; Palmer and Strobeck 1997; Pomiankowksi 1997; Whitlock and Fowler 1997), as heritability estimates are known to be specific to characters, populations, and environmental circumstance. Therefore, comparing studies that differ in all these factors may not be valid (Markow and Clarke 1997). However, Møller and Thornhill (1997b) claim that they do not assume that the heritability of two species is the same quantity and in order to review the literature comparisons must be drawn between studies. In this respect, Møller and Thornhill's analysis is useful in that it presents so much of the literature on this subject.

Palmer and Strobeck (1997) have also criticised the meta-analysis from the perspective of many of the statistical considerations that we indicated in Chapter 2. As measurement error is likely to inflate fluctuating asymmetry values, Palmer and Strobeck (1997) suggest that Møller and Thornhill (1997a) should have excluded all studies that did not examine the relative influence of measurement error. Additionally, Palmer and Strobeck argue that the artificial inflation of asymmetry by inaccurate measurements may have led to spurious significant heritability estimates if there were large among-observer differences in measurement error in the studies reviewed. Unfortunately, most manuscripts do not report such details and hence it is extremely difficult to assess the applicability of this criticism. Conversely, Pomiankowski (1997) remarks that measurement error will increase the residual variance term in the estimate of asymmetry heritability and that measurement error is relatively large for asymmetry (compared with trait size), hence heritability estimates may tend to be low due to measurement error.

In a similar vein, both Leamy (1997) and Palmer and Strobeck (1997) have remarked that Møller and Thornhill (1997a) did not distinguish sufficiently between studies that tested for the presence of directional asymmetry and antisymmetry from those that did not. This is a valid and important consideration. However, previous studies that did not test for the statistical properties of fluctuating asymmetry did not appear to differ in effect size from the more recent studies that have been more rigorous. Møller and Thornhill (1997b) have subsequently indicated that developmental instability is still significantly heritable when their analysis is restricted to the most conservative studies (mean(s.e.) = 0.162(0.063); t_5 = 2.59; P = 0.049). This average effect size is also significantly different from zero (mean(s.e.) = 0.203(0.065); t_4 = 3.12; P = 0.036). Therefore, the statistical considerations raised by Palmer and Strobeck (1997), although justifiable and applicable, may not have influenced the overall result demonstrated by Møller and Thornhill's (1997a) original analysis.

Another general criticism to have been discussed in the literature is that the studies reviewed by Møller and Thornhill did not control for the influence of trait size in the same manner (Palmer and Strobeck 1997). Additionally, as the most popular way of controlling for the effect of size is attained by dividing asymmetry by average trait size (i.e. $(|L–R|)/\frac{1}{2}(L + R)$), heritability estimates for asymmetry are confounded by the trait size heritability. Møller and Thornhill (1997b) accept that this is a general problem of most studies of heritability because the characteristics of any trait that heritability is calculated for are likely to be related to the size of another character, as many traits will be influenced by common developmental mechanisms and circumstance. Additionally, linkage disequilibrium and pleiotropic effects will also result in characters being phenotypically correlated (Falconer 1989). This does not necessarily invalidate the study of heritability and there is no particular reason why studies of the heritability of developmental instability, more than any other character, should control for the influence of trait size (Møller and Thornhill 1997b).

Despite the criticisms discussed in the literature (see Houle 1997; Leamy 1997; Markow and Clarke 1997; Palmer and Strobeck 1997; Pomiankowksi 1997; Swaddle 1997b; Whitlock and Fowler 1997), most of these researchers (with the exception of Markow and Clarke (1997)) agree that developmental instability does have an additive genetic component in many characters and populations. However, this does not mean that the heritability of developmental instability of all traits in all species is equal (or indeed significant); although every character will display some degree of heritability (Lewontin 1983). It is likely that many different genes are involved in determining developmental stability. If there are a large number of these loci, there will be genetic variation in developmental instability and relatives will tend to resemble each other with respect to the stability of their phenotype. Therefore, it may not be too surprising to expect developmental instability to have an additive genetic component (Møller and Thornhill 1997b).

Some of the more positive products of the open criticism of Møller and Thornhill's meta-analysis are the suggestions made for future studies of the heritability of developmental instability. Pomiankowski (1997) and Markow and Clarke (1997) suggest that comparisons among traits that differ in the intensity of selection and their relative fitness values may prove fruitful. This is echoed by Møller and Thornhill (1997b), who also point out that measurement error is relatively smaller compared with asymmetry in exaggerated traits and hence the considerations raised by Palmer and Strobeck (1997) would be minimised. Swaddle (1997b) has suggested that asymmetry values can be successfully transformed by Box–Cox transformations (see also Palmer and Strobeck 1986; Swaddle *et al.* 1994) to fit the normality assumptions of linear regression in order to obtain more accurate heritability estimates. Additionally, as asymmetry values often do not correlate among traits on the same individual (section 2.3.4), assessments of the heritability of developmental instability should incorporate asymmetry measures from a number of traits. Swaddle

(1997b) has suggested that this could be approached from a multivariate perspective in which additive genetic variance–covariance matrices, that include measures of developmental instability from a number of traits, are constructed and compared.

It may be significant that fluctuating asymmetry appears to have a relatively low level of genetic heritability. This is in accord with Fisher's fundamental theorem (Fisher 1930), which suggests that traits most closely associated with fitness have little additive genetic variation, as selection for individuals with low asymmetry will lead to fixation of the best-adapted genotypes. This low level of heritability probably also arises because fluctuating asymmetries result from a vast number of genetic and environmental influences during development (as detailed in the remaining sections of this chapter and Chapter 6), and the traits studied are often subject to directional selection which will act to decrease developmental stability (Houle 1992). This low level of heritability also helps to confirm the assumption that fluctuating asymmetry can be used as an indicator of environmental conditions (section 6.10). If fluctuating asymmetry was highly heritable, it would provide a less reliable bioassay. Møller and Thornhill (1997a) suggest that monitoring traits that are subject to intense directional selection may prove the most sensitive bioassay technique.

Whitlock (1996) has explored the heritability of developmental stability and fluctuating asymmetry in a mathematical model. In this approach he treats fluctuating asymmetry as the phenotypic expression of developmental stability, and as developmental stability is a reflection of 'noise', any one particular level of developmental stability could give rise to a range of asymmetry values. Hence, expression of asymmetry is merely a pinpoint on the distribution of available asymmetry scores that could have occurred for a particular trait and individual. If developmental stability of an individual can give rise to a normal distribution of asymmetry scores, Whitlock provides a convincing mathematical argument which indicates that the heritability of fluctuating asymmetry is equivalent to the repeatability of the asymmetry estimate multiplied by the heritability of developmental stability. Hence, low levels of heritability for fluctuating asymmetry may not mean low heritability of developmental stability because asymmetry measures are error-prone and generally have low repeatability (section 1.10.2). Therefore, the low levels of heritability reported above may underestimate the true heritability of developmental stability in many systems.

Overall, it would appear that fluctuating asymmetry in traits of many species does have a heritable component. This should also be expected given that germline mutations affect the level of individual asymmetry. So why has this genetic variation for developmental instability been retained if there is selection against instability? There are a number of reasons why this could have occurred. First, directional selection will tend to increase levels of fluctuating asymmetry as development becomes destabilised (see section 5.3.4). This may occur in traits subjected to sexual selection, or in systems where there is continuous selection due to host–parasite interactions (Hamilton 1982,

1986). Second, incorporation of mutant genes into the genome will increase levels of fluctuating asymmetry (see section 5.3.5). In systems where there is selection for novel genotypes, such as in the case of host–parasite interactions, similar effects may arise. Third, there may be directional selection for developmental stability, as symmetric individuals perform better (Chapter 7), and this directional selection pressure may, ironically, tend to generate a loss of developmental stability (see Chapter 3). Fourth, there may be intrinsic developmental costs in the production of a perfectly symmetric phenotype (section 7.5). Fifth, in many systems there will be fluctuating selection pressures, as there is a range of environmental conditions across the occupied habitat. As genotypes do not perform equally well in all environments, the genetic variation for developmental stability will be retained if there is gene flow among environments. Finally, if fluctuating asymmetry is lowest in heterozygous individuals (see section 5.3.2.3), the genetic variation for developmental stability may be maintained by heterozygote advantage.

5.3 Genetic factors

The genetic factors that result in fluctuating asymmetry can be summarised into at least five categories: loss of genetic variation, degree of protein heterozygosity, hybridisation that disrupts co-adapted gene complexes, episodes of directional selection (refer to Chapter 3), and invasion of the genome by mutant genes. The following sections (5.3.1–4) discuss these topics.

5.3.1 *Loss of genetic variation*

Inbreeding has two major effects on the genomic structure of a population. First, it reduces genetic variation and hence individuals in a population become more similar. Second, inbreeding increases homozygosity, as the number of unique alleles at each locus is reduced. This increase in homozygosity, often coupled with an increased expression of deleterious recessive alleles, often leads to reductions in fitness and viability (see Phelan and Austad 1994 for a review; also section 5.3.2). We expect inbred strains to be more developmentally unstable as they are more homozygous and so possess a reduced range of enzymatic products, and hence are less capable of resisting developmental upsets in a broad range of environments than more heterozygous populations. Additionally, if there are specific dominant genes that influence developmental stability (Falconer 1981), increased expression of recessive alleles may lead to decreased stability as there is a relative reduction in expression of genetically dominant alleles in inbred populations.

Reduced genetic variation in inbred strains of laboratory rats *Rattus norvegicus* has resulted in greater individual asymmetry in bilateral osteometric characters than in control rats (Leamy and Atchley 1985). This relationship has also been demonstrated in other taxonomic groups, for example in a laboratory experiment, Mather (1953) crossbred two long-established inbred homogenic strains of *Drosophila* (Oregon and Samarkand) to produce F_1 and F_2 hybrids, and therefore generations of increasing genetic variation. By measurement of fluctuating asymmetry of sternopleural chaetae for 20 males and 20 females from 25 cultures of parent, F_1 and F_2 stocks, Mather demonstrated that the population variance in fluctuating asymmetry was higher in the original inbred strains than the crossbred hybrids. These results were subsequently corroborated by the findings of Dobzhansky and Wallace (1953) and Reeve (1960). Leamy (1984) and Bader (1965; see Fig. 5.1), independently, indicated that the degree of inbreeding is positively related with fluctuating asymmetry in mice. Leary *et al.* (1985a) have reported a similar phenomenon in rainbow trout *Salmo gairdneri*, whilst Clarke *et al.* (1986) have indicated that inbreeding increases fluctuating asymmetry in thoracic leg segments in the marine harpacticoid copepod *Tisbe holothuriae*. Beardmore (1960) has also provided experimental evidence for inbreeding increasing levels of fluctuating asymmetry; two inbred strains of *Drosophila melanogaster* had greater asymmetry than their subsequent F_1 hybrids. However, it should be stated that inbreeding does not necessarily lead to a rise in fluctuating asymmetry in all cases (see Table W5.2 on the Web site).

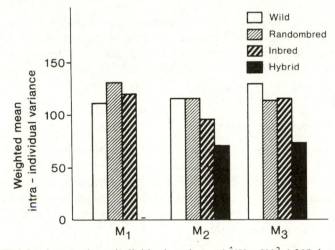

Fig. 5.1 Weighted mean intra-individual variance $(\hat{A}(X_1 - X_2)^2 / 2N)$ in mm for each molar tooth (M_1, M_2, and M_3) for each population (wild, random-bred, inbred, hybrid) of mice. Where X_1 and X_2 are measurements from left and right sides and N is the number of pairs measured. Levels of asymmetry are higher in inbred than hybrid strains. Adapted from Bader (1965).

It has also been shown that the cheetah *Acinonyx jubatus*, a species that has been subjected to a massive population bottleneck followed by intense levels of inbreeding, shows elevated levels of fluctuating asymmetry in cranial morphology compared with similar species of wild cat (Wayne *et al.* 1986). This indicates that the high intensity of inbreeding and lack of biochemical genetic variation in the cheetah has dramatically decreased levels of developmental stability.

However, Willig and Owen (1987) have criticised this comparative study of felid fluctuating asymmetry, mainly on the basis of the methodology that Wayne *et al.* (1986) employed. This inspired Kieser and Groeneveld (1991) to conduct a similar study that incorporated all the remarks made by Willig and Owen. This revealed that the inbred cheetah does not display any greater levels of fluctuating asymmetry than two closely related, but genetically more diverse, African wild cat species. Perhaps in the case of the cheetah, inbreeding is not so closely associated with decreased developmental stability as first thought. However, the absence of elevated asymmetry in cheetahs may be expected if the high level of inbreeding has resulted in intense selection against deleterious alleles which has eventually led to reduced asymmetry. Therefore, asymmetry may be expected to be low after a severe period of selective mortality, similar to that experienced by the cheetah. This same explanation may account for the lack of a relationship between inbreeding and developmental stability in haplo-diploid organisms such as honey bees *Apis mellifera* (Clarke *et al.* 1986, 1992).

Overall, our review of published reports that relate the effects of inbreeding on developmental stability indicates that the loss of genetic variation associated with inbreeding destabilises developmental processes in a wide range of taxa (Table W5.2). In our survey of 77 published reports from 29 species, approximately 78% of studies indicated that inbreeding is related to unstable developmental pathways; whereas a minority of studies (approximately 9%) displayed the opposite relationship. Therefore, there is convincing evidence to suggest that inbreeding results in an increase in developmental instability.

5.3.2 *Protein heterozygosity*

Protein heterozygosity is an indicator of genetic variability (see later), therefore we expect populations and individuals that are more heterozygous to be better able to cope with a broader range of environmental conditions than homozygotes. Hence, heterozygotes are predicted to show greater developmental stability, as they should possess a larger buffering capacity against environmental perturbations.

Lerner (1954) first hypothesised that there should be a negative relationship between developmental homeostasis (phenotypic variation) and heterozygosity; and subsequently many researchers have found such a relationship. However, a significant number of studies have failed to find this predicted negative relationship. In this section we review the evidence that documents the possible relationships between heterozygosity and fluctuating asymmetry and report a meta-analysis, the results of which indicate that this relationship may not always be straightforward.

5.3.2.1 *Heterozygosity and fitness*

The fitness advantage of heterozygosity has been proposed to be the primary device that maintains genetic diversity in natural populations (Dobzhansky

1955; Lewontin 1974; Charlesworth and Charlesworth 1987), yet the mechanism behind this advantage is still poorly understood and is a topic of much debate (e.g. Smouse 1986; Houle 1989, 1994; Mitton 1995). From an empirical point of view, many researchers have indicated a positive relationship between protein heterozygosity and suggested fitness indicators, such as growth rate (Baker and Manwell 1977; Johns *et al.* 1977; Makaveev *et al.* 1978; Singh and Zouros 1978; Bottini *et al.* 1979; Zouros *et al.* 1980; Fujio 1982; Pierce and Mitton 1982; Cothran *et al.* 1983; review in Mitton 1995). Heterozygosity has also often been associated with increased viability (e.g. Mitton and Koehn 1975; Koehn *et al.* 1976; Allard *et al.* 1977; Zouros *et al.* 1983; Farris and Mitton 1984), high feeding rates (e.g. Garton 1984; Garton *et al.* 1984), feed conversion rates (Mitton *et al.* 1994) and high fecundity (Koehn unpublished data, cited in Mitton and Koehn 1985). Therefore, there is convincing evidence to indicate that heterozygotes have a general fitness advantage over homozygotes, and also that the link between heterozygosity could be causal as enzyme heterozygosity increases metabolic efficiency (Koehn and Shumway 1982; Garton 1984; Garton *et al.* 1984).

For example, Cothran *et al.* (1983) discovered a complex combination of relationships involving heterozygosity, weight, fecundity, and foetal growth for the white-tailed deer *Odocoileus virginianus*. Adult females with high levels of enzyme heterozygosity weigh more than homozygous individuals; they also enjoy an advantage in fecundity (Johns *et al.* 1977) and have a higher rate of twinning. The weight of females is positively correlated with the weight of their foetuses; and foetal growth rates are positively correlated with foetal heterozygosity.

Experiments investigating the relationship between heterozygosity and growth rates in the American oyster *Crassostrea virginica* have found that the oxygen demand of heterozygotes is significantly lower than that of homozygotes (Koehn and Shumway 1982). This relationship has also been found in the tiger salamander *Ambystoma tigrinum* (Mitton, Carey, and Kocher, unpublished data reported in Mitton and Grant 1984). Danzmann *et al.* (1986) also indicated that there is a difference in metabolic efficiency between heterozygote and homozygote marine invertebrates, which gives rise to an enhanced growth rate in individuals that are more heterozygous (Garton 1984; Garton *et al.* 1984). Therefore, if all individuals have the same energetic input, highly heterozygous (and perhaps symmetric) individuals' lower routine metabolic costs should leave more energy to invest in growth, reproduction, or any other activity (Mitton and Grant 1984). Berger (1976) has reviewed several other biochemical mechanisms whereby the enzymes of heterozygous individuals can function more efficiently than those of homozygotes. Other models suggest that enzyme activities of heterozygotes at one locus will be similar to the activity of one homozygote or intermediate between a pair of homozygotes that differ in activity (Kacser and Burns 1981; Hilbish and Koehn 1985). Therefore, when the effects are summed across a large number of loci, heterozygotes are expected to be metabolically more efficient than homozygotes. The applicability of this hypothesis may differ greatly among taxonomic groups and may also be dependent on the nature of the enzymes (e.g. monomers, multimers) studied.

If heterozygosity at any locus is consistently associated with faster developmental rate, then it is probable that the enzyme products of this locus directly influence metabolic flux leading to differences in the rate of development. If, however, certain loci show strong positive associations between developmental rate and heterozygosity in some strains, but negative associations in other strains, then it is likely that such loci are marking chromosomal segments which carry other genes that have a strong effect on the rate of development.

Heterozygosity has been associated with increased fecundity in *Drosophila melanogaster* (Seradilla and Ayala 1983; Bijlsma-Meeles and Bijlsma 1988), the blue mussel *Mytilus edulis* (Rodhouse *et al.* 1986), brine shrimp *Artemia franciscana* (Gajardo and Beardmore 1989), and guppies (Beardmore and Shami 1979); and also increased mating success in *Colias* butterflies (Watt 1979, 1983; Watt *et al.* 1983, 1985, 1986). However, there have also been a number of reports where researchers have failed to find links between heterozygosity and putative measures of fitness (Zink *et al.* 1985; Zakharov 1987; Booth *et al.* 1990; Patterson and Patton 1990; Hartl *et al.* 1991; Clarke *et al.* 1992; Livshits and Smouse 1993b; Whitlock 1993). These discrepancies may be accounted for if the number of loci contributing to overall symmetry and fitness is large (Palmer and Strobeck 1986), as some studies may not measure enough loci to gain an accurate picture of overall heterozygosity.

Ryabova *et al.* (1995) investigated the relations between allozymic heterozygosity at two *LDH* (lactose dehydrogenase) loci, the *PGM* (phosphoglucomutase) locus, and the *EST* (esterase) locus with fitness traits in adult stellate sturgeon *Acipenser stellatus*. This analysis revealed an unpredictable relationship between fitness and heterozygosity, as heterozygosity at *LDH3*, *PGM1*, and *EST2* loci was negatively related to female fecundity, but heterozygosity at *LDH4* was positively related to the same measure. Additionally, heterozygosity at the *LDH3* locus was associated with high growth rates and increased juvenile survival in a hatchery population of sturgeon, but also appeared to be negatively related to female fecundity in a natural population. Nonetheless, the overall balance of the published literature indicates that heterozygosity is often associated with increased fitness in a broad range of taxa.

In general, it is not clear how heterozygotes appear to gain fitness advantages over more homozygous individuals. It may be related to differences in metabolic and enzymatic efficiency, as described above. Mitton (1993, 1995) has suggested that it may be related to differential rates of protein turnover (Hawkins *et al.* 1989) between homozygous and heterozygous individuals. Many of the proteins involved in metabolic pathways have a short half-life and need to be replaced daily (Koehn 1991). This persistent replacement of proteins is energetically costly and is reduced in heterozygous blue mussels compared with homozygotes (Mitton 1993). This may 'free-up' the energy that is not spent in protein renewal for other purposes, such as growth and maintenance of developmental stability, that may manifest itself in increased fitness for individuals that are more heterozygous. The mechanism behind the fitness advantage of heterozygosity is not clear, and it could differ between species. However, what is clear is that, in many cases, this fitness advantage is manifest.

5.3.2.2 *Heterozygosity and developmental stability*

There have been numerous findings to indicate that individuals that are more heterozygous at enzyme loci show less deviation from the phenotypic mean (e.g. Fleischer *et al.* 1983; Livshits and Kobyliansky 1984) and are usually less asymmetric than more homozygous individuals from the same randomly mating population (e.g. Leary *et al.* 1984; Wolff 1987; Zouros and Foltz 1987). It was Lerner's (1954) initial model that first predicted an inverse relationship between heterozygosity and phenotypic variability. He surmised that heterozygotes can produce a greater variety of biochemical products and hence can canalise their development better under a wide range of environmental conditions. As fluctuating asymmetry represents a measure of developmental homeostasis, and heterozygotes have a canalisation advantage, we may expect there to be a negative relationship between heterozygosity and fluctuating asymmetry.

This predicted negative relationship between heterozygosity and reduced fluctuating asymmetry has been shown in a wide variety of animal species, both within and among populations (see Table W5.3 on the Web site). However, there have also been a considerable number of reports that have found no relationship, or even a positive relationship, between heterozygosity and fluctuating asymmetry in an equally broad range of taxa (see Table W5.3). It is possible that some of the reports that have failed to find a negative relationship between fluctuating asymmetry and heterozygosity may have suffered 'outbreeding depression', that is, crosses between two populations may have resulted in the separation of co-adapted gene complexes (section 5.3.3). This segregation of co-evolved pieces of the genome may have disrupted developmental stability, so giving rise to the sometimes observed positive relationship between fluctuating asymmetry and heterozygosity.

In some studies, the relationship between heterozygosity and asymmetry appears straightforward and clear-cut. For example, Blanco *et al.* (1990) examined that relationship between five polymorphic enzyme loci and fluctuating asymmetry in three meristic characters (gill rakers on the first branchial arch; pectoral fin rays; pelvic fin rays) between two natural and two cultured stock populations of Atlantic salmon *Salmo salar*. In all populations, and in all characters, heterozygotes displayed lower asymmetry. Also, as the number of heterozygous loci increased, fluctuating asymmetry decreased. This effect appeared to be fairly robust, as it was detected by studying a small number of loci; although there was no one particular locus that explained the negative relationship. Blanco *et al.* (1990) also noted that between their populations of salmon, the populations with the highest proportion of heterozygous individuals had the lowest levels of population fluctuating asymmetry.

However, in some studies the relationship between asymmetry and hetero-zygosity is clearly unreliable. For example, 27 meristic and 9 metric characters were assessed for fluctuating asymmetry in the brown hare *Lepus europaeus* and compared with heterozygosity at 13 polymorphic loci (Hartl *et al.* unpublished data, cited in Mitton 1993). In this study, a negative relationship between fluctuating asymmetry and heterozygosity in the meristic traits is reported, but no relationship amongst the metric characters. Suchentrunk (1993) has reported this same phenomenon in brown hares: a positive correlation between hetero-zygosity and fluctuating asymmetry for molar teeth, but a negative correlation for non-metric skull traits. Therefore, reduced levels of allozymic heterozygosity may influence developmental stability and, hence, fluctuating asymmetry in different kinds of traits in very different ways, and it is not necessarily a phenomenon that affects all morphological systems.

The observed relationship between heterozygosity and fluctuating asymmetry may also differ greatly with membership of a particular taxonomic group. It has been proposed that fluctuating asymmetry is more likely to reflect genetic variability in poikilotherms rather than homeotherms (Handford 1980; Wooten and Smith 1986; Novak *et al.* 1993). The ability of an individual to maintain stable developmental trajectories in fluctuating environmental conditions is related to its genetic variability (Teska *et al.* 1991). Homeotherms generally experience very controlled, invariant environmental conditions early in devel-opment, e.g. in mammals due to the constancy of intra-uterine conditions, whereas poikilotherms experience greater environmental fluctuations (Novak *et al.* 1993). Mammals also have a relatively longer period of early controlled development than poikilotherms. These differences may result in fluctuating asymmetry being much lower in homeotherms than poikilotherms. Conversely, poikilotherms may be better adapted to environmental fluctuations and hence homeotherms may in fact be more sensitive to varying developmental condi-tions than poikilotherms. However, there is circumstantial evidence to indicate that heterozygosity and fluctuating asymmetry are more readily related in poikilotherms (e.g. Soulé 1979; Vrijenhoek and Lerman 1982; Leary *et al.* 1983, 1984, 1985a, c) than homeotherms (e.g. Handford 1980; Smith *et al.* 1982; Wooten and Smith 1986). In a meta-analysis of studies reporting a relationship between heterozygosity and asymmetry, Vøllestad *et al.* (1997) investigated the differences in this relationship between poikilotherms and homeotherms. They found that there tended to be a positive association between asymmetry and heterozygosity in homeotherms, whereas this relationship tended to be negative in poikilothermic animals. This suggests that developmental stability is more strongly associated with protein heterozygosity in poikilotherms, perhaps because these organisms are more susceptible to temperature fluctuations which may influence biochemical pathways associated with developmental processes. Hence heterozygous poikilotherms may possess biochemical path-ways that are more buffered against temperature fluctuations than individuals that are relatively more homozygous. There are many other differences between poikilotherms and homeotherms which may also give rise to this apparent difference between these taxonomic groups. These differences can be related to genetic, physiological, or even environmental differences between organisms from these two taxonomic groups.

We must also point out that it is uncertain whether measuring heterozygosity at a small number of loci is a reliable indicator of the heterozygosity of the genome as a whole. In general, it is believed not to be (see Chakraborty 1981; Palmer and Strobeck 1986), but there may be certain conditions in which it could be a good predictor of overall heterozygosity. These conditions are:

(i)   When there is a large amount of linkage disequilibrium (i.e. when the genome is structured into blocks of tightly linked genes) (Leary *et al.* 1984, 1985c).

(ii) When there is a high degree of inbreeding, as heterozygosity at specific loci indicates that an individual has not been subject to inbreeding and so is more likely to be heterozygous at other loci (Weir and Cockerham 1973).

(iii) When there is non-random mating due to a small population size (Chakraborty 1981; see Palmer and Strobeck 1986 for discussion).

It is also noteworthy that these three factors will vary between populations and between species; therefore, even if the same loci are measured in all studies, the validity and accuracy of the estimate of genomic heterozygosity will vary. In practice, researchers have not measured the same loci, and so the problems of interpreting single-locus heterozygosity in terms of genomic heterozygosity are even greater.

Mitton (1993) has argued that heterozygosity at a few polymorphic loci is not likely to represent overall genomic heterozygosity, as the number of loci sampled are far too small (see also Chakraborty 1981, 1987). Mitton (1993) proposes that heterozygosity at certain key loci are likely to be related to developmental stability. His examples of such loci are the malate dehydrogenase (*MDH*) locus in honey bees and lactate dehydrogenase (*LDH*) locus in killifish *Fundulus heteroclitus*. Genotypic variance at the *MDH* locus creates variation in respiration rates and oxygen consumption (Coelho and Mitton 1988). However, Clarke and Oldroyd (1996) claim there is no evidence of heterosis at the *MDH* locus in honey bees, and the asymmetry of only one in five of Mitton's (1993) characters was influenced by heterozygosity (a relationship that becomes non-significant once Bonferroni adjustments for multiple comparisons (Rice 1989) have been applied). In fact, heterozygotes at this locus are intermediate between homozygotes in a number of respects, such as thermostability (Cornuet *et al.* 1994) and oxygen consumption during hovering flight (Coelho and Mitton 1988). Heterozygosity at the *LDH* loci generally increases the amount of oxygen supplied to the muscles during swimming (Mitton 1993), as LDH is closely associated with the amount of ATP in red blood cells (Powers *et al.* 1979). As the *LDH* locus is related to haemoglobin-oxygen efficiency, it may not be too surprising that alterations to gene expression at this locus will influence developmental stability. It is possible that studies that report a significant negative relationship between heterozygosity and fluctuating asymmetry at single gene loci (e.g. Allendorf *et al.* 1983; Leary *et al.* 1983, 1984, 1993; Clarke and McKenzie 1987; McKenzie and Clarke 1988; McKenzie *et al.* 1990; McKenzie and O'Farrell 1993) have targeted the 'correct' loci. This hypothesis could be tested in the literature by comparing the loci screened in 'successful' and 'non-successful' studies. At present, the effect of single-locus heterosis remains unclear; Leary *et al.* (1984) and Clarke and Oldroyd (1996) suggest that variation at *cis*-acting regulators or chromosomal segments in linkage disequilibrium with specific loci may give rise to heterotic effects in some cases.

Overall, genomic heterozygosity may not necessarily be the best indicator of genetic variability, and generally underestimates it (Crow and Kimura 1970). Livshits and Kobyliansky (1985) present evidence to indicate that heterozygosity is linked between loci in humans, the loci are not completely independent. As the genome develops, a number of genetic units co-adapt with each other into a definable complex (Shapiro 1970). These polygenic complexes appear to buffer developmental processes from an external stressor (discussed in section 5.3.3), which when split up, or invaded by deleterious genes, may lower the buffering capacity resulting in developmental asymmetries. There is empirical support for this hypothesis in that humans that suffer major genetic disorders incur greater fluctuating asymmetry (e.g. Sofaer 1979; Barden 1980; Malina and Buschang 1984) as do other animals (e.g. Leary *et al.* 1984). It may be more fruitful to look at the relationship between fluctuating asymmetry and other more reliable measures of genomic variability, such as those based on mitochondrial or nuclear RFLP analysis (see Novak *et al.* 1993). Otherwise, perhaps the number of loci studied should be maximised, or specific loci should be targeted in all studies.

As the mechanisms that determine the fitness advantages of heterozygosity are poorly understood, the mechanisms underlying the relationship between heterozygosity and developmental stability are also contentious. Several genetic models have been proposed to explain an observed relationship between heterozygosity and fluctuating asymmetry: multiple-locus heterosis (Smouse 1986); additive gene action (Chakraborty 1987); and genetic overdominance or because dominance hides deleterious recessives in the heterozygotes (e.g. Soulé 1979; Kat 1982; Livshits and Kobyliansky 1984; Leary *et al.* 1985b; Nei 1987; Blanco *et al.* 1990). Overall, this association between heterozygosity and reduced fluctuating asymmetry may be due to either homozygosity of deleterious recessive alleles or heterozygous advantage (Berger 1976; Makaveev *et al.* 1978; Soulé 1982; Watt *et al.* 1983; Mitton and Grant 1984; Allendorf and Leary 1986; Mitton 1993); or, more specifically, the heterozygous advantage of increased parasite resistance (see Thornhill and Gangestad 1993). Individuals at the geographical extremities of a population are most likely to be more homozygous than those nearer the mean (Møller 1993d), and there is evidence (although not fully convincing) that these individuals have higher levels of fluctuating asymmetry (Soulé and Cuzin-Roudy 1982; Møller 1993d). However,

Zakharov (1989) argues that there is no evidence that individuals with marginal genotypes have greater asymmetry than those with modal genotypes. His data from mice and fruit flies, along with the vertebrate data from Leary *et al.* (1985a), Parker and Leamy (1991) and Soulé and Cuzin-Roudy (1982), indicate that when the asymmetry between left and right sides gets as large as mean trait differences among individuals then more marginal phenotypes are created. There is some evidence to indicate that individuals with these increased levels of asymmetry are pushed, by arithmetic, into the more marginal phenotypes; but there is a general lack of evidence to suggest that marginal genotypes themselves create increased levels of asymmetry.

In a re-evaluation of Soulé and Cuzin-Roudy's (1982) original theory that phenotypically marginal outgroups should display greater asymmetry, Clarke (1995) found new evidence to support some of Soulé and Cuzin-Roudy's statements. Clarke's (1995) re-analysis was centred on identifying outgroups in terms of deviations in the size of left and right sides of a bilateral trait independently of each other, rather than identifying outgroups from the average size of both sides $((L + R)/2)$. This meant that the left and right character values rendered separate outgroups and ingroups. Clarke adopted this approach as Soulé and Cuzin-Roudy's method tended to artificially overestimate the asymmetry of the ingroup, and underestimate the coefficient of variation, due to mathematical artefacts (see Soulé and Cuzin-Roudy 1982; Clarke 1995). Clarke found that in 50% of cases (of data from four insect species *Lucilia cuprina, Apis mellifera, Chrysopa perla,* and *Musca vetustissima*) outgroups displayed greater asymmetry; although there was no convincing evidence to indicate that this effect was due to increased homozygosity in outgroups. Clarke suggests that developmental noise will invariably swamp the asymmetry determined directly by the underlying genotype in non-fitness traits due to their lower buffering capacity. Hence, non-fitness traits will be sensitive to environmental perturbations and display larger environmentally induced asymmetries. Therefore, Clarke concludes that homozygosity will not be reliably related to asymmetry in most characters; although it is possible that homozygosity of specific loci could directly influence developmental stability.

Crusio and van Abeelen (1993) have proposed that there should never be a simple relationship between heterozygosity and fluctuating asymmetry because of the different selection pressures that may act within natural selection. They argue that although natural selection may, in part, act to canalise development, it may also act to keep it 'malleable' so that developmental processes can respond to environmental change. By crossing inbred strains of mice, they have demonstrated that the links between heterozygosity and developmental stability may not be straightforward, and will depend on the nature of selection pressures acting on the population. This hypothesis is further supported by findings that splitting up co-adapted gene complexes and episodes of directional selection also act to disrupt developmental stability (see section 5.3.3 and Chapter 3).

In an effort to clarify the debate over heterozygosity and fluctuating asymmetry, Vøllestad *et al.* (1997) have conducted a meta-analysis of published and unpublished investigations that report a relationship between these two variables. In this analysis, Vøllestad *et al.* utilised composite asymmetry scores (i.e. sum of asymmetry scores for individual traits, or number of asymmetric characters) where possible; but some studies rendered only single-trait asymmetry values. Although the number of loci screened by biochemical analysis and the selection of these loci differed greatly among studies, Vøllestad *et al.* chose to use sample size as the weighting factor in their meta-analysis. To reduce any possible phylogenetic biases, as the majority of studies have been performed on fish species, Vøllestad *et al.* calculated a weighted mean effect size for each species, with their weighting variable being the number of individuals sampled in each study. This approach allowed them to investigate the relative size of effects in both within-population and among-population studies.

Overall, the total mean effect indicated a negative relation between heterozygosity and fluctuating asymmetry ($z_r = -0.113$, $N = 35$); although the effect differed between type of study ($z = -2.01$, $P = 0.027$). Within-population studies indicated that there was no relation between heterozygosity and asymmetry ($z_r = 0.004$, $P = 0.41$), but there was a non-significant trend for a negative relation between heterozygosity and fluctuating asymmetry among populations ($z_r = -0.15$, $P = 0.089$). Both of the P values were generated from one-tailed tests of significance, therefore the relationships between heterozygosity and asymmetry may be weaker than Vøllestad *et al.* reported. Vøllestad *et al.* also indicated that effect sizes differed between poikilothermic and homeothermic animals ($F_{1,100} = 14.39$, $P < 0.001$), and that there was a significant interaction between type of study and thermoregulatory group ($F_{1,100} = 5.27$, $P = 0.024$). In homeotherms, there was a positive relation between asymmetry and heterozygosity both within and among populations. Whereas in poikilotherms, there was no relation within populations, but a negative relation between fluctuating asymmetry and heterozygosity among populations.

Temperature fluctuations are likely to influence the functional efficiency of enzymatic pathways, and the study of enzyme kinetics indicates that allozymes vary in their response to temperature (Zera *et al.* 1985). Therefore, heterosis may be more prevalent in local environments that experience greater fluctuations in temperature (Houle 1989). As the enzymatic pathways of poikilotherms are liable to be under the influence of greater variations in temperature than those of homeotherms, we may expect heterosis to be stronger in poikilotherms, as indicated by the analysis of Vøllestad *et al.*

Therefore, there is some evidence to indicate that there is a negative relation between heterozygosity and fluctuating asymmetry among populations, especially in poikilotherms. But there is no general relation between heterozygosity and asymmetry within populations. In general it could be concluded that enzymes catalyse metabolic reactions, and so altering the genetic variation at enzyme-coding loci is likely to have some influence on developmental processes and homeostasis. Therefore, heterozygosity can influence developmental stability, although the strength of this association may alter among loci, taxa, and thermoregulatory group.

5.3.3 *Hybridisation*

The proper functioning of a biological system is dependent on the cooperation of its component parts. In physico-chemical processes this entails cooperation between interacting gene products, and so indirect cooperation between gene loci. Therefore, optimal biological performance is related to the interaction of a number of gene products. This has led to the evolution of interacting, cooperative groups of loci, otherwise known as co-adapted gene complexes. Populations that have completely, or nearly completely, become reproductively isolated may possess different co-adapted gene complexes. Combining such different genomes by hybridisation may upset development because the new allelic combinations in the hybrids have not been subjected to natural selection; this would result in greater fluctuating asymmetry in hybrids (Soulé 1967; Soulé and Baker 1968; but see Jackson 1973a). It has also been proposed that hybridisation actively disrupts co-adapted complexes of suppresser genes that travel around the genome and induce mutations at new sites (Thompson and Woodruff 1978). This would lead to a disruption of developmental stability within the genome. In support of this, there is evidence that mutation rates rise significantly following hybridisation (Thompson and Woodruff 1978).

Hybridisation has been shown to increase levels of fluctuating asymmetry in *Drosophila* (Tebb and Thoday 1954) and many species of fish (Graham and Felley 1985; Leary *et al.* 1985b). For example, Graham and Felley (1985) investigated morphological asymmetry in hybrid and 'pure' populations of banded sunfish *Enneacanthus obesus* and blue-spotted sunfish *E. gloriosus*. Within traits, the hybrid populations tended to be more asymmetric and also possess more overall asymmetry than the pure populations (Fig. 5.2). Graham and Felley (1985) also noted that heterozygosity was positively associated with asymmetry, indicating that the disruption of co-adaptation within the genome has more of a deleterious effect on developmental stability than the possible stabilising influence of increased heterozygosity. Hybrids between yellow-shafted flickers *Colaptes auratus auratus* and red-shafted flickers *C. a. cafer* show greater asymmetry in the length of their malar patch than the original pure species (Graham, Chamberlin-Graham, and Moore, unpublished data reported in Graham 1992). Leary *et al.* (1985b) showed that, in the wild, hybrids between artificially introduced brook trout *Salvelinus fontinalis* and endemic bull trout *S. confluentus* display greater asymmetry and are sterile. Also, hybrids between two species of fire ant (*Solenopsis invicta* and *S. richteri*) display greater fluctuating asymmetry than pure species (Ross and Robertson 1990).

However, in a number of cases, there is a decrease of fluctuating asymmetry in hybrids from inbred strains. This discrepancy can be explained by the separation of co-adapted gene complexes causing a decrease in the buffering capacity but not outweighing the positive effects of increased heterozygosity (Dobzhansky 1970; Leary *et al.* 1985c). Alibert *et al.* (1994) have demonstrated that there can be a decrease in fluctuating asymmetry in a hybrid zone, in this case between subspecies of the house mouse (*Mus musculus domesticus* and

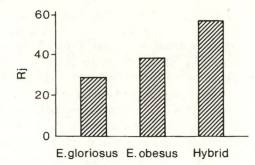

Fig. 5.2 Standard deviations of asymmetry values for bilateral morphological characters against population from which the measurements were taken. R_j is the sum of the populations' ranks for asymmetry variance across all characters. The characters measured were: lateral line scale number, pored lateral line scale number, pectoral fin ray number, number of scales above lateral line, eye diameter, cheek depth, and distance from the posterior edge of the orbit to the posterior edge of the opercle. Hybrids were more asymmetric than the original populations. Adapted from Graham and Felley (1985), which includes more explanatory details.

M. m. musculus). These studies indicate that the links between hybridisation and fluctuating asymmetry are not necessarily straightforward. In some systems, hybridisation will not separate co-adapted gene complexes and the benefits of increased heterozygosity may help to increase developmental stability and hence reduce fluctuating asymmetry. However, these cases are relatively infrequent in the literature. For example, Zakharov and Bakulina (1984) found that asymmetry was increased in hybrid outcrosses from three inbred *Drosophila* populations compared with the initial inbred lines themselves. This indicates that separation of gene complexes disrupts developmental stability, which outweighs the heterozygous advantage that the outbred lines may encounter, an advantage that could have acted to restore developmental stability.

Parker *et al.* (1985) used sperm from 10 species of fish (Centrarchidae) to fertilise eggs of Florida large mouth bass *Micropterus salmonides floridanus*. They observed an increased occurrence of phenodeviants and other developmental anomalies as the genetic distance between the Florida bass and paternal species increased. These observations further support the hypothesis that co-adaptation of gene complexes is related to developmental stability, as hybrids between closely related species, which are likely to have similar gene complexes, exhibit greater developmental stability than hybrids between species that are evolutionarily more distant and hence have gene complexes that are more disparate. The similar gene complexes of closely related species appear to be able to mix without an increase in developmental instability.

It may be significant that re-coadaptation between gene complexes can take place relatively quickly in natural populations of hybrids, and hence developmental stability will be restored in a relatively short period of time (Felley 1980). Thus, some studies that do not report an increase in fluctuating asymmetry in hybrid zones may have had sufficient time to evolve 'new' co-adapted gene complexes (e.g. Jackson 1973a, b; Felley 1980; see Graham and Felley 1985). Hybrids between the frog species *Hyla cinerea* and *H. gratinosa* do not exhibit greater fluctuating asymmetry, which is especially surprising because they are recent hybrids (Lamb *et al.* 1990). This pattern (or lack of pattern) has also been observed by Jackson (1973a, b) in lizards *Sceloporus woodi* and *S. undulatus*; but these lizards may have re-evolved co-adapted gene complexes, as explained above. Also it has been demonstrated that there is little genetic difference between the two lizard species studied, indicating that there should be little cost to hybridisation (Joswiak *et al.* 1985). There are other examples where researchers have failed to uncover a relationship between hybridisation and fluctuating asymmetry. For example, Vrijenhoek and Lerman (1982) observed no loss of stability in hybrids of *Poeciliopsis monada* and *P. lucida*. At least in the case of the latter study, the lack of stability loss may be related to the high levels of heterozygosity seen in these hybrids. So, with the exception of Lamb *et al.*'s (1990) investigation, most studies suggest that hybrids do suffer a decrease in developmental stability, although this may be partly compensated by increased heterozygosity or recent re-coadaptation.

5.3.3.1 *Genetic balance*

The apparent association of increased fluctuating asymmetry and hybridisation has also been viewed in terms of 'genetic balance'. Levels and expression of fluctuating asymmetry are believed to be, at least, in partial control by polygenic complexes (e.g. Mather 1953; De Marinis 1959; Rasmuson 1960; Ferguson 1986; Livshits and Kobyliansky 1989), although the heritability of fluctuating asymmetry for individual traits has been reported as low (section 5.2). Soulé (1982) proposed that fluctuating asymmetry may result from additive effects of polygenic loci and that this index of developmental stability may result from a measure of 'genic balance'. However, there are a number of researchers who believe that fluctuating asymmetry is mainly a non-additive phenomenon, relating to epistasis or dominance within the genome (Goldschmidt 1955; Thoday 1958; Livshits and Kobyliansky 1985, 1989, 1991; Clarke and McKenzie 1987; Clarke *et al.* 1992). However, recent meta-analysis indicates that there is significant additive genetic variance, although this does not preclude the effects of epistasis (Møller and Thornhill 1997a; section 5.2).

Splitting up co-adapted gene complexes by hybridisation may decrease developmental stability through a number of ways. Whitt *et al.* (1977) have demonstrated that genes can be expressed at inappropriate stages of development (i.e. too late or too early) in hybrids of largemouth bass *Micropterus salmoides* and green sunfish *Lepomis cyanellus*. This may have arisen for a number of reasons which include cytoplasmic incompatibility of mitochondrial and nuclear DNA (see Whitt *et al.* 1977). It is also possible that disruption of the genome by hybridisation may alter regulatory loci that code for enzymes in the major amino acid synthesis pathways. Burton (1990) has observed amino acid imbalances in hybrids of the copepod *Tigriopus californicus*, which may disrupt developmental stability.

Hybridisation may also increase the likelihood of genomic and somatic mutations. Some elements of the genome are mobile, and changing the composition of these elements could induce large rates of random transposition into crucial parts of the genome (Gvozdev 1986). If the parts of the genome affected are directly involved in development, then developmental stability may be affected in some way. Alternatively, transposable elements might disrupt co-adapted gene complexes (section 4.3.3).

5.3.4 *Mutant genes*

Introducing a mutant gene will usually change the course of development and so upset developmental stability. Developmental stability can only be restored after the genotype has had time to readjust to the new developmental pathways. In other words, incorporation of a mutant gene dictates that the balance achieved by the genome is no longer valid, so morphogenesis becomes unstable and asymmetries arise during developmental processes. There is evidence that incorporation, by natural selection, of a new mutant influences levels of fluctuating asymmetry. For example, the Australian sheep blowfly *Lucilia cuprina* evolved resistance to the insecticides dieldrin in 1957 and diazinon in 1967 (Clarke and McKenzie 1987; McKenzie and Clarke 1988). Flies susceptible to either pesticide and those resistant to diazinon have similar but significantly lower levels of fluctuating asymmetry than those resistant to dieldrin alone. Dieldrin is an insecticide to which they have not been exposed for over 30 years. Furthermore, incorporation of the diazinon resistant allele into a new genetic background through repeated backcrossing markedly increased fluctuating asymmetry. When the diazinon resistant allele was first incorporated into natural populations (20 years ago), it adversely affected development and increased levels of fluctuating asymmetry. Subsequently, selection has apparently acted to restore co-adaptation by favouring alleles at other loci that reduce harmful developmental effects of the diazinon resistant allele, and has returned fluctuating asymmetry to the levels now seen in wild-type flies (Fig. 5.3). In a study of the effects of mutant genes on human developmental stability, Shapiro (1983) has provided evidence that chromosomal mutations may alter the gene product dosage and hence alter biochemical pathways and disrupt developmental homeostasis in Down's syndrome patients.

By artificially inserting a mutant gene for muscular dysgenesis (the arrest of myogenesis and degenerative alterations to skeletal structures) into mice, Atchley *et al.* (1984) demonstrated that the developmental stability of traits related to skeletal muscle production can be disrupted (Table 5.1). Fluctuating asymmetry of other traits did not appear to be affected. This indicates that alterations to specific genes alter developmental pathways and stability of specific traits, and not developmental pathways as a whole. Goldschmidt (1940), in his contentious theory, proposed that mutations can induce major phenotypic changes (i.e. create a monster), the bearers of which may be able to

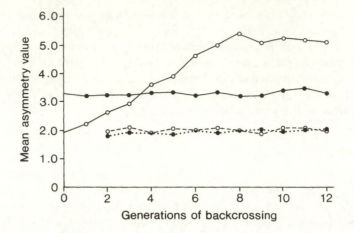

Fig. 5.3 Mean asymmetry of sheep blowflies in resistant heterozygotes and in susceptibles at each generation of backcrossing of *Rop-1/Rop-1* and *Rdl/Rdl* to M₁5. Resistance in these flies to the insecticides is mainly determined by these two unlinked genes (diazinon, *Rop-1*, chromosome IV; dieldrin, *Rdl*, chromosome V). Open symbols represent *Rop-1/+*; closed symbols are *Rdl/+*. Broken lines represent the wild-type flies from *Rop-1/Rop-1* (open symbols) and *Rdl/Rdl* (closed symbols) backcrosses. Range of s.e., 0.06–0.10. Adapted from Clarke and McKenzie (1987).

Table 5.1 Variance estimates for the difference between left and right sides of the mouse mandible. Variances are within-group mean squares from an ANOVA between left and right sides. $+/+$, homozygotes; $+/mdg$, heterozygotes with introduced muscular dysgenesis gene; F, variance ratio of $(+/+)/(+/mdg)$ with associated P value. Adapted from Atchley *et al.* (1984), which contains further details.

Trait	$+/+$	$+/mdg$	F	P
Posterior mandible length	0.1456	0.1652	1.14	>0.10
Anterior mandible length	0.2457	0.1799	0.73	>0.10
Height at mandibular notch	0.1813	0.3371	1.86	0.05
Height at incisor	0.3869	0.8006	2.07	0.03
Concavity	5.5520	4.4971	0.81	>0.10
Height of ascending ramus	0.4799	0.3841	0.80	>0.10
Condyloid width	0.9778	1.6844	1.72	0.08
Condyloid length	2.1389	1.3473	0.63	>0.10
Coronoid height	1.7448	3.5612	2.04	0.03
Coronoid area	7.3324	11.4142	1.56	>0.10
Angular process length	1.4804	1.3841	0.93	>0.10
Tooth-bearing area	0.3256	0.5796	1.78	0.06
Superior incisive process curve	6.4838	8.8047	1.36	>0.10
Inferior incisive process curve	3.8864	2.7898	0.72	>0.10

survive in novel conditions or perform new tasks ('hopeful monsters'), reproduce, and hence be subjected to further selection that will hone their adaptations. This theory of evolutionary processes is not generally supported by
modern evolutionary biologists, although there is evidence that there are some
mutations that can induce major phenotypic changes (e.g. indirect effects of
early lethals) and hence alter many developmental pathways simultaneously
(discussion in Maynard Smith 1975).

In conclusion, there is convincing evidence that incorporation of novel genetic
elements into the genome, by both artificial and natural selection, results in
decreased developmental stability; although it is likely that the identity of the
introduced mutant specifies which developmental pathways are destabilised.

5.4 Summary

- In this chapter we review the evidence that indicates there is a significant
  heritable component to developmental stability.
- We also describe the genetic factors that can lead to a reduction in
  developmental stability.
- Inbreeding often leads to unstable development.
- Heterozygosity at enzymatic loci can be associated with increased developmental stability among populations, especially in poikilotherms; however,
  there is no general relation between heterozygosity and stability within
  populations.
- Hybridisation the separates co-adapted gene complexes can give rise to
  unstable development.
- Additionally, incorporation of novel mutants into the genome can disrupt
  developmental stability.
- Many of these genetic factors are not independent of each other and, hence,
  genetic changes in natural populations may produce unpredictable changes
  in developmental stability. Therefore, the observations reported here must be
  applied with caution when interpreting and predicting the relationship
  between genetic factors and developmental stability in uncontrolled conditions.
- These studies are useful in that they illustrate the intimate ways in which
  developmental stability can reflect aspects of the genetic make-up.

6

Causes of developmental instability. II. Environmental factors

6.1 Introduction

A large number of environmental factors can result in a reduction of developmental stability and increased levels of fluctuating asymmetry in bilateral traits. The list of factors is almost endless and the review given here is not exhaustive but merely a representation of the kinds of factors that researchers have thought important in explaining the variations in fluctuating asymmetry that they have observed. These may be tied to the natural ecology of the species studied, or limited to the laboratory equipment through which the investigator can manipulate the developmental conditions of the study organism.

When an organism is exposed to a novel set of environmental conditions, or even a novel element within the same environment, developmental processes become unstable and levels of fluctuating asymmetry appear to rise. What we hope to do in this chapter is to give accounts of the types of environmental elements and changes that can produce these increases in asymmetry, and also give some indication of how these environmental influences may give rise to the observed disruptions of developmental stability. It is important to note that the environmental factors detailed in this chapter are far from independent in their effects; in fact many of them will interact at very obvious levels. In the experiments that are based on observations from organisms *in vivo*, many of the environmental factors are confounded. It is only in the carefully designed laboratory based experiments that the influence of one environmental variable can be separated from another.

Here we present evidence to indicate that a range of external factors can disrupt developmental stability. Abnormal ambient temperatures have been shown to increase developmental asymmetries in flies, rats, mice, snakes, fish, and lizards (section 6.3). Nutritional stress appears to be directly related to developmental stability, as birds subjected to increasingly harsh food deprivation regimes develop increasingly asymmetric wing feathers during their moult. This seems not only to be an effect of food deficiency but also the unpredictability of food sources (section 6.4). In section 6.5 we review the relation between chemical factors and fluctuating asymmetry. Most of these cases relate to environmental pollution of waterways, but there have been some experimental laboratory analyses of the effects of chemical agents. Rats exposed to an hallucinogenic drug and grunion growing in water polluted with DDT all exhibit greater levels of fluctuating asymmetry. High population density has been shown to be positively related to asymmetry in field populations of shrews and laboratory populations of blowflies, fruit flies, rainbow trout, and chickens (section 6.6). Similarly, increased audiogenic stimulation appears to induce asymmetries in the development of laboratory rats and other rodents (section 6.7). The influence of parasitic infection on developmental stability will be discussed in Chapter 9.

In section 6.8 we indicate that certain traits may be more susceptible to environmental factors than others. These appear to be ornamental traits that are subject to directional sexual selection. There may also be stress-specific phenotypic responses, which are detailed in section 6.9. We also discuss the nature and interaction of external environmental factors and how they can be viewed to act synergistically with internal genetic factors. Developmental stress could be viewed in terms of the efficiency of energy transfer mechanisms within cells, which may provide a useful framework under which we can interpret the developmental effects of both environmental and genetic influences (section 6.10).

As developmental instability can be caused by factors acting from the environment, fluctuating asymmetry can be used as a sensitive indicator of environmental conditions during morphogenesis (section 6.11).

6.2 Stress and energetic expenditure

In section 4.2.1 we discussed stress in terms of increased energetic expenditure as a consequence of exposure to suboptimal habitats. We also indicated that under stressful conditions there may be specific costs endured by organisms, such as increased levels of corticosterone or increased metabolic rates (Selye 1974). Therefore, if an organism, or population of organisms, experiences suboptimal environmental conditions energetic efficiency is reduced and the stability of developmental pathways will be disrupted as energy is diverted away from maintaining developmental stability. This will lead to greater phenotypic variance (section 4.3.5) and to the production of asymmetric phenotypes in suboptimal environmental conditions (cf. Parsons 1994a). In sections 6.3–7 we review the evidence which indicates that developmental instability can be caused by an number of agents acting from the environment, and that these factors result in greater asymmetries due to their disruptive effects on development.

6.3 Adverse temperatures

Skinnes and Burås (1987) examined the developmental stability of one Mexican and six Norwegian strains of wheat *Triticum aestivum* in response to a range of constant temperatures during seed set and seed development. The coefficient of variation of a range of agronomic characters increased at temperatures both above and below the optimal developmental conditions. Additionally, there were differences among the strains in their buffering capacity to abnormal temperatures; the Mexican strain tended to be more stable in aberrant temperature conditions than any of the Norwegian strains. This may have resulted from the breeding strategy of the Mexican strain, which is found at both high elevation (i.e. low temperatures) and at sea level (i.e. high temperatures) in Mexico; whereas the Norwegian strains have a more limited natural range.

Increased temperature has led to more asymmetry, especially with increasing maternal age, in *Drosophila* (Beardmore 1960; Parsons 1961, 1962; Heiskanen *et al.* 1984; but see Wakefield *et al.* 1993). Thoday (1955) has observed the same phenomenon in *Drosophila*, which appears to become further exaggerated under the additional effect of inbreeding. In a similar fashion, rearing temperatures above and below normal conditions have been observed to increase levels of fluctuating asymmetry in Australian sheep blowflies *Lucilia cuprina* (Clarke and McKenzie 1992; McKenzie and Yen 1995; Fig. 6.1). Additionally, thale cress *Arabidopsis thaliana* exhibits more stable growth trajectories at medium to low temperatures (Griffing and Langridge 1963). Beardmore (1960) demonstrated that strains of *D. melanogaster* exhibit a loss of developmental stability when reared in incubators that had fluctuating temperature conditions. Developmental pathways appeared to be more stable, and hence more traits were more symmetric, when the ambient temperature was kept constant. These early observations have since been repeated; *D. melanogaster* reared in a constant 25°C environment showed lower levels of fluctuating asymmetry than flies raised in a fluctuating environment from 20°C to 29°C (Bradley 1980). So even though the average ambient temperature did not differ, the temperature changes in the fluctuating environment appeared to result in unstable development. This may be due to the fluctuations themselves or the highest (or lowest) temperature experienced.

Increased water temperatures correlate with increased fluctuating asymmetry in the fins of rainbow trout (Leary *et al.* 1992). A similar relationship has been demonstrated in the Siberian sturgeon *Acipenser baeri* (Ruban 1992). High temperatures also decrease developmental stability during growth processes in rats (Riesenfeld 1973). At the other end of the temperature scale, Fox *et al.* (1961) found that low temperatures increased the asymmetric embryonic development of garter snakes *Thamnophis elegans*. Similarly, Siegel and Doyle (1975b) have reported increases in fluctuating dental asymmetry for mice born and raised in cold environments. The same investigators have also indicated that cold stress increases asymmetry in humeri of *Peromyscus floridanus* and *P. gossypinus* (Siegel and Doyle 1975c). Zakharov (1989) demonstrated that ambient temperatures above and below normal incubation temperatures increase levels of scale fluctuating asymmetry in two species of sand lizard *Lacerta* spp. Gest *et al.* (1983, 1986) have indicated that two groups of neonatal

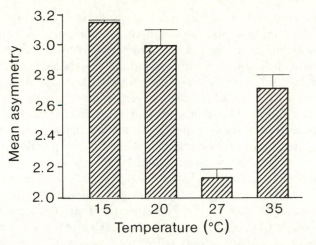

Fig. 6.1 Mean (+ s.e.) asymmetry value versus rearing temperature (C) in Australian blowflies. Mean asymmetry value is a composite measure of left–right asymmetries for three meristic traits (refer to Clarke and McKenzie (1992) for more details). Values are means of replicates. Adapted from Clarke and McKenzie (1992).

rats, one stressed with heat and the other exposed to cold conditions, both exhibit greater femoral asymmetries than unstressed control rats. Similar experimental treatments have demonstrated that both heat and cold stress can lead to thinner cortical bone in mice and rats (Brandt and Siegel 1978) and greater asymmetry in the parietal bones of rats (Mooney *et al.* 1985) than unstressed control animals. Therefore, it would appear that high, low, and fluctuating temperatures can all act to decrease the stability of developmental pathways.

6.4 Nutritional stress

Kirpichnikov (1981) stated that temperature and food supply are the most common environmental factors influencing the frequency of phenodeviants within fish populations. There have been several experimental investigations of nutritional (or energetic) stress and its relation with developmental stability that indicate that this form of environmental variable can have a marked effect on developmental pathways. Increased levels of nutritional stress have been shown to increase developmental instability in *Drosophila* (Parsons 1964), rats (Sciulli *et al.* 1979), mice (Erway *et al.* 1970), and European starlings *Sturnus vulgaris* (Swaddle and Witter 1994). Between-habitat differences in nutritional stress have been implicated as a cause of inter-population fluctuating asymmetry differences in the Montana grizzly bear *Ursus arctos horribilis* (Picton *et al.* 1990). European nuthatches *Sitta europaea* were observed in induced moult to grow fourth retrices that tended to be more symmetrical when provisioned with food supplements during winter than when the nuthatches relied on the naturally available (lower) food abundance (Nilsson 1994); however, this result was not significant. Nilsson also observed that the induced feathers (developed during the winter) were more asymmetric than those grown during natural moult in the summer. Food availability was higher in the summer compared with the winter and hence Nilsson (1994) concluded that nutritional status was related to feather asymmetry. However, this correlation does not indicate that energy intake (or seasonality) is causally related to levels of fluctuating asymmetry, as the process of feather plucking and induced growth (that only occurred in winter) may have disrupted developmental stability through an unidentified stress-related mechanism. A more formal investigation into the relation of nutritional stress to feather asymmetry was performed by Swaddle and Witter (1994).

In their study, Swaddle and Witter (1994) monitored levels of fluctuating asymmetry in primary feather lengths during the moult of captive European starlings. These birds were randomly assigned to four treatment groups that experienced varying levels of nutritional stress. The first group was a control group and received *ad libitum* food and water. The second group of birds had their food source removed for four hours at 'dawn', and were termed 'morning-deprived' birds. The third group had their food removed for four hours beginning four hours after dawn, and so were referred to as 'afternoon-deprived' birds. The last group of birds randomly received either a 'morning' or an 'afternoon' food deprivation, so the birds could not predict at what time their food source would be removed. This group of birds were termed 'variable-deprived'. Swaddle and Witter found that birds in the food deprivation groups exhibited greater fluctuating asymmetry in their primary feathers at the end of moult than controls (Fig. 6.2a). Within the three deprivation groups, variable-deprived birds showed greater asymmetries than both morning- and afternoon-deprived groups; there was no difference in primary asymmetry between morning and afternoon birds. This indicates that the nutritional stress of food deprivation induces greater asymmetries during moult, and that unpredictability of the food deprivation creates relatively more developmental stress than either morning or afternoon fixed times. Therefore it is not only energy intake that is important, but also the predictability of this energy intake. These conclusions are further supported by the observation that levels of primary feather asymmetry were negatively related to subcutaneous fat levels, once differences due to the experimental treatments were controlled for (Fig. 6.2b). The most symmetric individuals stored the most fat, which would act as a buffer against the nutritional stress of the food deprivations. This experiment implies a causal role for nutritional or energetic stress in the production of feather fluctuating asymmetries. Birds under greater nutritional stress exhibit greater levels of fluctuating asymmetry.

6.5 Chemical factors

A number of different chemical factors have been shown to increase developmental instability in both laboratory and field based studies. In laboratory

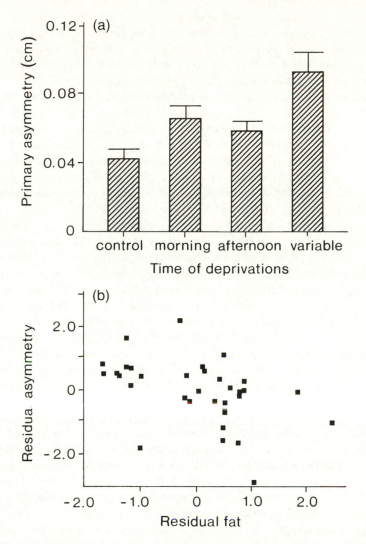

Fig. 6.2 (a) Mean (+ s.e.) primary feather asymmetry against food deprivation treatments in the European starling. Control, no deprivation; morning, food-deprived in morning; afternoon, food-deprived in afternoon; variable, deprived randomly either in morning or afternoon. Deprivations affected primary asymmetry (Kruskal-Wallis, $H =$ 11.51, d.f. = 3, $P = 0.01$) so that food-deprived birds were more asymmetric than controls ($W = 64.5$, $P = 0.017$), and the effect was stronger with variable deprivations ($W = 196.5$, $P = 0.025$). (b) Residual primary asymmetry against residual mean fat score, controlling for the effects of food-deprivation treatments and individual dominance. There is a significant negative relation between asymmetry and fat. Adapted from Swaddle and Witter (1994).

conditions, Siegel *et al.* (1977) dosed pregnant rats with concentrations of an hallucinogenic drug. They found that drug-treated rats gave birth to young with more dental asymmetries; although surprisingly within the treatment levels, rats exposed to the lower concentrations produced offspring that were more asymmetric than mothers that had been treated with a higher level of the drug. Graham *et al.* (1993b) treated larval *Drosophila melanogaster* with various concentrations of lead and benzene in a laboratory experiment. They observed that the concentrations of lead and benzene did not influence the number of adults that emerged from these larvae or the total number of sternopleural bristles found on each adult. However, both lead and benzene increased the asymmetry in number of these bristles. This particular experiment also indicates, as initially proposed by Yablokov (1986), that fluctuating asymmetry could be used as a sensitive indicator of environmental perturbations before it directly affects viability of a population (see section 6.11).

Many of the field observations relating chemical factors to fluctuating asymmetry have been conducted in areas where there is a high level of chemical pollution, and hence have been driven by the ecological incentives of monitoring habitat degradation. For example, high levels of fluctuating asymmetry have been found in species of fish inhabiting waters around industrial centres and in ponds containing high concentrations of mercury and low pH values (Ames *et al.* 1979; Zakharov 1981). Fish species inhabiting Swedish waters contaminated with heavy metals and arsenic showed an increased frequency of phenodeviants compared with fish from less polluted areas (Bengtsson *et al.* 1985). Also, fluctuating asymmetry levels of marine and freshwater invertebrates have been shown to be greater in an area polluted by a fertiliser manufacturing facility (Clarke 1993a). In a separate study, Clarke (1993b) demonstrated that the fly *Chrysopa perla* situated near an agrochemical factory exhibited increased frequencies of phenodeviants compared with individuals residing in more distant areas, although there was no apparent difference in levels of fluctuating asymmetry. Brook trout inhabiting acidified lakes exhibit greater fluctuating asymmetry than those in more neutral lakes (Jagoe and Haines 1985). Fox *et al.* (1991) have reported a far greater occurrence of bill abnormalities and asymmetries in double-crested cormorants *Phalacrocorax auritus* from the Great Lakes area of North America where there are high incidences of polyhalogenated aromatic hydrocarbons (e.g. PCBs) than in birds from relatively unpolluted prairie localities. Grey seals *Halichoerus grypus* inhabiting polluted regions of the Baltic are more asymmetric than those from less polluted areas (Zakharov and Yablokov 1990). Similarly, grunion *Leuresthes tenuis* from populations off the southern coast of California, where pollution is believed to be high, are more asymmetric than those from Baja California, which is relatively less polluted (Valentine *et al.* 1973). In a subsequent laboratory investigation Valentine and Soulé (1973) manipulated levels of a toxic chemical, *p,p'*-DDT (1,1,1-trichloro-2,2-bis(*p*-chlorophenyl)ethane), in a flowing water experiment. They placed 400 fry in each of nine increasing concentrations of *p,p'*-DDT and measured the resulting variance of left minus right values in number of pectoral fin rays (V_a) for each population. They found a positive relation between V_a and strength of toxin concentration (Fig. 6.3). Grunion that developed in the most polluted water had increased pectoral fin ray asymmetry. Valentine and Soulé also performed a static water experiment, within the same study, to examine congenital differences in asymmetry. They collected grunion eggs from the same sites studied by Valentine *et al.* (1973) and found that fry developing from eggs from the least polluted areas were the most symmetric. This result was attributed to low adult and egg DDT burdens in the less polluted areas.

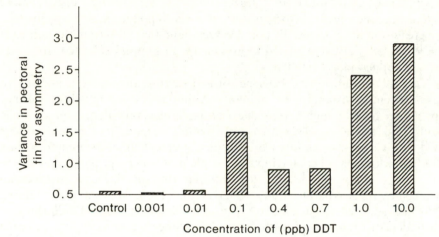

Fig. 6.3 Variance in pectoral fin ray asymmetry (left minus right) values, V_a, of grunion raised in varying concentrations of DDT (ppb). Adapted from Valentine and Soulé (1973).

These types of studies have not been restricted to measuring the developmental impact of aquatic pollution. Clarke and Ridsdill-Smith (1990) found that Australian bush flies that develop in dung from cattle that had been treated with the pesticide Avermectin B grew wing veins that were more asymmetric than flies developed in dung from untreated cattle. Also, increased heavy metal concentrations have been implicated in causing high fluctuating asymmetry levels in the common shrew *Sorex araneus* (Pankakoski *et al.* 1992). Additionally, barn swallows *Hirundo rustica* that have been exposed to increased levels of atmospheric radiation, due to the Chernobyl reactor leak, show greater asymmetries than those from unpolluted areas of the Ukraine (Møller 1993b).

In a similar way to Møller's (1993b) study of the impact of pollution in the Ukraine on developmental asymmetry in the swallow, Graham *et al.* (1993a) conducted a survey of plant asymmetry at points of increasing distance from the sources of chemical pollution. In both species that they studied, an annual plant *Convolvulus arvensis* and the black locust tree *Robinia pseudoacacia*, leaf asymmetry was greatest in areas closest to the source of pollution. This study was expanded to include other plant species, such as a herbaceous umbel *Aegopodium podagaria* and a perennial herb *Epilobium angustifolium* at sites in northern Russia (see Freeman *et al.* 1993). Similar patterns were observed; plants closest to the epicentres of the chemical pollution were the most asymmetric. Freeman *et al.* (1993) also monitored levels of bilateral symmetry in three populations of brown alga, *Fucus furcatus latifrons*, off the Washington coastline. The three populations experienced varying levels of environmental pollution. In accord with their Russian studies, they found that algae growing in the most polluted areas exhibited the largest asymmetries. Additionally, Kozlov *et al.* (1996) have found that leaves of birch trees *Betula* display more asymmetry when situated close to sources of aerial pollution, that asymmetries are greatest in the most heavily polluted areas, and that nickel concentrations in the leaves are positively related to asymmetry in samples collected from sites around copper–nickel smelters.

A series of studies have been performed linking alcohol consumption to developmental instability in humans. Although alcoholic adults do not appear to have a higher frequency of phenodeviants than non-alcoholic adults (Gualteri *et al.* 1982), it has been demonstrated that alcoholic mothers tend to give birth to children that display greater dental asymmetries than children born to mothers having a lower alcohol consumption (Kieser 1992). This latter study did not include children that displayed foetal alcohol syndrome. In a separate study, Qazi *et al.* (1980) showed that children who had suffered foetal alcohol syndrome had an increased frequency of phenodeviants compared with controls. Wilber *et al.* (1993) compared dermatoglyphic asymmetry of children who had suffered foetal alcohol syndrome with those who displayed only foetal alcohol effects. Asymmetry was seen to increase from control children through the foetal alcohol effect group and was greatest in the foetal alcohol syndrome group. Therefore, the asymmetry increased with the severity of the alcohol's effect. Alcohol is not the only chemical substance that humans frequently enter into their bloodstream; in a study conducted within university departments. The influence of smoking on developmental stability is not straightforward, as Kieser and Groeneveld (1994) have shown that parental smoking has no effect on levels of asymmetry of foetuses or neonates.

6.6 Population density

In a cyclic population of the common shrew in Siberia, it was observed that population density was positively correlated with fluctuating asymmetry values of offspring born that year (Zakharov *et al.* 1991). Hwu and Thseng (1982) reported that the coefficient of variation of numerous agronomic characters among strains of rice *Oryza sativa* increased with increasing plant densities. Overcrowding the larvae of the Australian sheep blowfly, increasing larval competition, has the effect of increasing subsequent adult fluctuating asymmetry (Clarke and McKenzie 1992), but this effect has only been observed in wild type and not in resistant flies that already exhibit high levels of asymmetry (McKenzie and Yen 1995). Clarke and McKenzie (1992) performed this experiment in laboratory rearing conditions and were interested in using fluctuating asymmetry as a quality control indicator for the rearing conditions used by many entomologists. They obtained blowfly eggs from replicate colonies and placed them at varying densities on a common rearing substrate. These eggs subsequently hatched and the larvae fed on the available substrate and pupated. A random selection of adults that emerged from these pupae were then assessed for asymmetry in three meristic traits: number of bristles on the frontal head stripe, number of bristles on the outer wing margin, and the R_{4+5} wing vein. A composite asymmetry value was calculated and this was compared between the treatment groups. They found a significant effect of larval density on fluctuating asymmetry, which appeared to follow a linear relationship. Adults that were raised as larvae in more crowded conditions exhibited greater levels of fluctuating asymmetry (Fig. 6.4).

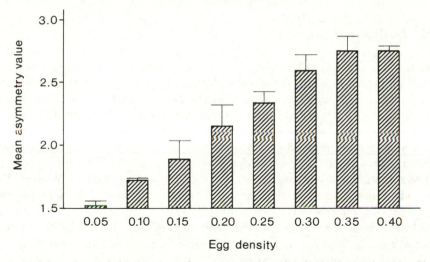

Fig. 6.4 Mean (+ s.e.) asymmetry value versus rearing density in Australian blowflies. Mean asymmetry value is a composite measure of left–right asymmetries for three meristic traits (refer to Clarke and McKenzie (1992) for more details). Values are means of replicates. Adapted from Clarke and McKenzie (1992).

Rasmuson (1960) observed that the frequency of phenodeviants increased with larval density in *Drosophila melanogaster*, and also that the frequency of these phenodeviants was positively related to both asymmetry and the coefficient of variation. Similarly, population density and variance asymmetry were positively correlated for one meristic character (pectoral fin ray number) but not in an additional two characters (number of gill rakers on outer arch, pored lateral line scale number) in bluegill sunfish *Lepomis macrochirus* (Wiener and

Rago 1987). Leary *et al.* (1991) demonstrated that asymmetry of eight meristic characters increased with population density in rainbow trout *Oncorhynchus mykiss*, but not significantly. Møller *et al.* (1995b) raised farm breeds of chickens *Gallus gallus* at three different population densities (20, 24, and 28 chickens per square metre). They observed an increase in two composite measures of relative asymmetry with density. Chickens raised in the most crowded conditions developed asymmetries 30% larger than those at lower population densities. Rettig *et al.* (1997) planted even-aged clones of *Populus euramericana* trees at three population densities (0.167, 0.5, and 2 per square metre) and two levels of competition from weeds, in a factorially designed experiment. They observed that leaf asymmetry increased significantly with population density, and also with level of competition from the weeds. Interestingly, there was also an interaction effect of population density and interspecific competition in that plants at high densities with an abundance of weeds were the most asymmetric.

Therefore, evidence from a range of taxa indicate that increasing population density results in decreased developmental stability.

6.7 Audiogenic stress

Siegel and Smookler (1973) demonstrated that rats exposed to audiogenic (noise) stress for a period of 21 days prenatally and 21 days post-partum exhibited greater fluctuating asymmetry in the mandibular first molar teeth than unstressed control animals. Siegel and Doyle (1975a) proceeded to demonstrate a similar effect induced by audiogenic stress for either of these two periods of development. In this latter experiment, they exposed two groups of pregnant laboratory rats (Sprague-Dawley) to different audiogenic stress regimes. The first group was stressed from the day of conception to the day of birth (21 days). Whereas the second group was stressed from the day of birth for the next 21 days. During the stress periods, the rats were exposed to three sessions of high-intensity noise per week (500–4000 Hz, 1007 ± 2.3 dB), with each session lasting for four hours (randomised 2.5 hours of sound and 1.5 hours of silence). A third, control, group was raised under normal laboratory conditions. All animals were sacrificed and measured 21 days post-partum. They found that both mandibular and maxillary first molar teeth exhibited greater fluctuating asymmetry in the stress treatments compared with the controls. A further investigation was performed to see if these effects were present in the developmental pathways of traits where the presumed genetic component is lower than that of the dental system. In this experiment, Siegel and Doyle (1975b) investigated the effects of audiogenic stress on long bone morphology in rats. Here they found no effect of audiogenic stress on asymmetry. This may have been due to the plastic nature of bone development (Moss and Young 1960; Riesenfeld 1966), and hence some adaptive remodelling (i.e. compensation for asymmetry) may have occurred during the growth processes. However, in a separate experiment rat litters exposed to audiogenic stress between birth and the time of weaning developed greater asymmetries in mass per unit length of long bones than control litters (Doyle *et al.* 1977). The apparent differences between dental and skeletal traits observed by Siegel and Doyle (1975b) may be related to dental development being more genetically predetermined; and so perhaps the developmental pathways associated with the production of dental traits are less able to compensate for any environmentally induced upsets during morphogenesis.

Subsequent investigations have provided further evidence that noise stress can induce developmental instability during morphogenesis. Siegel and Doyle (1975c) indicated that audiogenic stress can induce asymmetries in humeri of *Peromyscus floridanus* and *P. gossypinus*. Brandt and Siegel (1978) demonstrated that noise-stressed rats develop thinner bones. Additionally, neonatal rats exposed to audiogenic stress exhibit greater asymmetries in femurs (Gest *et al.* 1983, 1986) and parietal bones (Mooney *et al.* 1985) compared with unstressed controls. The effect of noise stress increasing dental asymmetries has been further demonstrated by Siegel and Mooney (1987) in laboratory rats. Offspring of stressed individuals showed greater asymmetries of dental calcium in their molars than offspring of unstressed controls.

6.8 Trait-specific susceptibility to environmental factors

In a series of experiments, Møller (1992c, 1993b) has indicated that some traits may be more susceptible to environmental factors than others. The secondary sexual ornaments of swallows (i.e. male outer tail feathers) appear to have a lower buffering capacity than other morphological traits in males, and analogous traits in females. This effect has been documented in response to both ectoparasitic infection (Møller 1992c) and proximity to a source of radioactive contamination (1993b). In this latter experiment, Møller (1993b) compared morphology and mate choice behaviour of swallows from the contaminated Chernobyl area with swallows from a control area south-east of Chernobyl that was unaffected by the radioactive leak. A number of morphological characters were measured (wing length, outer tail length, inner tail length, bill length, bill height, bill width, tarsus length, and body mass), several of which exhibited bilateral symmetry and so rendered measures of fluctuating asymmetry. The effect of fluctuating asymmetry and aberrant morphology on mate choice was assessed by the relation between breeding date and length of the outermost tail feathers, as it is known that breeding date is strongly negatively related to tail length (see Møller 1994b). Secondary sexual ornaments appeared to be affected to a greater degree by the radioactive contamination than normal morphological traits. Outer tail feathers of male swallows exhibited the greatest levels of fluctuating asymmetry and the most frequent rates of aberrant feather morphology (Table 6.1). These deviations in tail morphology influenced mate choice as they appeared to delay the date of pair formation and hence the dates of breeding and, subsequently, breeding success.

Table 6.1 Mean (± s.e.) values of fluctuating asymmetry in barn swallows from difference periods and areas in the Ukraine. All values are expressed in mm. The Chernobyl radiation leak occurred in 1986. Adapted from Møller (1993b).

	Period			
	Pre-1986		Post-1986	
Character	Males	Females	Males	Females
Chernobyl				
wing asymmetry	0.63 ± 0.20	0.52 ± 0.19	0.82 ± 0.14	0.48 ± 0.10
outer tail asymmetry	0.73 ± 0.33	2.00 ± 0.25	5.25 ± 2.27	1.70 ± 0.30
inner tail asymmetry	0.09 ± 0.09	0.12 ± 0.09	0.12 ± 0.05	0.15 ± 0.05
Control area				
wing asymmetry	1.07 ± 0.23	1.11 ± 0.12	0.54 ± 0.07	0.52 ± 0.10
outer tail asymmetry	0.93 ± 0.51	0.25 ± 0.25	1.84 ± 0.49	1.49 ± 1.13
inner tail asymmetry	0.13 ± 0.09	0.11 ± 0.11	0.01 ± 0.01	0.03 ± 0.02

In a similar study, Dufour and Weatherhead (1996) reported asymmetry values for nine morphological characters in the male red-winged blackbird *Agelaius phoeniceus*. They too found that a sexually selected trait, epaulets, displayed greater asymmetry than the non-ornamental traits, although epaulet asymmetry were a poor predictor of overall asymmetry across traits. This may further indicate that ornamental traits are more susceptible to environmental perturbations, although environmental factors were not explicitly examined in this study.

The difference in susceptibility to environmental factors may be due to directional selection, which presumably acts on secondary sexual traits whilst stabilising selection acts on normal morphological traits. The intense directional selection could act to destabilise the genome through selection against genetic modifiers (Møller and Pomiankowski 1993a; Chapter 3). Additionally, such traits may be more susceptible to upsets during morphogenesis due to the increased physiological requirements and complex developmental pathways involved in producing secondary sexual ornaments (Pomiankowski *et al.* 1991).

These observations raise the important issue that fluctuating asymmetry may be trait specific. Even within the same individual, genetic factors can be different between traits, and these differences can influence the effect of environmental factors on fluctuating asymmetry production. This issue may particularly relate to the use of fluctuating asymmetry as an environmental bioassay technique. There may be some traits that are particularly sensitive to external factors due to the differential intrinsic genetic stress that their developmental processes experience. Identification of these traits may be extremely useful when employing fluctuating asymmetry as a bioassay technique of monitoring habitat quality as these traits will be very sensitive indicators of environmental conditions (see section 6.11).

In a recent meta-analysis of the relations between environmental stress on developmental stability, Leung and Forbes (1996) have indicated that the reported difference in susceptibility of sexually selected versus non-sexually selected traits may not be as strong as Møller reports in the barn swallow. Analysis of sexually selected traits from 11 species and non-sexually selected traits from 37 species indicated that both categories of trait display a similar relation of developmental stability with environmental stress, i.e. increased asymmetry in more stressful conditions (Leung and Forbes 1996). However, as the number of published studies reporting a relationship between environmental conditions and developmental stability of sexually selected traits are still relatively low, Leung and Forbes remark that there is still a need for more studies to be performed using ornamental traits. Leung and Forbes (1997a) have also generated a mathematical model for the production of fluctuating asymmetry based on the opposing factors of developmental noise and developmental stability. This model indicates that asymmetry in traits with low developmental stability (i.e. large asymmetries), such as ornamental traits, is not more related to stress or quality than asymmetry in highly stabilised traits. Therefore, there are theoretical reasons to believe that trait type may be relatively unimportant in assessing environmental stress levels. However, as the model of Leung and Forbes (1997a) is fairly simplistic and there is some empirical evidence to suggest that ornamental traits may be more revealing, we suggest that it would be interesting for researchers to perform direct comparisons of asymmetry–stress relations in ornamental and non-ornamental traits within species.

6.9 Stress-specific responses

The *Drosophila* experiments indicate that individuals that are susceptible to one kind of stress appear to be susceptible to other varieties of stress (Hoffmann and Parsons 1989). Therefore, there may be some kind of general stress resistance that is indicated by fluctuating asymmetry. For the vast majority of environmental factors this is likely to be the case, as the factors are very general in their nature. However, it could be possible that some specific factors may not influence fluctuating asymmetry, or only affect certain traits. For example, in a study of two populations of mice from a hazardous-waste dump site and a waste-free area, there were no inter-population fluctuating asymmetry differences although there was inter-population morphometric variation and evidence of chromosomal damage in the polluted population (Owen and McBee 1990). Also, it has been demonstrated that phenylthiourea (PTU) induces a developmental stress that leads to decreased body weight in *Drosophila*, but it has little effect on fluctuating asymmetry (see Parsons 1990b). It could be that most factors, such as ambient temperature or nutritional stress, are generalised and non-specific, and so influence the basic physiological and metabolic pathways of the entire organism. These generalised factors may be expected to have multiple consequences and affect growth patterns of most traits, whereas a chemical compound such as PTU, which is a specific chemical inhibitor, may be a more specialised factor and influence far fewer metabolic pathways, perhaps only influencing the development of selected traits. Therefore, attention to the nature and intensity of the factor may be required in some situations, especially in controlled laboratory conditions where stresses are more likely to be specific. In the field, this type of situation may be rare as factors are often likely to interact and, on the whole, be general in their nature.

6.10 Environmental stress

The point of presenting such a list of examples is to demonstrate the vast range of different factors in the environment that can disrupt developmental home-ostasis (see Table W6.1 on the Web site). Fluctuating asymmetries have many different potential causes and in some field-based studies it may be extremely difficult to ascertain which of these effects is causing the asymmetry (e.g. Hershkovitz *et al.* 1993). In many cases it may only be through factorial manipulations of specific factors that we will discover the role of each type of factor in fluctuating asymmetry production. This is most easily performed in the laboratory where developmental conditions can be controlled and regulated. There is the additional problem that environmental factors are likely to be non-additive and interactive in their nature. Also, there will not often be discrete boundaries and distinct definitions of each form of stress. For example, population density may influence all the forms of environmental stress that we have listed above. In terms of nutritional stress, there is variability in the competition for food resources. High population densities may also increase ambient temperature and levels of audiogenic stress, as organisms are in close proximity to each other. This close proximity may also increase the chance of cross-contamination of parasitic infection and also chemical stress due to excretory products. In uncontrolled field studies, it will be very difficult to tease apart the relative effects of each of these factors.

Additionally, the list of examples we have presented in this chapter is by no means exhaustive. Environmental stresses can come in practically any guise. For example, in a pilot survey of the developmental stability of soybeans *Glycine max*, Turner and Freeman (unpublished data reported in Freeman *et al.* 1993) demonstrated that fluctuating asymmetry can be increased by proximity to high- voltage overhead cables. It would appear that the electromagnetic field generated by the power lines disrupts developmental stability and, hence, the largest asymmetries develop in soybeans that are grown directly under the cables, and asymmetries decrease with distance from the power source. Alternatively, these asymmetries may relate to heavy metal pollution in the areas close to the power cables. As these stresses can, potentially, come from very obscure environmental sources, the stress may be very difficult to detect and quantify. This is one of the reasons why fluctuating asymmetry has been suggested as a useful bioassay technique for monitoring environmental conditions (section 6.11). Through monitoring of developmental stability it is possible to detect an environmental factor without knowing its origin. In theory, this could provide an early warning system of environmental change and habitat degradation.

Another important point to remember at this stage is that the genetic factors described in Chapter 5 may influence the susceptibility of individuals and populations to the environmental factors listed above. That is, there will be a genotype-by-environment interaction. For example, Jokela and Portin (1991) have shown that incorporating an extra Y chromosome into the genome of *Drosophila melanogaster* raised at two different temperatures can have unpredictable effects on sternopleural bristle fluctuating asymmetry. At 18°C, the extra Y chromosome appeared to increase asymmetry, whereas at 25°C the extra chromosome was associated with low levels of bristle asymmetry. Jokela and Portin postulate that the insertion of the Y chromosome may shift the optimal developmental temperature for *Drosophila melanogaster*.

Parsons (1992) views stress as 'an agent placing an organism at a disadvantage requiring the continued expenditure of excess energy (assessable by metabolic rate) which ultimately threatens survival'. He sees the energetics of the adenine nucleotide pathway as the crucial measure of stress (see discussion in section 4.2.1). Organisms under stress require more energy to perform the same functioning as an unstressed organism. This may be a useful way to conceptualise the effects of adverse environmental conditions and is eminently testable. The various stimuli from the environment that influence the adenine nucleotide pathway can almost be limitless, and can certainly interact with each other. The important element is that they all affect the pathway and hence induce developmental stress upon the organism. Within this framework, we can also include the genetic factors that were described in Chapter 5. Internal alterations to the genetic make-up of the organism, and the organism's cells, will also influence the efficiency of the adenine nucleotide pathway. Hence, we may have a 'catch-all' framework for interpreting the effects of developmental stresses, whether the stimuli that exert these stresses originate from internal genetic processes or external environmental influences. As metabolic efficiency is central to this conceptualisation of developmental stress, it may be that certain species and certain traits within species are more prone to responding to stress. Species with high metabolic rates may exhibit larger responses and be more sensitive to developmental stress (Parsons 1992). It has been suggested that small passerine birds, due to their high metabolic rates, may be good indicators of developmental stress (Root 1990). Similarly, behaviourally active mutants of *Drosophila melanogaster* may be extremely sensitive to stress during their development (Trout and Hanson 1971; Parsons 1991a). Additionally, certain species may be more susceptible to specific factors than other species, for example, poikilotherms may be more sensitive to temperature induced stresses than endotherms.

6.11 Environmental monitoring by fluctuating asymmetry

Fluctuating asymmetry is sensitive to factors from the environment. If it is assumed that genetic diversity is not great between populations of the same species (but see later), then any differences in fluctuating asymmetry between populations are most likely to be due to relative differences in environmental conditions. Populations developing in the most adverse environments will exhibit the greatest levels of fluctuating asymmetry. Therefore, fluctuating asymmetry could be used as a bioassay technique for monitoring habitat quality (e.g. Valentine *et al.* 1973; Leary and Allendorf 1989; Clarke 1993a; Freeman *et al.* 1995; Tracy *et al.* 1996).

Previous techniques used in bioassays of environmental conditions have relied on variables that measure phenotypic quality of individuals within the habitat, such as life-time reproductive success, fecundity, survivorship, development time, fat content, muscle mass, and relative body mass; these are all time-consuming and relatively difficult to measure, and are not without problems of interpretation (Witter and Cuthill 1993). One of the attractions of fluctuating asymmetry is that it is quick, simple, and cheap to measure. Studies can often be performed with relatively small sample sizes, as long as the sample is not biased. Also, there is no need for complicated or cumbersome equipment, and the same techniques can be employed in the laboratory and the field, hence allowing direct comparison between studies. Measurement of fluctuating asymmetry can also be non-invasive and permit continual longitudinal reassessment of populations, hence allowing the researcher to monitor the effectiveness of any action taken and measure trends over time. It should also be possible to detect traces of prior developmental instability in preserved specimens and fossils from their extent of bilateral asymmetry. This may indicate changing environmental conditions over evolutionary time (cf. Møller and Pomiankowski 1993b). Fluctuating asymmetry also has the significant advantage over other measures of individual condition in that the optimum value is known *a priori* (i.e. zero asymmetry), whereas this is often difficult to predict for measures such as fecundity, survivorship, or body size.

However, the use of fluctuating asymmetry as a biomonitor of habitat quality can be criticised from two perspectives. First, as a relatively severe stress is required to induce notable asymmetries in experimental conditions, it could be expected that the detection of environmental effects under field conditions would be difficult, and perhaps may only be detectable in ecologically marginal conditions (Parsons 1990c). However, as stated above, in the field many of these environmental effects are confounded with each other and will probably be multiplicative. In experimental conditions, one kind of stress is isolated from the others, therefore a large level of an individual stress may be required in the laboratory, whereas seemingly low levels of stress can induce large asymmetries in the field. Therefore, we would expect differences in fluctuating asymmetry to be detectable among populations in the field, and that fluctuating asymmetry will be sensitive to small changes in habitat quality. Additionally, most laboratory studies provide subjects with food and water *ad libitum*, whereas in the field this will rarely occur. As it is known that nutritional stress increases developmental instability (section 6.4), we would expect developmental processes to be less stable in field conditions and hence more susceptible to additional environmental variables that may disrupt developmental stability.

The second major criticism of this approach is the assumption of genetic homogeneity among populations, which can often be violated. Populations could differ in their relative levels of inbreeding, heterozygosity, hybridisation, mutation rates, and intensity of directional selection. Therefore, the use of fluctuating asymmetry as an indicator of environmental conditions must be performed with caution (Soulé 1967; Saunders and Mayhill 1982; Hallgrímsson 1993). In the laboratory, where gene flow can be controlled, this approach may work extremely effectively and accurately. But in the field, where there is often little knowledge of the genetic make-up of individuals or populations, interpretation of fluctuating asymmetry data may be complicated as the relative influences of genetic and environmental factors could be confounded. In these situations, asymmetry differences may be more attributable to intrinsic developmental noise rather than influences of the environment. To separate the relative influences of genes and environment, it is necessary to gain some information about the relative genetic stresses to which the populations are subject. This could be achieved by molecular techniques, but these are expensive and extremely time-consuming, and may not give information concerning the nature of the selection pressures and evolutionary histories of the populations under study. Alternatively, genetic effects could be controlled by introducing a population of known genetic background into the different habitats. If a transgenic population is introduced, it is unlikely to outcompete the natural population (see Parsons 1992) and so have no deleterious effect on the resident species. Importantly, it may provide a method of monitoring the stress response of a known genotype to the prevailing environmental conditions. If the transgenic species is selected carefully, or several species are used in the same study, meaningful comparisons of relative levels of stress can be made between habitats. Alternatively, the fluctuating asymmetry of an asexual clonal species could be monitored, as the genetic diversity of sexuality is then removed and populations are more likely to be homogeneous.

Perhaps the most appropriate use of fluctuating asymmetry in biomonitoring is in repeated-measures applications. In other words, measurement of asymmetry in the same populations or individuals over time. Therefore, each population can serve as its own genetic control, as long as the populations are reasonably stable (cyclic populations often exhibit large changes to their genetic structure over time (Graham *et al.* 1993a)) and do not have large levels of immigration and emigration. Longitudinal changes in levels of fluctuating asymmetry over time, in these situations, will most likely be due to environmental changes, and so fluctuating asymmetry will provide a sensitive indicator of changing environmental conditions.

There may be some species that are more sensitive to suboptimal environmental factors than others. Species with high metabolic rates may be more sensitive to environmentally induced perturbations than those with lower metabolic rates (e.g. small passerine birds), as stress can be defined as a metabolic response (see Parsons 1992). Additionally, within species there may be certain traits that are more susceptible to increased fluctuating asymmetry than others. Møller (1992c, 1993b) has provided evidence to indicate that traits under directional selection, such as secondary sexual ornaments, are more sensitive to environmental stresses than ordinary morphological traits. Conversely, Dufour and Weatherhead (1996) suggest that mechanically functional traits may be more revealing. However, Leung and Forbes' (1996) meta-analysis suggests that there is little variation among classifications of trait in their susceptibility to environmental stressors. Where possible, we suggest that several traits, ornamental and biomechanically functional, should be measured on the same individuals and populations. However, in cases where few traits can be measured, ornamental traits may be more useful for the reasons already stated plus the additional point that asymmetry in these traits is often larger, and hence measurement error should be relatively smaller than in non-ornamental traits. As measurement error should be minimised (section 1.10.2), this may mean that ornamental traits may render more accurate measurements of developmental stability in many situations.

There are a couple of additional considerations to make when selecting a suitable study species. The species must, obviously, be widespread and cosmopolitan. Also, to minimise the probability that the study species has already adapted to the environmental stress, it may be best to select a species with a relatively long generation time. Obviously, traits also have to be developed within the period of study, and so to allow comparisons within populations over time, the same trait has to be developed repeatedly (e.g. annual moulting of avian feathers). Therefore, if the species and traits are chosen with care, a very sensitive bioassay of environmental conditions can be constructed.

6.12 Summary

- We review the various environmental factors that can give rise to decreased developmental stability and hence increased asymmetry. These include adverse temperatures, nutritional deprivations, chemical pollutants, high population density, and audiogenic stress.
- Traits that are already subject to genetic stresses, for example secondary sexual characters that are under directional selection, may be more sensitive to environmental factors than ordinary morphological traits.
- Species with high metabolic rates may be more sensitive to environmental factors than species with lower metabolic rates.
- Certain species may be more susceptible to specific stresses than other species.

- It is possible that asymmetry can be used as a biomonitor of environmental conditions, providing that genetic differences between populations and samples are accounted for. In this respect, clonal and asexual organisms may prove to be particularly suitable models.
- The most appropriate application of fluctuating asymmetry as a bioassay of environmental conditions may be in longitudinal monitoring of stable populations over time.

7

Developmental instability and performance

7.1 Introduction

In this chapter we review the evidence that links asymmetry with decreased performance, particularly in relation to locomotion. In other words, we investigate the functional costs of asymmetry. We start by explaining why symmetry is the ideal phenotypic expression in many cases of performance (section 7.2). This is because symmetry keeps a body in a state of stable equilibrium, whereas asymmetry disrupts this state and so requires energy to maintain the status quo. This expense of energy may reduce overall performance and hence decrease efficiency; the reduction in performance does not just relate to locomotor performance but also any other costly activity which will be in a trade-off for energetic resources (e.g. growth, reproduction, immune defence). The relation between asymmetry and locomotor performance has been illustrated by a negative correlation between morphological asymmetry and handicap rating in flat-racing thoroughbred horses *Equus caballus*, between running speed and asymmetry in racing dogs *Canis familiaris*, and between asymmetry and gait quality in chickens *Gallus gallus*. However, these studies only reveal a correlation between asymmetry and performance; these relationships could be due to asymmetry directly affecting locomotory performance, or that asymmetry reflects intrinsic properties of the individual that are correlated with poor performance. Manipulation experiments that disentangle these effects indicate that morphological asymmetry can directly influence locomotory ability in birds and also that intrinsic properties of an individual that relate to developmental stability effect performance (section 7.4).

We focus on the particular example of bird aerodynamics in section 7.3 as most of the relevant experimental investigations have been performed using avian taxa, and birds represent a large body of the recent literature on developmental stability. Theoretical accounts of the differential effects of wing and tail asymmetry predict that asymmetry will be particularly costly in terms of aerial manoeuvrability and agility. As a result of the differences in aerodynamic functioning of wings and tails, it is also predicted that asymmetry should be less costly at larger trait sizes in wings, but more costly in tails. This example illustrates the different influences that asymmetry can have in different traits.

Experimental tests of the predictions made in section 7.3 illustrate that asymmetry does have a large influence on turning performance in birds (section 7.4). Both wing and tail asymmetry reduce the turning ability of birds. In wings, there does not appear to be an interaction effect of primary length and primary asymmetry, i.e. the costs of asymmetry do not alter at different primary lengths (section 7.4.1). In birds' tails, the costs of asymmetry appear to increase with tail length (section 7.4.2). These results largely agree with the developed aerodynamic theory, although we do discuss ways in which the theory could be developed.

In section 7.5 we indicate that the developmental stability of a character may be related to its functional importance. We propose that traits with high functional significance may be constrained into symmetrical development, whereas the developmental stability of less functional traits can be relaxed at little cost.

The functional significance of a trait may also play a role in the maintenance of a certain degree of developmental instability within a population (section 7.6). The functional costs of asymmetry probably accelerate as a trait becomes more asymmetric, and so minor deviations from symmetry incur little cost. However, the intrinsic developmental costs of producing perfect symmetry are likely to decelerate with asymmetry. We suggest that the different trajectories of these competing costs may lead to a trade-off between functional and developmental aspects, so that a small degree of asymmetry is maintained in the population. The functional importance of a trait may alter the accelerating trajectory of the functional costs, so that more functionally important traits will have steeper trajectories. Therefore, it is possible that the trade-off between functional and developmental costs may result in functionally important traits being more developmentally stable than less functional traits.

7.2 Asymmetry and performance

Symmetry or regularity is often a feature of stable structures and is a fundamental engineering principle. A single component of poor quality can result in the breakdown of a complex structure. Notwithstanding this, asymmetry does not necessarily result in instability; for example, the undulating sutures of mammalian skulls prevent parts of the skull from sliding apart during fights by rams *Ovis aries*. However, on the whole, irregularity often results in low stability and poor performance. A revealing example concerns a rather uncommon hair disease in roe deer *Capreolus capreolus* (Herzog *et al.* 1983). Roe deer with this disease lose parts of their hair, therefore they suffer from problems of thermoregulation that eventually may lead to death. This parakeratosis was originally believed to arise because of the activity of ectoparasites, but careful analysis revealed that it was caused by physical damage of poorly constructed hair. Mammalian hair consists of regularly arranged scales that can have coronal or imbrical patterns; however, the hair of the diseased deer consisted of irregularly sized and shaped scales with no imbrical pattern. This lack of regularity to the hairs' component parts gave rise to structural weaknesses, breakages, and major hair loss.

In most cases of performance and locomotion, symmetry is the ideal phenotype. Symmetry keeps a body in stable equilibrium, whereas asymmetry makes the body unstable and will make performance less efficient. After all, if birds had very asymmetric wings they would fall out of the sky; and if dogs had asymmetric legs, they would be forever chasing their tails. This may seem obvious, but to shed more light on the situation it may be useful to think of morphological asymmetry directly resulting in suboptimal performance in the following way. Force is a product of mass and acceleration. Therefore, if a structure is asymmetric, mass distribution around the axis of symmetry will be uneven, and the force generated will be asymmetric even if the intended acceleration is symmetric. Concomitantly, there is likely to be an asymmetric turning moment, as a turning moment is a function of force and distance from the plane of symmetry. Therefore, if forces are acting at different distances from the plane of symmetry on the left and right sides, a moment will occur even if the forces generated by both sides are equal. This moment will tend to rotate the body about the plane of symmetry and will have to be counteracted by a moment in the opposite direction to keep the body in a state of equilibrium. The correction of this turning moment will expend energy and decrease overall performance.

Symmetry is also believed to minimise the variability of load on an organism (Alexander *et al.* 1984). As a result of the reduction in load variability, pressures, stresses, and strains will also be minimised and hence performance can be made more efficient as fewer resources need to be directed toward resisting these biomechanical constraints. Domestication of animals such as pigeons and hens has increased asymmetry, perhaps as a reflection of the decreased need for efficient flight and locomotion that has accompanied domestication (Parsons 1990a) in addition to increased directional selection (Møller *et al.* 1995c; Chapter 3). Therefore the selection pressures for minimised load variability have been lessened, or removed, and developmental stability has been relaxed at little functional cost. This is a plausible explanation of the observed correlation, but this effect is also confounded with increased homozygosity, as domestication has led to intensive inbreeding in most species. Homozygosity and inbreeding can give rise to a decrease in developmental stability (Chapter 5), as can be observed in the increased asymmetry of colour patterns in many domesticated animals (section 8.4.1).

The overwhelming predominance of symmetry in nature would tend to support these statements. Practically all taxa display symmetry to some extent, especially in external morphology. Internally, large asymmetries can often occur. One example of gross internal asymmetry is the alimentary canal of mammals. None of the organs in the digestive tract are paired and in most cases there are extreme directional asymmetries in positioning of these structures. However, externally, symmetry is by far the most common phenotype, of which bilateral symmetry appears to occur most frequently. The difference between internal asymmetry and external symmetry implies that symmetry is not a necessary consequence of development, it is there for a reason. One of these reasons is the locomotor performance advantage that symmetry entails. This, in turn, implies that sessile animals and plants should have relatively high asymmetry due to the absence of locomotor costs, which is supported by the observation that plants do appear to possess high levels of asymmetry (Møller and Eriksson 1994).

One example of a correlation between performance and symmetry appears to occur between handicap ratings and morphological asymmetry of flat-racing thoroughbred horses. Manning and Ockenden (1994) measured 10 paired characters (six on the head and four on the forelegs) from 73 racehorses and correlated asymmetry in these characters with handicap rating. Handicap ratings are assessed after each race that the horse runs and, as it sounds, determine the weight that the horse must carry in its next race. Better horses are given larger handicaps (i.e. more weight to carry) than poorer horses so, in theory, making races more even and exciting for the spectator. Of course horse trainers and owners know this, as do the jockeys that ride the animals. So handicap ratings may not offer the best measure of a horse's true performance at any one time, as trainers and jockeys can successfully manipulate the handicap (at least to some extent) so that their horse hits peak physical form at a time when it is not carrying its largest handicap. However, average handicap over a long period of time (or a season) may be a reliable indicator of how well that horse has done in the past. Manning and Ockenden found that all of their measures of asymmetry correlated negatively with handicap rating, although only one of these was significant (the distance between the cheekbone and the mouth, see Table 7.1). As all their traits were negatively associated with performance, a composite index of asymmetry across these traits was highly negatively related to performance. Handicaps are also altered with age of the horse, as older horses are believed to be able to run faster than younger horses. This means that four-year-olds carry more weight than three-year-olds and in turn, three-year-olds carry more than two-year-olds. Once these weights were added to the initial handicaps, the negative relationships between rating and asymmetry became stronger. Now height asymmetry of the first pair of upper incisors was also significantly negatively related with age-adjusted handicap rating (Table 7.1). This appeared to be due to age being negatively related with asymmetry. Older horses were more symmetric. This could have occurred for three reasons. First, asymmetry could decrease with age due to wear and damage of tissues, as teeth and soft tissues would be liable to abrasion. However, this is not likely to be the case, as damage would normally lead to an increase in asymmetry with time as physical abrasion tends to occur asymmetrically (Swaddle *et al.* 1996). Second, morphological asymmetry could actually decrease as the horse got older, due to regrowth of soft tissues or adaptive remodelling of bones. This could occur, as many of the traits measured here include soft tissue structures. Third, asymmetric horses could be selected against during training, and so only the more symmetric horses go on to compete as three- and four-year-olds, as they are better performers and are more likely to win money for their trainers and owners. This could also occur in this system. If symmetric horses are better performers, we may expect to see this negative relationship with age, which also implies that trainers and owners are already selecting their horses, indirectly, in terms of symmetry.

Table 7.1 Pearson product-moment correlations between character asymmetry and handicap ratings (plus age allowance) of flat-racing thoroughbred horses *Equus caballus*. The most significant correlations between asymmetry and performance arise from asymmetry estimates of facial characters. Adapted from Manning and Ockenden (1994).

Character	Ratings	P	Ratings + age	P
Elbow–knee	−0.15	0.19	−0.15	0.21
Knee–ergot	−0.05	0.65	−0.06	0.64
Knee thickness	−0.15	0.21	−0.24	0.03
Coronet band	−0.19	0.11	−0.22	0.06
Ear height	−0.10	0.40	−0.02	0.89
Incisor height	−0.21	0.07	−0.38	0.001
Incisor width	−0.12	0.31	−0.14	0.24
Nostril width	−0.22	0.06	−0.20	0.09
Cheekbone–ear	−0.04	0.72	−0.06	0.59
Cheekbone–mouth	−0.35	0.0027	−0.33	0.0047
Overall mean asymmetry	−0.43	0.0002	−0.48	0.0001

Perhaps the most surprising element of this study was that asymmetry of foreleg features was not significantly related to performance, only facial features were. If morphological asymmetry is directly related to performance, perhaps we would expect the asymmetry of the locomotor apparatus to exhibit the strongest relationship. However, the hindlegs are more important in determining the speed of a horse than the forelegs and, unfortunately, these were not measured.

In a similar investigation, Møller *et al.* (1997) compared gait quality with asymmetry in chickens *Gallus gallus*. Møller *et al.* evaluated the gait of chickens on a six-point scale (0–5; 5 representing a lame individual) and found that gait was significantly positively related to asymmetry in all three characters that they measured (length of tarsometatarsus, width of tarsometatarsus at spur, width of upper joint of tarsometatarsus). In both these examples, increased asymmetry was associated with decreased locomotory performance.

Overall asymmetry may be negatively related to performance as overall asymmetry is a good indicator of the genetic and environmental 'health' of the individual. So it may be that asymmetry does not directly cause reduced performance; it may be a correlate of reduced performance as it is a general indicator of poor genotypic and phenotypic condition. In general, it is not possible to assess the accuracy of this statement without performing manipulation experiments where the functional role of asymmetry can be disentangled from the intrinsic properties of the individual. We report a number of such experiments in sections 7.4.1 and 7.4.2, the results of which indicate that asymmetry can directly cause reduced performance but there may also be an influence of intrinsic properties, as fluctuating asymmetry before the experiment influences performance after asymmetry manipulations have been performed (e.g. Møller 1991). The direct effects of asymmetric morphology and the indirect correlates of poor developmental stability are not mutually exclusive and can operate at the same time. If both of these systems do operate, they will tend to polarise the performance difference between individuals of varying developmental stability.

7.3 Avian asymmetry and aerodynamics

As we have described above, the moment about the plane of symmetry increases with trait size, if the force is applied at greater distances from the plane in larger traits. Therefore the mechanical costs of an absolute asymmetry increase with trait size. Evans and Hatchwell (1993) have stated that this will be independent of other functions of the trait, as this principle is a mechanical property that applies equally to sexual ornaments under directional selection as to normal morphological traits under stabilising selection. Generally, this assumption will be true. However, in one interesting case this relationship of mechanical costs between trait size and asymmetry breaks down. This is the case of birds' wings where costs appear to decrease with trait size. The details of these investigations are given in this section.

Perhaps the one area of performance-related asymmetry research to receive the most attention over recent years has been in the field of avian aerodynamics. This is primarily due to the large behavioural and ecological interest in birds as study species. Avian aerodynamics also provides an intriguing example of the influence of asymmetry on locomotion, as the interaction of asymmetry and trait length is predicted to have different functional costs in the wings of birds compared with their tails. This difference appears to have been overlooked by many ornithologists, but it is an important theoretical point that needs to be tested empirically. We hope that this section will act to both inform the reader of the current state of the literature and to stimulate functional approaches to the study of avian developmental stability. This is a comparatively new field of interest and offers exciting opportunities for novel and innovative research. Insects and insect flight might also be interesting in this context.

We shall start by briefly reviewing a recent theoretical model of the aerodynamic performance of birds' tails (Thomas 1993a). The literature on avian wing performance is not reviewed here; the interested reader could refer to one, or both, of the books by Pennycuick (1989) and Norberg (1990). After we have explained how birds' tails are functionally different from their wings, we shall consider the influence of asymmetry, trait size, and most importantly, the interaction of trait size and asymmetry on aerial performance.

7.3.1 *Avian tail aerodynamics*

Adrian Thomas (1993a) has recently produced a theoretical model of avian tail aerodynamics. The main assumption behind this model is that of invariant flow over the cross-section of the tail, and that this flow is comprised of the vector sum of flow due to flight velocity and the flow velocity induced by the wings. This may be a fair assumption, as Spedding (1987a, b) has shown that flow is fairly invariant over the wings during gliding and cruising flight in the kestrel *Falco tinnunculus*. However, flow over the centre of the tail must be disturbed by the body, even if flow from the wings is uniform across the span. Given that this assumption is likely to be accurate, the tail can be modelled as a separate aerofoil acting independently of the wings.

As a bird's tail has much lower aspect ratio (span squared divided by area, and so is a relative measure of aerofoil length) than a wing, flow over the surface of the aerofoil is more likely to be dominated by vortices at the aerofoil tips and will be almost two-dimensional across the span. Therefore, lifting surface theory must be used to model a tail aerodynamically instead of lifting line theory that is normally applied to birds' wings. This basic modelling difference creates some important implications for the influence of asymmetry on the relative performance of wings and tails, as we discuss below. Thomas adopts the techniques used by Jones (1946, 1990) and models the tail as a slender lifting surface that has low aspect ratio and is, essentially, a thin flat delta-shaped wing.

Two important findings from the Thomas model are:

(i) There are sharp peaks in the spanwise distribution of pressure differences at the leading edges of the tail, so the edges of the tail have a large influence on its aerodynamic performance (Fig. 7.1a). Hence, fluctuating asymmetry in the outer feathers may have a disproportionately large effect on tail performance, as asymmetries in leading edges could result in large rolling moments that reduce aerodynamic performance of the tail (Thomas 1993a, b).

(ii) The area behind the point of maximum continuous span only generates drag and no lift, as it is dominated by the wake of the forward section (Fig.

7.1b). This implies that the section behind the point of maximum continuous span can become exaggerated and ornamented at little functional cost to the bird, as drag will always be an order of magnitude smaller than the lift produced. This is probably why birds' tails are often so elaborately ornamented compared with other traits.

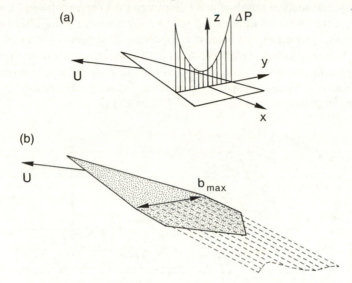

Fig. 7.1 (a) Spanwise pressure difference distribution across a theoretical triangular tail of a bird. The pressure difference peaks at the edges indicating the importance of the leading edges of the tail. (b) The area behind the point of maximum continuous tail span only contribute drag and no lift, as it in the influence of the wake of the forward section. Adapted from Thomas (1993a).

In general, Thomas (1993a) found that the forces generated by the tail are influenced by tail width and tail area, at any given angle of attack. Tail width (maximum continuous span) determines lift, tail moment, and induced drag, whereas tail area determines profile drag. Hence, the aerodynamic performance of the tail must vary with tail shape and spread of the tail feathers. Thomas (1993b) then went on to consider the effects of fluctuating asymmetry on the aerodynamic performance in both birds' wings and tails by invoking classical aerodynamic theory for explanations of wing asymmetry and his model (Thomas 1993a) for tail asymmetry.

7.3.2 *Wing asymmetry and aerial performance*

By adopting Norberg's (1990) reworking of Pennycuick's (1975) momentum jet model, Thomas investigated the qualitative relations between wing asymmetry, wing length, and aerodynamic costs in the barn swallow. The reasons for this choice of species are obvious given the large literature on this bird (see Møller 1994b), and Thomas claims that this choice should not qualitatively affect the results of his analysis. Thomas models asymmetry as an asymmetry in wing length (and no change in mean wing chord), and assumes that asymmetric birds compensate for asymmetry by flexing the longer wing and hence reduce overall wing span.

The calculations of the influence of asymmetry on the power required for flight indicate that the costs of absolute wing asymmetry decrease with wing length and aspect ratio; but the costs of constant relative asymmetry do not alter with wing span. Overall, asymmetry can be costly, especially at slower flight speeds. A 5% relative asymmetry can require more than a 10% increase in power required for flight at slow flight speeds. At higher speeds (above about 12 m s^{-1}) asymmetry is weakly negatively related to flight costs, whereas below this figure asymmetry can be strongly positively associated with flight costs (Fig. 7.2).

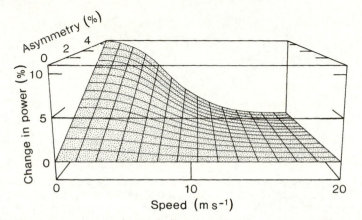

Fig. 7.2 Change in power required for flight caused by wing asymmetry of a bird. Adapted from Thomas (1993b).

As birds generally have a lift to drag ratio of 10:1, the lift asymmetry is more important than an asymmetry created in drag. The asymmetry in lift (L) generated by left and right wings produces a rolling moment (M_a) which is proportional to the asymmetry (a):

$$M_a = La / 2.$$

Thomas (1993b) assumes that this moment is independent of wing length and only depends on the force generated by the wings and the asymmetry. This is not necessarily a valid assumption, as the moment will be proportional to the distance from the plane of symmetry through which the force is acting. Therefore, Thomas' (1993b) model will be prone to under estimating the aerodynamic cost of asymmetry at longer wing spans.

As an alternative to flexing the longer wing, a bird may attempt to counteract asymmetry by adjusting the distribution of lift across the wings, perhaps by altering the angle of attack of the two wings independently. The lift differential required to balance the effects of a given level of absolute asymmetry is inversely proportional to wing span. Therefore the costs of absolute asymmetry decrease with wing length. An asymmetry in lift distribution will reduce flight performance. If a bird can compensate for this asymmetry, then it would be optimal to do so. In steady flight this may be possible as wings will operate well below their maximum lift coefficient value, and the compensation will most likely be through increasing the lift coefficient of the shorter wing. However, during manoeuvring and accelerating flight this would not be possible. The rate at which a bird can turn and the tightness of this turn are limited by the difference between its maximum lift coefficient and that needed to support the bird's weight. As asymmetry results in an increase in the lift coefficient needed for the bird to support its own weight, the difference between the 'maximum' and 'body-weight' lift coefficients is reduced. This dramatically reduces the turning performance of an asymmetric bird (Thomas 1993b). The yawing moment of drag is much less than the rolling moment of asymmetric lift, and will be compensated for by the measure outlined above. For example, if lift is increased on the shorter wing, the induced drag produced by that wing is increased and drag is balanced symmetrically.

The predictions from this analysis are that wing asymmetry is costly in that it increases the power required for flight at slow speeds and reduces turning performance. The costs of a given absolute asymmetry are likely to decrease with wing length.

7.3.3 Tail asymmetry and aerial performance

The effect of tail asymmetry on aerial performance depends on the angle of spread of the tail (Thomas 1993b). When the tail is furled the effects will be minimal, but at large angles of spread the effects could be large. Therefore tail asymmetry is only likely to influence slow flight and manoeuvring flight. As the aerodynamic forces generated by the tail depend on the square of its maximum continuous span (Thomas 1993a), tail shape will influence the effects of tail asymmetry on flight. Asymmetry in forked tails may have more influence on flight than asymmetry in graduated and pintails; although this relationship is confounded by highly aerial species usually possessing deeply forked tails. Balmford *et al.* (1993) have found that relative fluctuating asymmetries of outer tail feathers are generally larger in species with graduated and pintails than in those with deep forked tails. Also, asymmetry in central feathers will be far less costly than the same absolute asymmetry in outer feathers, due to their relative moments about the plane of symmetry.

Tail asymmetries can be counterbalanced by either (i) a differential lift and thrust from the wings or (ii) by tilting the tail at an angle to generate the necessary balance in roll or yaw moments (Thomas 1993b). In the first of these scenarios, the force differential (k_t) can be expressed as:

$$k_t = 4M_t/bf$$

where M_t is the moment created by the tail asymmetry, b is the wingspan, and f is either the total lift or total drag of the bird. As the force produced by the tail asymmetry must be less than that produced by the wings, and the moment arm of the tail asymmetry will be less than the wingspan, k_t must be small. In the second example, tail asymmetry could be counteracted by tilting the tail at an angle. Tilting the tail in order to obtain a straight flight path will impair the turning performance of the bird, as the tail is tilted to initiate and control manoeuvres. Thomas (1993b) suggests that compensation of asymmetry through differential lift and thrust of the wings is less costly, and therefore more probable in nature.

As tails have a lower lift to drag ratio than wings, asymmetry in sections of the tail that do not generate aerodynamic lift will also be costly as they induce drag. This drag will tend to create a yawing moment, and will be related to the distance from the centreline of the bird to the centre of the drag-inducing section of the tail. Obviously, this distance (Z_a) will again depend on the length of feathers (R) and the relative spread of the tail (D):

$$Z_a = R \sin(D/2).$$

Therefore, the moments in both lift and drag created by an absolute tail asymmetry increase with tail length (Fig. 7.3a) and tail spread. Also, asymmetry in outer feathers will impose larger flight costs than the same asymmetry in more central feathers. At theoretically infinite tail lengths, the costs of an absolute asymmetry will asymptote. The costs of a relative asymmetry will rise exponentially with tail length (Fig. 7.3b). However, the overall costs of tail asymmetry are far less than the equivalent asymmetry in wing length (Thomas 1993b). This is supported by observations that asymmetries are much smaller in wings compared with tails (Møller and Höglund 1991; Balmford *et al.* 1993). Both manoeuvrability (Fig. 7.3c) and agility (Fig. 7.3d) are adversely affected by tail asymmetry, and these effects increase with increasing tail length.

Another recent inspection of the function of tail feathers has revealed that tail asymmetry may incur even larger costs than Thomas has predicted. Åke Norberg (1994) investigated the aerodynamic function of elongated outer tail feathers in barn swallows by analysis of high-speed film of swallows capturing insect prey in the air. He also investigated the bending and twisting behaviour of these feathers under different loads and analysed the aerodynamic performance of tail feathers in a wind tunnel. The angle of attack and angles of tail spread

---

Fig. 7.3 (a) Absolute and relative asymmetry reduce the lift a bird's tail can produce, and this increases with tail length. (b) Asymmetry can have large effects in the non-aerodynamically functional part of a tail. The thrust differential required to balance asymmetry increases with tail length. (c) Asymmetry decreases aerial manoeuvrability, and this effect is maximised at longer tail lengths. (d) A similar effect is seen in terms of aerial agility. The notation aa represents an absolute asymmetry of 4mm, and ra represents a 5% relative asymmetry. Adapted from Thomas (1993b). (e) Cross-section of a theoretical tail. The suction force created by the vortices on top of the tail is perpendicular to its surface, therefore the flap tilts the force outward, but also forward because of the flap's forward directed aspect. Adapted from Norberg (1994).

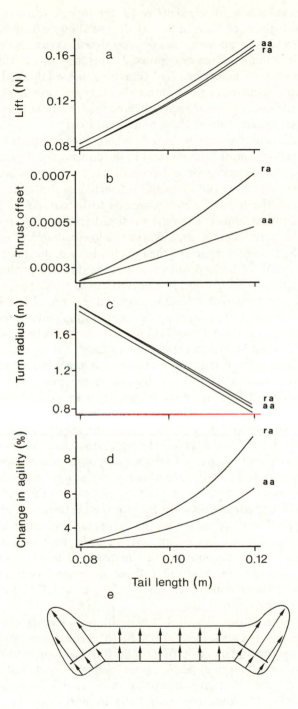

Tail length (m)

were manipulated by wire attachments to fix the base of feathers in position. Analysis of the high-speed films indicated that swallows can spread their tail feathers by up to 120° during aerial manoeuvres. Feathers can also be twisted by up to 90° and be either elevated or depressed. Therefore, these feathers can be manipulated in all three dimensions. The films also showed that tail feathers are usually oriented at positive angles of attack and so generate aerodynamic lift. Inspection of tail feather performance indicated that the outermost feathers are capable of large rotations about the longitudinal axis.

The wind tunnel observations of the tail performance of a male barn swallow indicated that the outermost feathers yield in the direction of the incident air, in other words they bend backwards and inwards when the bird is head-on to the wind. This also occurs if the end of the tail is lowered, i.e. set at a positive angle of attack, so that the tail end bends upwards to lie parallel to the air flow generated by the wind tunnel. The outermost tail feathers also rotate at their base so that the outer edge of each feather is lower than the inner. As the outermost feathers are overlapped by the next feathers in, this rotation causes a bending in the profile of the aerofoil section of the tail. The profile of a cross-section of the tail is no longer flat; it is curved down at the tips, resembling a Kruger type aircraft wing with a leading-edge slat (see Fig. 7.3e). This form of wing profile maximises lift to drag ratios in a delta wing, and can maintain high lift at relatively large angles of attack (Thomas 1993a; Norberg 1994). As well as this alteration of wing profile increasing lift to drag, it may also influence air flow over the inner part of the wings. Hence Norberg's interpretation of tail aerodynamics has more ecological validity and applicability, as the tail is not considered totally independent of the wings. As a swallow manoeuvres, the wings are fully extended perpendicular to the body and the tail is spread to its maximum (approximately 120°). At this point, the leading edge of the tail is close enough to the trailing edge of the wings that the tail may help with downwards deflection of the air over the rear part of the wings. Norberg (1994) suggests that the tail could act in an analogous way to a trailing-edge wing flap and hence increase lift at higher angles of attack without the bird stalling. In other words, it would greatly increase the turning performance of the bird.

Norberg suggests that, as length of the elongated outer tail feathers of a swallow influence tail profile shape, an asymmetry in these elongated feathers will incur increased costs beyond those estimated by Thomas (1993b). Asymmetry in these outermost feathers will also create asymmetry in induced drag that would have to be compensated for by changes in wing attitude and so would probably increase total drag.

Both the investigations of Thomas (1993a, b) and Norberg (1994) indicate that tail asymmetry is costly in terms of flight performance, particularly in relation to slow-speed and manoeuvring flight. Birds with asymmetric tails will suffer a reduction in turning performance and this cost will increase at increasing tail lengths. Also, an asymmetry in outer tail feathers is likely to be more costly than the equivalent asymmetry in more central plumes. Interestingly, asymmetry is larger (not smaller) both in relative and absolute terms in the outermost tail feathers (Møller 1994c). Hence, tail asymmetries may not be severely constrained by biomechanical costs.

7.3.4 *The difference between wings and tails*

As the mechanisms that generate lift in wings and tails are different, the relationship between asymmetry and trait length can be different. The cost of an absolute asymmetry decreases with wing length. Whereas the cost of an asymmetry will increase with tail length in the aerodynamically functional part of a bird's tail (Fig. 7.3a). Similarly, the flight cost of an absolute asymmetry in an aerodynamically non-functional part of a tail (perhaps a tail ornament) will increase with ornament size (Fig. 7.3b). In all cases, larger asymmetries incur larger costs. This relationship of costs of asymmetry with trait length make it even more curious why tail ornament length appears to be negatively related to ornament asymmetry in some species (Møller and Höglund 1991) and not others (Balmford *et al.* 1993). These differences may relate to the relative signalling and aerodynamic functioning of the tails in the different species that these two surveys measured (see Chapter 8).

There is one important issue to come from these theoretical analyses of wing and tail asymmetry on aerodynamic performance that has not been made clear, as yet. As a tail is spread it increases in aspect ratio, as it becomes relatively longer due to the angle of the feathers from the plane of symmetry. This may mean that the tips of the outermost feathers play less of an aerodynamic role in lift production in extremely spread tails, especially in those with very deeply forked profiles. Therefore, asymmetry in the outer tail feathers of deep forks may be relatively unimportant in vortex formation, and so developmental stability could be relaxed at little functional cost. This could be modelled theoretically and tested experimentally using the kinds of approaches adopted by both Thomas and Norberg.

Although we do not want to dwell on the limitations of the aerodynamics models we have described above, as they are intended to be simplifications of biology, we would like to raise one point that seems relevant at this stage that may stimulate future research. This is the problem of modelling wing asymmetry as an asymmetry in wing length. Very few developmental stages will actually give rise to an asymmetry in wing length, as the wingtip is most often comprised of a number of feathers. Only in species with extremely pointed wings will primary feather asymmetry actually influence overall wing length, and hence alter wing length asymmetry (see discussion in Swaddle *et al.* 1996). In these rare cases it will only be the asymmetry of the most distal primary that influences wing length. However, Thomas' way of modelling wing asymmetry is perfectly acceptable, given the state of the aerodynamic literature. Aerodynamicists model the wing of a bird as a solid aerofoil and do not take account of changes to the dimensions of individual feathers within the wing. Primary feathers exhibit fluctuating asymmetries and these asymmetries will lead to subtle size and shape differences between the left and right wings that cannot be accounted for at present by aerodynamic theory (Swaddle *et al.* 1996). What is needed is a very detailed 'strip' analysis of the influence of wing morphology on aerodynamic performance. Each of these strips could represent the size and asymmetry of an individual feather in the wing and need not be limited to the primaries; the secondaries could be included. This analysis would be far from straightforward, but it would yield an aerodynamic theory that would provide incredibly valuable insights into many facets of ornithology and have numerous applications. For example, it could be applied to the functional significance of avian wingtip shape (Lockwood *et al.* 1997) and the aerodynamic costs of stages and patterns of avian wing moult (Swaddle and Witter 1997a), as well as investigations of primary feather developmental stability.

The recent theoretical considerations of the cost of asymmetry in birds' tails (and in fact any biomechanical structure) is further complicated by observations of asymmetry in musculature associated with swallows' tails. Møller and Moreno (1997) found asymmetries in the muscles that rotate, elevate, and depress the outermost tail feathers, but not other tail feathers, of the barn swallows, and these muscular asymmetries may act to lessen the aerodynamic costs of feather asymmetry. The extent of muscle asymmetry and muscle size was positively correlated with feather asymmetry and feather size. Hence, Møller and Moreno concluded that the aerodynamic costs of a tail of given length and asymmetry cannot be accurately inferred from external morphology alone. However, musculature is not the only internal factor that can display asymmetry and so other internal variables such as skeletal asymmetries, muscular excitation asymmetries, innervation asymmetries, and vascularisation asymmetries should also be considered. This form of reductionism may be unhelpful to theoreticians who want to simplify and predict the costs involved with asymmetric structures; but it also suggests that the most appropriate method for assessing the performance costs of asymmetry are to adopt whole-organism approaches that directly quantify the costs in live specimens (cf. Møller 1991; Evans *et al.* 1994; Swaddle *et al.* 1996).

7.4 Experimental investigations of asymmetry

In this section we review the experimental evidence that asymmetry reduces performance. All of the experiments reported here are on avian species, which further justifies our focus on the theoretical issues of avian asymmetry in the previous section. First we shall give some details of the only study of flight performance and wing feather asymmetry and then review the two published experiments that have investigated the influence of tail asymmetry on flight.

7.4.1 *Wing asymmetry and performance*

Swaddle *et al.* (1996) investigated the effect of primary feather asymmetry on two aspects of flight performance in the European starling. They focused on flight parameters that were likely to have a large influence on the ecology of this species, namely escape take-off ability and aerial turning performance. Escape take-off was assessed by releasing the birds from a perch in a long narrow aviary with the simultaneous sounding of a loud vocal startle stimulus. The escape flight was filmed and the video was subsequently analysed to obtain measures of flight speed and angle of trajectory. Turning performance was assessed in an aerial obstacle course that consisted of padded wooden poles suspended from the ceiling of a flight aviary. The tips of the birds' wings were dipped in a small, standardised volume of washable ink and the number of pole contacts that the birds made through the course was recorded from the resulting ink marks left on the obstacles.

In observations of birds with naturally occurring levels of wing feather asymmetry, Swaddle *et al.* (1996) found that asymmetry reduced take-off performance, but asymmetry was confounded with reduced primary length as many of the instances of asymmetry occurred in birds with feather damage. Analysis of only those birds with intact, undamaged plumage indicated that asymmetry was not related to take-off performance. By experimental manipulation of primary feathers, in a factorial design for both primary length and asymmetry, they deconfounded length and asymmetry from intrinsic properties of individuals. They found that primary length (and not asymmetry) decreased take-off performance. In contrast, primary asymmetry (and not length) negatively affected aerial turning performance (Fig. 7.4), as predicted by Thomas (1993b). In a subsequent investigation, Swaddle (1997c) has shown that within-individual changes in starling wing feather asymmetry (due to natural moult) are related to flight performance. In this case, increased asymmetry was associated with decreased take-off and level flight performance even though the asymmetry differences were very small (approximately 0.5% of trait size).

The investigations of Swaddle *et al.* (1996) indicate that primary asymmetries mainly affect aerial manoeuvrability and agility, whereas primary lengths appear to influence the speed of flight in take-off. There did not appear to be any interaction of primary length and asymmetry in terms of the flight performance measures that Swaddle *et al.* made. This may be an artefact of the parameters that they measured, or it could be that the costs of asymmetry do not change at different wing feather lengths. Thomas (1993b) predicted that the costs of asymmetry would decrease with wing length, whereas Evans and Hatchwell (1993) predicted the opposite. Evans and Hatchwell's prediction was based on the assumption that a given amount of asymmetry will always create greater turning moments when a further distance from the plane of symmetry, i.e. at longer wing lengths. The results of Swaddle *et al.* do not give strong support to either hypothesis. More detailed and sensitive investigations of flight behaviour and wing kinematics may reveal which hypothesis holds more truth.

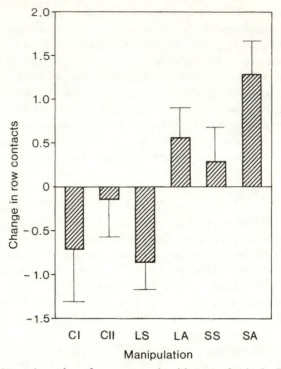

Fig. 7.4 Mean (+ s.e.) number of contacts made with rows of poles by European starlings *Sturnus vulgaris* flying through a manoeuvrability course. LS-long symmetric primaries; LA-long asymmetric primaries; SS-short symmetric primaries; SA-short asymmetric primaries; CI, CII-control groups. Asymmetrically manipulated individuals made contact with more poles than symmetrically manipulated individuals. Adapted from Swaddle *et al.* (1996).

7.4.2 *Tail asymmetry and performance*

In the first investigation of the effect of tail asymmetry on aerial performance, Møller caught swallows in mist nets and brought them into the laboratory (Møller 1991). There, they were randomly assigned to three treatment groups. Birds in the first treatment had 20 mm of either the left or right outermost tail feathers removed with scissors. In the second group, birds had 10 mm removed from either the left or right outermost tail feathers. The third group was a control treatment and birds were merely handled for an equal amount of time as the two previous groups. All birds were also measured, weighed, and ringed. The tips of their wings were then dipped in a small volume of black ink and they were then flown through an aerial manoeuvrability course (cf. Cuthill and Guilford 1990). In this course, birds had to negotiate baffles in order to fly through the maze into an open area. Collisions with these baffles were recorded by marks left by the inked wings.

The experimental treatments decreased aerial performance, as manipulated birds collided with more obstacles. Presumably this effect was due to increased tail asymmetry and not due to manipulation *per se*. Natural levels of tail fluctuating asymmetry and tail length also appeared to influence performance through the manoeuvrability course (Fig. 7.5). In an analysis of covariance, Møller indicated that performance after manipulation was influenced by the state of the bird's tail plumage before the manipulations were performed. Individuals that were naturally more asymmetric and those with naturally shorter tails tended to collide with more obstacles. This indicates that there may be intrinsic differences in flight ability between birds of varying tail length and tail length asymmetry. However, this experiment does not indicate the costs of asymmetry at different tail lengths, as length and asymmetry were not manipulated factorially, the most asymmetric birds also had the shortest mean feather length.

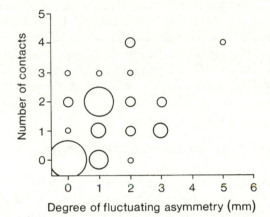

Fig. 7.5 Natural levels of tail feather asymmetry with manoeuvrability in male barn swallows *Hirundo rustica*. Manoeuvrability was assessed in an aerial obstacle course. Size of circles denote number of observations, which ranges from one to six. Adapted from Møller (1991).

A subsequent experiment has approached this same problem in a factorial design. Evans *et al.* (1994) manipulated both length and length asymmetry in the tail of male red-billed streamertails *Trochilus polytmus*. Evans and Hatchwell (1993) and Thomas (1993b) predicted that the cost of asymmetry will increase with trait size, and so long-tailed individuals may be constrained into stable, symmetrical development due to mechanical effects. Evans (1993) has even suggested that the reported negative relations of fluctuating asymmetry and size of secondary sexual ornaments has come about through these mechanical constraints and fluctuating asymmetry is not a reflection of individual quality, but merely a reflection of mechanical costs. Through their tail length and asymmetry manipulations, Evans *et al.* (1994) found that birds with shortened tails paid no manoeuvrability costs of asymmetry, whereas birds with asymmetric elongated streamer tails took longer and collided with more obstacles through the flight maze than symmetric long-tailed birds. Additionally, birds with naturally (i.e. before the manipulation) asymmetric tails and also those with naturally short tails took longer to negotiate the flight maze. These results support the case that the costs of asymmetry do increase with trait size, although asymmetry is not correlated with tail length in this species. However, it is not all together clear, if larger traits are constrained into being more symmetrical, why exaggerated secondary sexual ornaments display more asymmetry (not less) than the equivalent non-exaggerated trait (Møller and Höglund 1991; Møller and Pomiankowski 1993a) unless secondary sexual ornaments are non-functional. Thomas (1993a) has predicted that such traits are non-functional in birds; however, Norberg (1994) has provided analysis to indicate that the secondary sexual ornaments of barn swallows are at least partially functional. It is probable that secondary sexual ornaments do have some performance-related function, which may help to maintain their honesty as a signal, but this may not be as great as that of non-exaggerated traits under stabilising selection. This would mean that developmental stability could be relaxed in ornaments due to their decreased functional significance, but within a trait larger ornaments are constrained to symmetry because of mechanical constraints and intrinsic properties. These are not mutually exclusive explanations of the observed patterns.

7.5 Asymmetry and functional importance

The degree of developmental stability observed in a morphological character may be related to its functional importance. Traits that are functionally important, such as skeletal characters that are maintained throughout life, exhibit extremely stable developmental trajectories and hence very small levels of fluctuating asymmetry (Møller and Höglund 1991). Traits that are less functionally important, for example temporary traits like the feathers of birds that are replaced during moult each year, are less stable and exhibit relatively larger asymmetries. However, this pattern is bound to alter between species, as the relative functional importance of the same trait will alter from one species to another. Also, within a species, and potentially within an individual, the functional importance of a trait will alter with trait size. Most importantly the cost of any asymmetry of growth will also alter with trait size. Hence, it may be more accurate to relate the narrowness and height of the peak of the size–fitness function to the level of developmental stability and fluctuating asymmetry. If the size–fitness function is steep and tall, then development should be highly constrained to produce the ideal functional phenotype, i.e. symmetry, whereas if the peak is more broad and flat, then development will not be as stable, as variation in trait size will have little consequence to overall fitness and the trait is not particularly functionally important (Fig. 7.6).

   In these cases the degree of developmental stability may be directly related to the level of stabilising selection acting on the trait (Møller and Pomiankowski 1993a). Characters that are subject to intense stabilising selection could be highly canalised by genetic modifiers that regulate and prevent expression of

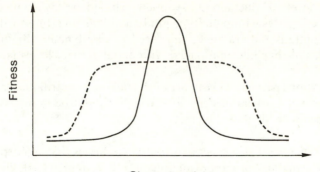

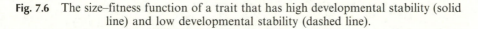

Fig. 7.6 The size–fitness function of a trait that has high developmental stability (solid line) and low developmental stability (dashed line).

extreme phenotypes. Traits under weaker stabilising selection are less developmentally stable and hence can develop a wider range of phenotypes and exhibit greater levels of fluctuating asymmetry. This may result in traits that are functionally important, and have highly constrained developmental pathways, exhibiting lower levels of fluctuating asymmetry. Traits under directional selection, such as secondary sexual ornamentation, will have less stable developmental trajectories, as increasing (or in some cases decreasing) size is selected for and genetic modifiers, which control development, are selected against (Møller and Pomiankowski 1993a). This results in sexual ornaments having far greater levels of asymmetry than traits under stabilising selection. There may also be a trade-off between size and symmetry in these ornaments (cf. Swaddle and Witter 1994), as individuals will have to optimise both size and symmetry to maximise their sexual advantage. Cheating is prevented as fluctuating asymmetry directly reflects the ability of an individual to buffer development against genetic and environmental stresses. Only individuals of high quality will be able to develop large secondary sexual ornaments that are also symmetric.

An example of the functional importance of a trait being related to its function is the difference in level of asymmetry displayed in birds' tails and wings. Wings are far more developmentally stable than tails (Møller and Höglund 1991; Balmford *et al.* 1993), this is because wings are far more aerodynamically important than tails (Thomas 1993a). Balmford *et al.* (1993) adopted this view of functional importance determining levels of asymmetry and generated some predictions based on Thomas' (1993a, b) aerodynamic predictions of the costs of flight feather asymmetry in birds. They then measured a selection of museum skins to test these predications. The predictions that they made were as follows:

(i)   Wing asymmetries should be lower in species where flight is increasingly important in their ecology. Specifically, they predicted that wing asymmetry should be lower in migrants than non-migrants.

(ii)   Relative levels of fluctuating asymmetry should be higher in the outer feathers of species with pintails or graduated tails than those with forked tails, as pintails and graduated tails are less aerodynamically functional, although tail shape is confounded with aerial activity levels (see section 7.3.3).

(iii) As projections past the widest part of the tail only contribute drag and no lift, the outer feathers of deep forked tails should be more asymmetric than those with shallower forks.

Balmford *et al.* (1993) also generated some hypotheses that were specifically related to the signalling properties of fluctuating asymmetry. If the level of asymmetry was primarily determined through sexually selected signalling criteria they predicted that:

(iv)   In species with sexually selected tails, relative asymmetry will be larger in the more-ornamented species due to the destabilising influence of sexual selection.

(v)   If asymmetry is related to phenotypic quality, absolute asymmetry should be negatively related to tail length in ornamented species (cf. Møller and Höglund 1991).

Through morphometric analysis of species that represented taxonomically independent groups, they found that wings were more symmetric than tail feathers. This is in agreement with the earlier study of Møller and Höglund (1991) and is concordant with Thomas' model (1993a). Balmford *et al.* also indicated that species that were relatively more aerial than terrestrial-based species tended to have lower levels of wing fluctuating asymmetry. In an analysis of migratory versus non-migratory species they report a small difference in asymmetry (males, one-tailed $P < 0.05$; females, one-tailed $P < 0.10$) with non-migratory males tending to display slightly more asymmetry than migratory males. These results tend to corroborate the predictions concerning the functional role of fluctuating asymmetry. Cuervo and Møller (1997c) have compared the flight morphology of resident and migratory populations where both populations are of the same species. They found that migratory populations have increased levels of wing length asymmetry, which they interpret as resulting from increased directional selection acting on migratory populations for increased wing length, as migratory species are known to have longer and more pointed wings (review in Lockwood *et al.* 1997). The directional selection acts against genetic modifiers and destabilises developmental processes in migratory populations (cf. Møller and Pomiankowski 1993a; Chapter 3).

Balmford *et al.* (1993) also report a trend for relative fluctuating asymmetry to be greater in the outer feathers of birds with graduated and pintails compared with those with forked tails. This is a more convincing test of the Thomas model, as the outer tails of forks will also be subject to more intense directional selection than those of graduates and pintails. This does appear to indicate that asymmetry will be minimised in the more aerodynamically functional parts of the tail.

In terms of their predictions in support of a signalling role for fluctuating asymmetry, their results differed from those obtained by Møller and Höglund (1991). Balmford *et al.* (1993) found no difference between the sexes in relative asymmetry of the most elongated tail feathers, although sexual dimorphism was absent or not great in many of the species. In species where there was pronounced sexual dimorphism there was also no effect. In this latter case the analysis was limited to just eight samples, so it may not be too surprising that they obtained a non-significant result. They did not test for differences in absolute asymmetry between the sexes. Finally, Balmford *et al.* did not find convincing evidence for a consistent negative relationship between trait size and asymmetry in the most exaggerated tail feathers. This conflicts with the previous study of Møller and Höglund (1991), who found consistent negative relationships, although different species were measured in both studies, and so we may expect interspecific differences in signalling strategies between the birds studied (see discussion in Chapter 8). It is also important to point out that testing predictions through non-rejection of a null hypothesis do not represent powerful tests. There may be non-biological reasons, which could be pure statistical artefacts, why a null hypothesis is not rejected.

In general, Balmford *et al.* (1993) propose that natural selection processes are important in determining levels of fluctuating asymmetry, and that fluctuating asymmetry is less likely to occur in functionally meaningful traits or parts of a trait that are closely related to performance. It can also be stated that the pattern of asymmetry manifest in a trait reflects a balance due to the reducing effects of natural and sexual selection and the increasing effects of directional and disruptive selection acting on the trait *per se* (i.e. not specifically on asymmetry) (refer to Chapter 3). Thus, it is possible to envisage a trait that is subject to intense natural selection for performance and yet have a large degree of asymmetry due to the intensity of the directional selection acting on character size.

7.6 Performance costs maintain asymmetry

It is possible that the relative intrinsic developmental costs and the external functional costs of asymmetry maintain and partially determine the developmental stability of traits. It is probable that the relative developmental costs of producing asymmetry decelerate as the phenotype strays from perfect symmetric form. These costs are likely to occur in terms of extremely sensitive feedback mechanisms between the left and right sides of a bilateral trait (Chapter 2). Costs are maximal when symmetry is produced and the cost of producing a slightly more symmetric phenotype is greater the closer the trait is to symmetry. On the other hand, the functional costs of asymmetry are likely to accelerate, as a minor asymmetry may not influence performance to any great degree but increasing asymmetry may lead to exponentially increasing functional costs (Fig. 7.7).

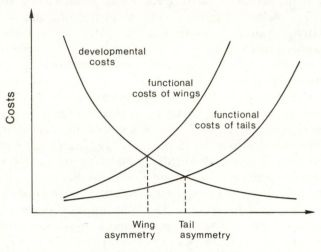

Morphological asymmetry

Fig. 7.7 Graph of decelerating developmental costs and accelerating functional costs of birds' wings and tails. Due to their functional importance, wings have lower levels of fluctuating asymmetry.

To illustrate this point we can consider the development of wings and tails of birds. At small asymmetries, the performance costs of both structures are likely to be small as the morphological and hence force and moment differences are minimal between left and right sides. As the difference between the two sides increases, the functional costs (in this case lift to drag ratio) are likely to rise exponentially, but at different rates for the two traits. Birds' wings are more functionally important than their tails (Thomas 1993a) and so the cost of asymmetry will be greater in wings than in tails. The intercept of the line between the decreasing developmental costs of symmetric production and the accelerating costs of functional performance gives two different values for asymmetry in birds' wings and tails (see Fig. 7.7). Wings should be more symmetric than tails, and indeed this relationship appears to be true (Møller and Höglund 1991; Balmford *et al*. 1993). Thus, there may be an optimum level of asymmetry due to a trade-off between developmental and performance costs of morphological asymmetry. This will tend to maintain a certain level of asymmetry within the population depending on the magnitude and exact pattern of these relative costs, which will vary between traits and species.

This sort of trade-off may occur in the example of racehorse performance and asymmetry that we described in section 7.2. It is probable that there is little variation in asymmetry of foreleg features compared with facial features due to the difference in relative functional costs of asymmetry between the two types of trait. This may make asymmetry difficult to assess in horse forelegs, as the asymmetry will be smaller in comparison with measurement error than in facial asymmetries. This may make the detection of a 'true' negative relationship between asymmetry and performance difficult to obtain in highly functional traits such as horse forelegs.

7.7 Summary

- In this chapter we review the evidence that links morphological asymmetry with performance. Asymmetric individuals suffer a reduction in performance.
- This relationship between asymmetry and performance can be both indirect (as asymmetric individuals may be of poor intrinsic quality and low-quality individuals do not perform well) and causal (as morphological asymmetry can directly cause reduced performance through mechanical principles).
- We pay particular attention to the role of asymmetry in avian aerodynamics, as this system provides an intriguing case where asymmetry is costly but the relative costs of asymmetry at varying trait sizes may be different between traits. In birds' wings the aerodynamic costs of asymmetry may decrease with trait size, but the costs may increase with trait size in birds' tails. Both the theoretical and empirical literature are reviewed.
- The developmental stability of a trait may be related to its functional importance. Characters that have high functional importance may be constrained into more symmetrical development.
- Finally, we discuss how the functional importance of a trait can both determine and maintain the level of instability during development. Functionally important traits display lower levels of asymmetry, but asymmetry could be maintained due to a trade-off with the intrinsic developmental costs of producing the perfect symmetrical phenotype.

8

Developmental stability and signalling

8.1 Introduction

Signals are either morphological structures or have a morphological basis. There are frequently considerable intra- and interspecific differences in signal design, particularly for those involved in sexual selection, but also for signals used in other contexts. Current differences in the efficiency of signals and large degrees of signal divergence during evolutionary history suggest that signals have experienced a recent history of consistent directional selection. The role of developmental stability in signalling theory in general and in sexual selection in particular is described (section 8.2).

Given that developmental stability of signals is presumed to be important in a signalling context, how could this situation have arisen? While it is possible that the symmetry of signals may have conveyed information on phenotypic quality from the very beginning of the evolution of a signal, this is not the only possibility. Most organisms and their parts demonstrate symmetry in one way or another, and signal perception may have evolved towards recognition of symmetry because all biological interactors such as mates, competitors, parasites, and predators, but also food demonstrate symmetry. Some recent analyses of evolution in simple computer models of neural networks have indicated that symmetry easily becomes entwined into the perceptual machinery. Psychological experiments have also demonstrated that symmetry is often more easily recognised than are other categories (section 8.3).

The various kinds of signals and the ways in which developmental instability may affect signalling are discussed (section 8.4). These signals include visual ones such as colour patterns and other morphological signals, but also signals in other sensory domains such as behaviour, including displays and vocalisations, pheromones, and electrical and tactile signals. The possibility is raised that asymmetry, which basically represents deviations from the norm, may also be important at levels below that of the individual. Possible examples include the major histocompatibility complex and signalling among gametes.

Sexual signals have been studied particularly intensively because asymmetries in secondary sexual characters are large which presumably makes them easy to detect. Several studies have investigated the relationship between developmental instability and sexual selection, and this extensive literature is reviewed. It is important to emphasise, as done already in Chapter 1, that much of the literature on fitness consequences of asymmetry is correlational, and it is therefore often impossible to know whether symmetry *per se* or a variable closely associated with it is responsible for a particular relationship. However, several experiments have revealed causal evidence for mate preferences of symmetrical partners by manipulation of the level of symmetry. Comparative studies have determined the relationship between sexual selection and developmental stability of secondary sexual characters involved in mate choice and intrasexual competition. These studies have generally investigated the amount of fluctuating asymmetry in sexual characters and compared this with asymmetry in non-sexually selected control traits. Furthermore, the relationship between the intensity of sexual selection and the relative level of asymmetry has been determined on two occasions. Finally, sexual selection in flowering plants depends on pollinator services, and observational and experimental studies have shown that pollination visitation rates are related to floral symmetry (section 8.5).

Given that developmental stability has an additive genetic basis, and that there is strong directional selection against asymmetry, there should be little variation in the level of genetically based developmental instability among individuals. This is a corollary of Fisher's fundamental theorem which posits that the rate of increase in mean fitness resulting from selection is equal to the standardized additive variance of fitness among genotypes (Fisher 1930). Two evolutionary processes may tend to maintain additive genetic variation in developmental stability of signals, and these are (i) an elevated mutational input due to intense directional selection and (ii) biased mutation (section 8.6).

8.2 General signalling theory

Signals convey messages to intended and unintended receivers, and the information content and truth or honesty of a signal can be classified with respect to the congruence of signal content with the state of the signaller. Signals are usually categorised as conventional signals which may refer to conventions as strategic correlates of quality, or as arbitrarily related to their message (Maynard Smith 1991; Guilford and Dawkins 1995). The first kinds of conventional signals are strategic choice handicaps where signallers have an evolutionary choice of signalling level at any quality level (Grafen 1990). A potential example is extravagant, costly tail feathers in males of a bird. Males can develop a range of tail lengths, but reliability is maintained because the cost of a given tail length is higher for low- than for high-quality individuals. Such signals could be faked, but generally are not at the evolutionary stable equilibrium. The second kinds of conventional signals are those designed to be arbitrarily related to the message. Well-known examples are status badges or warning colours where the costs depend on the target receiver. Non-conventional signals are either assessment signals that necessarily are connected with the underlying quality or that indicate quality directly (e.g. displays that signal body size) (Maynard Smith and Harper 1988; Johnstone and Grafen 1992), or those with designs closely determined by details of the message. In a number of cases there is no discrepancy of interests between signaller and receiver, and such signals can best be considered cost-free, conventional signals based on a strategic choice of the signaller (Guilford and Dawkins 1995).

Signals can be considered to be reliable or unreliable with respect to the state of the sender (Hasson 1994). Reliable signalling theory suggests that signals reliably reflect quality properties of individuals (Zahavi 1975; Grafen 1990). Reliability of signalling can, in cases where there is a conflict of interest between the two parties involved, be maintained by two conditions. First, they have to be costly to produce or maintain. Second, a given level of signalling has to be more costly for low- as compared with high-quality individuals. Cheating will be prevented only when these two conditions are fulfilled. In the following paragraphs we will discuss different kinds of signals in relation to developmental instability.

There are basically two ways of addressing the relationship between signals and their asymmetry. The first idea is that as signals become exaggerated, they become costly to produce and maintain. The assumption is that only signalling individuals in prime condition will be able to develop the most exaggerated versions of the signal and simultaneously produce a symmetrical signal (Møller 1990a). The second idea is that it is the intensity of directional selection that determines the level of developmental instability of a signal (Møller and Pomiankowski 1993a, b; review in Chapter 3). An elevated intensity of directional selection (as opposed to stabilising selection) will select against genetic modifiers that control the stable development of a character. If directional selection becomes particularly intense, the additive genetic variation in the signal may become depleted, and it may then only express a low level of developmental instability. The additive genetic variance of a secondary sexual character will depend on mutation–selection balance, and if a large number of alleles go to fixation due to intense directional selection, this will result in a reduction in the additive genetic variance.

Empirical tests are generally in accordance with the second explanation for the relationship between signals and their asymmetry, but at conflict with the first. For example, relative asymmetry in secondary sexual feather characters of birds is negatively related to the intensity of sexual selection as reflected by the mating system (Cuervo and Møller 1997a). However, there is no evidence that larger secondary sexual characters are disproportionately asymmetric. Further comparative tests of other signalling systems would be useful.

Reliable signalling theory assumes that cheating is prevented by the phenotype-dependent, differential cost of a given level of signalling. However, not all signals are reliable since examples of deception are widespread. These include flashing eyespots in moths, deceptive use of alarm calls in birds, and numerous cases of deception in primates (Trivers 1985; Munn 1986; Whiten and Byrne 1988). Intraspecific deception can only evolve to a low frequency in largely honestly signalling populations because such frequency-dependent selection requires a close fit of mimics to a particular model. This implies a high intensity of stabilising selection and therefore a low level of developmental instability once the mimicry is effective. Any major deviation of a mimic from a model will be strongly selected against because the signal will be inefficient in communicating a particular message. Furthermore, social control of deception may result from unreliable signallers being discerned as providing signals that are incongruent with their quality; lie-detectors use small, but predictable, deviations from the modal patterns of speech (Reid and Inbau 1966). Such deviations could be considered as context-dependent developmental instabilities.

Some conventional signals convey simple, uncostly information to a receiver that generally has common interests with the sender. The lack of conflict of interest selects for simple and uncostly conventional signals (or signs) that can readily be interpreted by all parties. A high level of stabilising selection is therefore involved in the evolution of such conventional signals. The level of developmental instability of such signals is therefore predicted to be very low.

Developmental instability generally increases with the intensity of directional selection (reviewed in Chapter 3). Directional selection thus may continuously generate a pool of inferior phenotypes with asymmetric signals. Two evolutionary processes may increase the level of developmental instability in signals, namely mutational input and biased mutation (see section 8.6). Mutations give rise to disrupted genomic stability, and a preference for symmetric signals by choosy individuals may allow screening of potential mates for deleterious mutations. Choosy individuals in a sexual selection context will thus benefit for either direct or indirect fitness reasons, since developmentally stable individuals will tend to perform better, but also provide genes for developmental stability to their offspring.

Costly signals themselves may generate stress which can increase asymmetry. Fluctuating asymmetry and other results of developmental instability may be disadvantageous (Chapter 7), though the disadvantage may be reduced by compensatory features. Direct disadvantages of asymmetry can potentially be offset by muscular or other adjustments that can reduce the direct impact of the asymmetry on performance. For example, fluctuating asymmetry in wing and tail feathers of birds has been predicted from aerodynamic theory to be costly for flight and manoeuvrability (Thomas 1993b). Thomas' study assumed that the muscular basis of flight was symmetrical, while the external feather morphology might vary with respect to asymmetry. A recent study of fluctuating asymmetry in tail feathers and muscles of the barn swallow has revealed clear positive covariation between asymmetry in feathers and asymmetry in certain muscles attached to these feathers. Møller and Moreno (1997) found that individual barn swallows had developed muscular compensations to asymmetries in a secondary sexual feather character (Fig. 8.1). The amount and direction of asymmetry in tail feathers was positively associated with amount and direction of asymmetry in tail muscle size. Asymmetries in muscles and feathers might be hypothesised to reflect the same underlying inability of the genotype to produce symmetrical traits. This is unlikely because feather asymmetry was only positively correlated with muscle asymmetry for the elongated, outermost tail feathers, but not other feathers in the tail. Furthermore, the relationship between muscle asymmetry and tail feather asymmetry was only found for muscles involved in tail manoeuvrability, but not for other kinds of muscles. Hence, the direct cost of the asymmetry in terms of performance appeared to some extent to be balanced by muscular adjustments. Similar compensation for deviations from perfect symmetry in morphology may be countered by anatomical adjustments in other signalling systems.

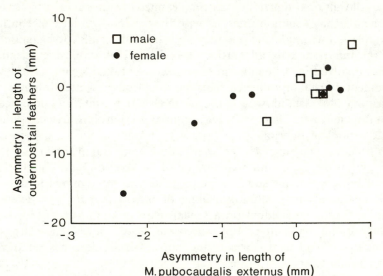

Fig. 8.1 Tail length asymmetry (mm) in relation to asymmetry in the length of the M. pubocaudalis internus muscle 7 (mm) among individual barn swallows *Hirundo rustica*. Adapted from Møller and Moreno (1997).

8.3 Origins of developmentally stable signals

Three different questions are central to every evolutionary phenomenon including developmentally stable signals; their origin, their selective history, and their maintenance. Although signals may provide information relevant for direct or indirect fitness benefits to a receiver, these benefits may not necessarily explain the origin or the selective history of a developmentally stable signal. Visual signals can be used as an example. Most animal and plant signals are symmetric in one way or another, for example the colour patterns of many animals and plants, the horns of antelopes and beetles, and the extravagant tails of many fish and birds. Such symmetric signals may have evolved because the brain which is used for processing visual signals may have particular properties that facilitate such processing. The perceptual neuronal network may be wired to symmetry because most animate structures demonstrate symmetry at one or more levels, and an ability to process sensory information with respect to symmetry thus may be highly beneficial (Møller 1992d). All food items, potential mates, competitors, predators, and parasites demonstrate symmetry of some kind, and any individual with a particular acuity to symmetrical features is likely to have experienced enormous fitness advantages. This may have resulted in a sensory bias at the evolutionary origin of a signal even before any mate preference for developmentally stable sexual signals evolved (Møller 1992d). Such biases in preferences have been put forward as a general explanation for the initial spread of extravagant signals (Ryan 1990). However, it is likely that as the perceptual wiring of the neural system evolved because of the advantages arising from acuity to symmetrical objects, this may simultaneously have selected for symmetrical sexual signals. Even though the sexual signals can be said to have evolved because of the pre-existing acuity of the sensory system, this acuity would also have provided certain individuals with direct or indirect fitness benefits during the selective history of a signal. For example, this could be the case if the possession of a specific sensory system resulted in an enhanced probability of finding food or avoiding being eaten by a predator.

8.3.1 *Symmetrical signals in neural networks*

The evolution of a simple neural system being exposed to a range of stimuli has been simulated in two simultaneously published studies of neural networks in computers (Enquist and Arak 1994; Johnstone 1994). In the first study, Enquist and Arak (1994) allowed a simple neural network to be exposed to a series of visual signals, and a reward system allowed particular signals to be chosen and used in the subsequent runs of the test, while the others were deleted. Random variation was added to these surviving 'progeny' to mimic genetic mutation. This cycle of selection and mutation was repeated a large number of times, and a network with a recognition ability eventually evolved. The model of Enquist and Arak allowed for the signal itself to evolve, and the simulation model thus mimicked the situation where a neural network co-evolves with the signal. Symmetry preferences in neural networks were shown to arise as a by-product of the need to recognise objects regardless of their position and orientation in the visual field (Fig. 8.2). The existence of sensory biases for symmetry may have been exploited independently by selection acting on biological signals (Enquist and Arak 1994), but other interpretations are possible as stated below.

In a second study, Johnstone (1994) showed that artificial neural networks readily evolved preferences for symmetry even in the absence of any link between symmetry and quality, but simply as a by-product of selection for mate recognition. In the recognition task of the model, the network is rewarded for detecting a signal from noise. This apparently gives rise to a preference for

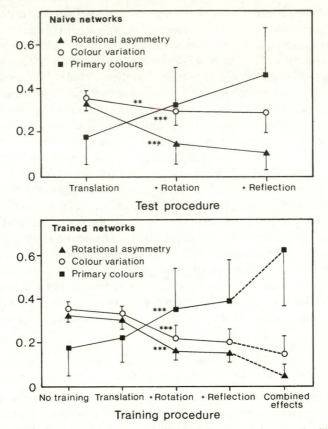

Fig. 8.2 Development of symmetry preferences in a neural network. Characteristics of preferred signals as a function of signal selection and network training procedures. The graph shows mean (+ s.d.) number of correct choices. Asterisks denote significant differences in comparison with preferred patterns generated by translation only. Statistical significance: **$P < 0.01$, *** $P < 0.001$. Adapted from Enquist and Arak (1994).

symmetry because the network appears to respond more easily to symmetric signals. Since the symmetrical patterns are closer to the population average of the training set, symmetrical traits will be preferred by the network because of averageness, not because of symmetry (Cook 1995). However, a preference for the average of a bilaterally symmetrical trait is also a preference for a symmetrical trait without training towards symmetry because the preference for symmetry evolves as an emergent property of the recognition system (Johnstone 1995).

The studies by Johnstone and Enquist and Arak assume that there is no cost involved for the specific preference for symmetry. This is unlikely to be the case because the assessment of any signal will have at least a time and possibly also an energy cost (see Pomiankowski (1987) for sexual signals and costs of mate choice). Any evaluation process takes time, and delay caused by assessment of features even as easily perceived as asymmetry is likely to have fitness consequences. For example, if an individual waits to mate until a partner exceeding a certain threshold level of signal has been found, this may result in a considerable delay, the fitness cost of which only can be offset by a balancing advantage due to direct or indirect fitness benefits. A second problem relates to the costs of assessment. If a signal is designed to allow rapid recognition of a mate of the correct species, it should stabilise at a uniform level that is small and uncostly. Most biological signals used in sexual contexts are large and costly, and therefore easy to assess, suggesting a different function of visual sexual signals. Selection for species recognition is likely to lead to the evolution of simple, unambiguous badges (Maynard Smith 1991), which must be stable, but should not show high levels of asymmetry within a population. Finally, as stated above, biases in the sensory and neural processing systems towards symmetry are associated with discrimination ability among biotic entities and inanimate objects (Møller 1992d).

The use of neural networks in the study of pattern recognition is an important step forward in our understanding of how the brain may work. However, the models used to simulate the effect of pattern recognition biases on signal evolution may be misleadingly simplistic, as suggested by Dawkins and Guilford (1995). Simple neural network models may often fail to capture essential properties of real biological recognition systems, and apparent hidden biases for symmetrical stimuli may be artefacts of the simplicity of the models. This criticism has important implications for our understanding of the models by Enquist and Arak (1994) and Johnstone (1994). Enquist and Arak suggested that symmetrical signals and inherent preferences for such signals have evolved as a consequence of the need to recognise objects in different positions and orientations in the visual field. In other words, a preference for symmetry has evolved as a solution to the problem of being able to recognise objects despite changes in their appearance because they are seen on different parts of, or orientations on, the retina. Such positional and rotational view invariance may arise because symmetrical signals look more similar than asymmetrical signals when rotated, and the neural network can therefore more easily learn symmetrical patterns. The network model is only able to solve the problem of signal recognition because it is not required to distinguish between orientations and recognise that the same stimulus is being presented in different orientations. Animals obviously are able to perceive the form of a signal independently of their position on the retina (Cook 1995). Enquist and Arak's model assumes that animals use solutions that allow them to solve view invariance problems at the cost of being unable to distinguish between individual presentations. This assumption needs to be tested in real animals.

Johnstone's (1994) neural network model is even simpler than that of Enquist and Arak because it is unable to recognise patterns if these are presented in different parts of or rotated on the artificial retina. The hidden preference in the network for symmetrical stimuli is produced by the use of the average set of training stimuli. Visual processing by real animals will obviously be able to recognise a slightly displaced or rotated signal as such, and the simplicity of the neural network model thus appears to give rise to the preference for the symmetrical signal. This is confirmed by recent computer simulations (Bullock and Cliff 1997). More realistic models that are able to recognise when the same signal is presented in different orientations and positions on the artificial retina are needed before we can reach any conclusions whether symmetrical signals have evolved as a by-product of the ability to recognise objects in different positions and orientations in the visual field.

8.3.2 *Developmentally stable signals and receiver psychology*

The psychology literature demonstrates that animals are perceptively sensitive to very small asymmetries (Schwabl and Delius 1984). Furthermore, animals ranging from bees to birds, dolphins, apes, and humans use symmetry as a category (Rensch 1957, 1958; Corballis and Roldán 1975; Delius and Habers 1978; Menne and Curio 1978; Barlow and Reeves 1979; Bornstein *et al*. 1981; Delius and Nowak 1982; Pashler 1990; von Fersen *et al*. 1992; Horridge 1996; Horridge and Zhang 1995; Giurfa *et al*. 1996). A number of studies have shown that animals are able to recognise symmetry as a concept, and that symmetric signals appear to elicit strong responses by receivers either initially or following training events. Recent studies have demonstrated that fluctuating asymmetry in skeletal traits in humans is directly related to the size of certain regions of the brain (Thoma 1996) and brain performance measured as IQ scores (Furlow *et al* 1997c). If this result proves to be general, it is likely that the ability to perform particular tasks as determined by the amount of neural tissue allocated will co-vary with individual levels of asymmetry. This may differentially affect the ways in which individual signallers and receivers are able to communicate.

The exact mechanism that allows symmetry detection, for example when a predator is confronted with the task of precisely locating a cryptic insect on a tree trunk, may utilise local feature detectors (Osorio 1996). In a mechanism that identifies edges and lines by classifying relative phases in spatial harmonics, an axis of bilateral symmetry resembles a line with no contrast. Performance of human symmetry perception is orientation sensitive, being best about vertical axes (Julesz 1971; Barlow 1980). It remains unknown whether this also applies to other organisms. Asymmetries arise as a consequence of the discrepancy between perceived lines and edges and physical intensity changes because the eye classifies local features by analysing phase relations in the spatial harmonics across the image. Points where the harmonics are in phase, or congruent, at zero-crossings ($0°$ or $360°$) in the sine functions are categorised as edges, whereas points of phase congruence at peaks and troughs ($90°$ or $270°$) are categorised as lines. For bilaterally symmetrical patterns spatial harmonics at the axis are at $90°$ and $270°$ (Delius and Nowak 1982), and the pattern therefore resembles a line of no specific contrast. Osorio (1996) developed a simple model demonstrating that arrays of filters operating locally across the visual image detect axes of symmetry by categorisation of spatial phase. Asymmetries will modify spatial phases and degrade the signal by making the mechanism inherently sensitive to fluctuating asymmetry and other deviations from perfect symmetry.

Asymmetry may make axes less detectable, or produce distortions analogous to blurring or kinking a physical line. Cells known to act as line or edge detectors are known from animals as diverse as insects and humans (Osorio 1996). For humans, bilateral symmetry is the most easily perceived of the transformational invariances in complex patterns (Julesz 1971), and finding axes of symmetry may be a task like that of judging the orientation of a row of dots (Barlow 1980). Osorio (1996) has pointed out that an animal which detects symmetry by phase congruence should have a bias for perfect symmetry irrespective of the information conveyed by the degree of asymmetry. Such a bias arises because asymmetries modify phase relations in spatial harmonics so that the axis is most detectable when symmetry is perfect. Other scientists, such as Enquist and Arak (1994), have argued that a preference for symmetry arises as a by-product of a mechanism that has evolved to recognise patterns irrespective of position or orientation, and perception should therefore be robust against variations that are a normal consequence of changing viewing angle and distance, and hence comparatively insensitive to small asymmetries. Studies of perception in organisms as diverse as bees, pigeons, and humans suggest that sensitivity to even small asymmetries is acute.

Psychological experiments on humans have generally shown that symmetrical patterns are more pleasing than asymmetrical patterns. Attneave (1954) found that symmetrical patterns were more accurately reproduced from memory, but only when the symmetrical patterns contained less information than the asymmetrical ones. The redundancy in the information of the symmetrical pattern was invoked to explain the apparent visual preference for such patterns.

The relationship between information redundancy in symmetrical patterns and attractiveness of patterns can be viewed in terms of the size of subsets of meaningfully related stimuli (Garner and Clement 1963). The simplest symmetrical pattern will have a subset of one meaningful stimulus, while the simplest asymmetrical pattern will have a minimum subset of two stimuli. Thus, when perceiving a reference pattern, the observer infers a subset of equivalent patterns, and the attractiveness of the reference pattern is inversely related to the size of the inferred subset. An experimental test revealed that attractive patterns do have small inferred subsets. This result provides support for the idea that symmetrical patterns are perceived as being attractive because they consist of a small subset of stimuli.

In addition to symmetrical objects containing a small subset of stimuli, they also possess desirable qualities *per se*. Szilagyi and Baird (1977) performed an experiment that allowed subjects to perform a design-production task in which an ideal pattern was created. The patterns created by the experimental subjects generally demonstrated much more symmetry than expected by chance. Hence, subjects found symmetrical patterns to be more ideal than asymmetrical ones.

Two experiments have addressed the question of whether organisms have an innate preference for symmetrical, visual signals. Horridge (1996) trained honeybees to use visual cues in a Y-maze choice apparatus for a food source. Choice experiments clearly provided evidence for faster learning when one of the patterns mimicking flowers was bilaterally symmetrical with a vertical plane of symmetry. Bees were able to discriminate between a bilaterally symmetrical form of the pattern and the same pattern rotated 90°. Interestingly, the choice of symmetrical patterns increased steadily from 50 to 80% during several hours of training (Fig. 8.3; Horridge 1996). At the start of the experiment there was no clear preference for the symmetrical signal but such a preference developed during the training session. This result provides clear evidence for honeybees learning a specific pattern rather than emerging with the ability to discriminate between symmetrical and asymmetrical signals. Thus there was no innate tendency for bees to choose bilaterally symmetrical signals in this particular experiment, whereas the innate tendency appears to be for faster learning of symmetrical patterns. Giurfa *et al.* (1996) used a different design to train bees on either symmetric or asymmetric models. After only seven trials honeybees demonstrated a clear preference for the learnt pattern even in novel stimuli. Bees showed a predisposition for learning and generalising symmetry as compared to asymmetry because they choose symmetry more frequently, approach it more closely, and hover longer in front of the novel symmetrical stimuli than the bees trained for asymmetry do for novel asymmetrical stimuli. Again, this experiment suggests that learning of symmetry is facilitated over learning of asymmetry.

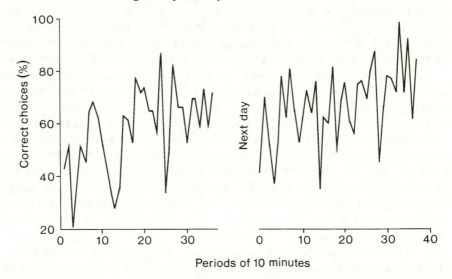

Fig. 8.3 Choice of a symmetrical pattern by honeybees *Apis mellifera* in the choice situation of a Y-maze provided with a food source. The choice of the symmetrical pattern increased steadily during a training session of several hours' duration. There is clearly no preference for symmetry at the start of the experiment (no significant deviation from 50%). Values are means. Adapted from Horridge (1996).

8.4 Developmentally unstable signals

Most mate recognition and sexually selected signals differ considerably among closely related taxa. That is the case for visual signals such as colour patterns, feather ornaments in birds, horns in antelopes, vocalisations of frogs, and pheromones of insects. These signals must have been subject to recent evolutionary divergence and hence a recent history of intense directional selection. Signals differ in the current prevailing patterns of selection. Some secondary sexual characters, such at the train of the blue peafowl *Pavo cristatus*, are subject to a current directional mate preference (Petrie *et al.* 1991), while other traits such as certain features of the song and song-producing structures of *Drosophila* are subject to stabilising selection (Carson and Lande 1984; review in Pomiankowski and Møller 1995). Persistent directional selection generally results in an increase in developmental instability, while stabilising selection has the opposite effect, as described in Chapter 3. We can therefore start investigating the properties of signals in terms of developmental stability and analyse these in terms of current selection patterns. In this section, we will briefly discuss a range of different signals and how the concepts of developmental stability relate to their structure and current function. Many extravagant morphological signals such as the feather ornaments of birds or the weaponry of male beetles, deer, and antelopes are currently subject to intense directional selection pressures and often demonstrate elevated levels of developmental instability. These and other examples are treated in greater detail in section 8.5.

8.4.1 *Colour patterns*

Colour patterns are common among a diverse group of organisms ranging from plants to animals such as annelids, molluscs, arachnids, insects, and various groups of fishes, amphibians, reptiles, birds, and mammals. Colour patterns generally are bilaterally symmetrical with a few notable exceptions such as zigzag stripes of snakes, the plumage of the wryneck *Jynx torquilla*, and the net pattern of the giraffe *Giraffa camelopardalis*. The near ubiquity of symmetry may suggest that symmetric patterns arose early during evolutionary history and thereafter have been lost secondarily only a few times. Alternatively, symmetric colour patterns may have arisen as an indirect consequence of basic body symmetry. Symmetric colour patterns may continuously be disrupted by mutation, and ubiquitous symmetry might be maintained by the advantages in terms of natural and sexual selection. These suggestions can be subjected to a preliminary test by use of data from evolution under domestication. As early as 1889 Wallace noted that domestic animals often lost the symmetric colour patterns of their wild counterparts, and he attributed this fact to species recognition. Asymmetric coat colour patterns have evolved independently a number of different times in mammals (Table 8.1). Rapid changes in the symmetry of colour patterns suggest that symmetry can readily be altered, perhaps because symmetric colour patterns are canalised traits with genes with major epistatic effects masking the genetic variance present (see Chapter 3). The high frequency of asymmetric colour patterns under domestication as compared to the relatively few cases under natural conditions may indicate that either the absence of one or more selection pressures under domestication may change the balance in favour of pattern asymmetry, or artificial directional selection may expose underlying genetic variation hidden by canalisation. The most likely selection candidate is the absence of any significant predation pressure under domestic conditions (Cott 1940). In this context, it is perhaps interesting that pattern asymmetry under domestication does not appear to be a reliable stress indicator (Woolf 1993). In conclusion, it is likely that symmetry in colour patterns is maintained by stabilising selection rather than being present due to a constraint.

Table 8.1 Cases of evolution of asymmetric coat colour patterns under domestication of mammals.

Rabbit *Oryctolagus cuniculus*

Deer Mouse *Peromyscus maniculatus*

Mouse *Mus musculus*

Rat *Rattus norvegicus*

Hamster *Mesocricetus auratus*

Guineapig *Cavia procellus*

Cat *Felis catus*

Dog *Canis familiaris*

Horse *Equus caballus*

Cow *Bos taurus*

Pig *Sus scrofa*

Goat *Capra hircus*

Sheep *Ovis aries*

Predator–prey signals belong to a group of signals that must be subject to intense stabilising selection. Individuals of many species that fall prey to predators have symmetrical facial or ear markings which may signal the direction of gaze. For example, this is the case in several hares and rabbits, antelopes and deer. Similarly, rump patches of conspicuous colour may provide information on the angle of escape to a potential predator by an individual of a prey species. For example, this is the case in several gallinule bird species, hares and rabbits, and deer. If facial markings signal the direction of gaze and rump patches signal the direction of escape, then asymmetric signals will appear to originate from a prey individual that has not spotted a predator, or that is not fleeing in a direction that most quickly brings it out of reach of a predator. If asymmetric colour patterns reflect poor phenotypic quality, as appears to be the case for asymmetric phenotypes in general, then predators will be able to capture asymmetric prey individuals more often than symmetric ones. Predators will on average be more successful in prey capture independent of whether signals are perceived as asymmetric due to prey not having spotted the predator, or prey having genuinely asymmetric colour patterns. Predators may therefore learn that prey capture success is directly associated with perceived signal asymmetry, and such a learnt response among predators will impose intense stabilising natural selection on symmetrical colour patterns of prey species. Interestingly, very asymmetric colour patterns are generally cryptic, at least for humans (Osorio 1994), and large degrees of asymmetry may render individuals less likely to become detected. However, these amounts of asymmetry greatly exceed what is the common level encountered for colour patterns demonstrating fluctuating asymmetry.

Warning or aposematic coloration has evolved a number of times among marine and terrestrial invertebrates and vertebrates. Batesian mimics evolve to closely resemble a distasteful model species, while several unpalatable Müllerian mimics converge in appearance, each species gaining protection from its similarity to the others. Aposematic coloration is mainly effective if there is a proper match between model and mimicking species. Mimics have evolved from ancestors that deviated considerably in appearance from models, and systems of mimicry must have arisen as a result of rapid evolutionary change, when the signal evolved. Once the mimics are established, the signal will be subject to intense stabilising selection in order to facilitate recognition by potential predators. We can thus predict that Batesian and Müllerian mimics will demonstrate low levels of developmental stability as suggested by a relatively high frequency of imperfect mimics in butterflies (Sheppard 1959). Developmentally unstable models will be at a selective disadvantage during encounters with potential predators because of the elevated risk of recognition errors, especially if pattern recognition is easier for symmetric patterns. The initial evolution of a mimic may therefore be facilitated by intense selection against asymmetrical phenotypes arising as a consequence of the rapid directional evolutionary change.

8.4.2 *Behavioural signals*

Behavioural signals may not seem readily assessed in terms of developmental stability. However, behaviour always has morphological correlates. This is the case for extravagant displays such as the mating dance of the three-spined stickleback *Gasterosteus aculeatus* (Tinbergen 1951), the stotting behaviour of gazelles *Gazella granti* (Caro 1986), and the threat display of moths with eyespots on their hindwings (Sargent 1976). Behavioural signals are frequently repeated, stereotyped actions. Behaviour has a morphological basis, and even slight differences in the morphology between sides due to developmental instability may become accentuated during display sessions. Alternatively, small deviations in the performance of a repeated behavioural trait may reliably reflect the smallest deviations in the underlying morphology, i.e. signal consistency, which ties in with the more recent signalling literature. Moreover, behaviour is always the output of psychological adaptations. Phenodeviant neural tissue may be expressed in behaviour, and the behaviour of such individuals may for that reason be considered deviant (see Markow and Gottesman (1993) for an example concerning schizophrenia in humans).

A potential example of a deviant behaviour concerns the song of the great tit *Parus major*. Male great tits, like most other birds, sing a stereotyped song a number of times. The song of the great tit consists of repeated notes of a specific frequency. When males have sung the same song a number of times, the frequency of the song usually starts decreasing, a feature called drift (Lambrecht and Dhondt 1988a, b). Males differ considerably in the number of songs that they are able to sing before drift affects the performance, and individuals are consistent in their ability to delay drift. It is easy to imagine that small phenodeviants in the syrinx or the muscles of the syrinx give rise to this drift. Females apparently prefer males that are able to sing many songs before signs of drift in frequency, and such males raise more offspring during their lifetime than others. These speculations are further supported by direct relationships between anatomical asymmetry in the vocal tract and individual differences in the calls of the oilbird *Steatornis caripensis* (Suthers 1994). Furthermore, the dominant frequency of calls of male jungle fowl *Gallus gallus* was negatively related to asymmetry in the length of a secondary sexual character (spur length) (Furlow *et al.* 1997b).

A second example involving calls concerns the song of males of the grasshopper *Myrmeleotettix maculatus* (Møller 1997b). Songs of grasshoppers are produced using a stridulation file on the tibia of the hind-legs, and stridulating males alternate between left and right leg when singing. The number of pegs in the file may differ between legs, and although most individuals are symmetrical with respect to peg number, some males display considerable asymmetry. Sonograms of songs clearly reveal that asymmetries in peg number are translated into 'asymmetric' songs. Experimental manipulation of the number of pegs by covering some with wax was used to generate 'asymmetric songs'. Females placed in a Y-maze choice chamber in which 'symmetric' and 'asymmetric' songs were played back from loud-speakers revealed a female preference for 'symmetric' songs. A similar relationship between song features and developmental instability has recently been reported for field crickets (Simmons and Ritchie 1996).

Animals construct a range of different structures such as the nests of many insects, fish, and birds, and actual display arenas such as the bowers of bowerbirds and the extravagant stone piles of black wheatears *Oenanthe leucura*. Bowers are highly elaborate structures that in different species may consist of alleys, maypole-like structures, or circular constructions of sticks (Gilliard 1969). All these structures demonstrate either bilateral or radial symmetry, and it is tempting to speculate that symmetry as well as the elaboration of such structures provide reliable information on the constructors.

Pheromone communication is widespread among many invertebrates such as insects and vertebrates such as amphibians, reptiles, and mammals. Pheromone chemical structures are extremely elaborate with enormous interspecific variability (e.g. Roelofs and Brown 1982; Birch *et al.* 1990). Elaborate pheromone molecules may exhibit production errors that render them slightly deviant with respect to the mode of the population, and such deviant pheromones may be slightly less efficient in eliciting a response by a receiver. Pheromone receptors are often extremely specific with only a single pheromone molecule of the right structure being able to elicit a response (Shorey 1976). Phenodeviant pheromones will obviously fit less well into a lock-and-key communication system. Finally, asymmetry in the fine morphology of pheromone detection systems such as antennae would make location of the source of pheromones errorprone.

Tactile and electrical signals are used by several widely separated groups of fish. Electrical signals are produced by electroplates or electroplaxes, which are aggregations of disc-like cells arranged so that they all face in the same direction, while the signals are received by the lateral line system or special receptor organs called mormyoblasts. The lateral line sensory system in fishes often demonstrates high levels of asymmetry (Kirpichnikov 1981), and this obviously may affect the ability of individuals to receive signals and behave appropriately in particular contexts.

8.4.3 *Cellular signals*

Fertilisation requires identification of and interactions between gametes of compatible mating types. If gametes are released externally in an aquatic environment, this requires long-distance signalling and assessment of species identity and potential quality properties of the gametes for gamete fusion to be non-random. The same may apply to signalling at the gamete level among species with internal fertilisation. Gamete signalling systems have evolved great levels of intricacy of design (Lee and Vaquier 1992; Levitan and Petersen 1995), that may involve reliable signalling (Pagel 1993). Minor phenotypic deviations from the species-specific norm for both signal and receptor may be considered to be cases of developmental instability, if the same genotypes are producing a range of these phenotypes, and they may form the basis of developmental selection against gametes with abnormal phenotypes (Møller 1997d).

Signalling at the cellular or subcellular level occurs continuously in unicellular and multicellular organisms. Most such signalling systems consist of relatively simple chemical substances and receptors. Problems of developmental instability at this level may be relatively small, since both signal and receptors appear to have remained unchanged during great evolutionary time spans. For example, this is the case for the immune system of vertebrates, as exemplified by the major histocompatibility complex (MHC) which is highly variable with an enormous ability to generate complex signals (Klein 1986). Phenodeviant signals will not only generate incorrect responses to parasites, but also to hosts. Equally complex signalling has evolved in the self-incompatibility system of plants that is responsible for pre- and post-zygotic female choice (e.g. Willson and Burley 1983; Seavey and Bawa 1986). Molecular crypsis by microparasites relies on a continuous change in the phenotypes of parasites and an inability of the host to recognise such change (e.g. Hyde 1990). The continuous, intense directional selection pressure among parasites and their hosts should generate a relatively high frequency of phenodeviant signals and receptors that should have important consequences for this co-evolutionary process.

8.5 Sexual selection and developmental stability

8.5.1 *Overview and experiments*

A number of studies have investigated the relationship between asymmetry and sexual selection (Table W8.1 on the Web site). Apparent asymmetry may be caused by factors other than unstable growth, and it is important to note that almost all studies have taken great care to exclude individuals with morphological asymmetries caused by damage. The relationship between developmental stability and sexual selection has been investigated in a meta-analysis based on 114 samples from 52 studies of 36 species of animals (Møller and Thornhill 1997). Eighty-six samples showed a negative relationship, 27 samples a positive relationship and one no relationship ($P < 0.001$). The overall effect size adjusted for sample size in the studies, which reflects the effect of developmental stability on sexual selection, was $r = -0.23$, $t = 7.53$, df $= 113$, $P < 0.001$. There was statistically significant heterogeneity in effect size among studies, and some of this heterogeneity could be accounted for by confounding variables. First, studies in which tests for the characteristics of fluctuating asymmetry were made and confirmed had a significantly larger effect size than studies without such tests (with test: $r = -0.36$, $N = 65$; without test: $r = -0.17$, $N = 39$, $P = 0.007$). Studies based on secondary sexual characters demonstrated a stronger effect than studies dealing with ordinary morphological traits (secondary sexual characters: $r = -0.38$, $N = 48$; ordinary character: $r = -0.24$, $N = 60$, $P = 0.03$). Finally, effect size was significantly larger in experimental than in observational studies, as expected if experiments remove the effects of confounding variables (experiments: $r = -0.51$, $N = 22$; observations: $r = -0.26$, $N = 85$, $P = 0.0028$). This meta-analysis demonstrates that developmental stability is associated with sexual selection.

The first experimental study to address the question of whether females prefer males with a symmetric phenotype was performed on the barn swallow (Møller 1992b). Males of this species have elongated outermost tail feathers that are the subject of a directional female mate preference. However, since long-tailed males also on average have very symmetrical tails (Møller 1990a), it is difficult to tell whether females prefer mates with long tails, symmetric tails, or both. (The same argument applies to other experimental studies of male phenotypes (such as for example, that of Andersson (1982)), because experimental manipulation of size simultaneously affects relative asymmetry.) Simultaneous experimental manipulation of tail length and tail asymmetry revealed that both affected the duration of the pre-mating period (Fig. 8.4; Møller 1992b). There was no effect on quality of offspring, but symmetric males produced more offspring per season than asymmetric males, mainly because their mates were more likely to lay a second clutch. The results suggest that females use ornament symmetry (or a correlate thereof) in their mate choice. Males with experimentally asymmetric and shortened tails provided a larger share of food for the nestlings (Møller 1994d), and the treatment thus did not appear to seriously impair the flight ability of males because males capture food for nestlings while on the wing. Therefore, female barn swallows did not choose mates with symmetric tails because of any greater efficiency in the providing of food by such males.

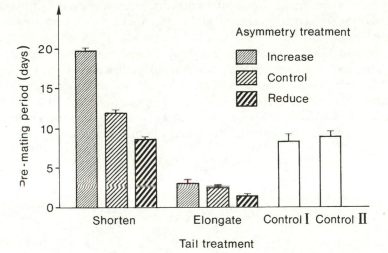

Fig. 8.4 Duration of the pre-mating period (in days) in relation to tail length and tail asymmetry of male barn swallows *Hirundo rustica*. Values are means (+ s.e.). Adapted from Møller (1992b).

The barn swallow experiment may have affected the appearance of the sexual signal, male flight behaviour, or both, and another experiment on the zebra finch *Taeniopygia guttata* tested whether a sexual signal *per se* affects a female mate preference (Swaddle and Cuthill 1994a). The study exploited the fact that females show preferences for particular leg ring colours (Burley 1981), and the coloration of rings thus becomes a window through which we can indirectly investigate the mind of birds. Males were randomly assigned rings of preferred colours placed symmetrically or asymmetrically. The measure of the female mate preference was assessed from the amount of time spent by females displaying courtship behaviour in front of a particular male. Females spent more time displaying in front of males with symmetric rings (Fig. 8.5). This experiment provided evidence for female preference for a symmetric signal *per se*. A subsequent study exploited the fact that birds see in the ultraviolet and investigated the effect of ultraviolet reflectance of leg rings placed symmetrically or asymmetrically (Bennett *et al.* 1996). There was a clear female preference for males with symmetric rings over males with asymmetric rings that had ultraviolet reflectance. Future studies may be able to utilise video images of animals with manipulated levels of asymmetry, since female spiders clearly respond to asymmetry even when displayed on a video-monitor (Uetz *et al.* 1996). Additional studies have confirmed the conclusions of these first experiments, that developmental stability plays an important role in sexual selection independently of whether the male character in question may have any direct effects on male behaviour (Table W8.1 on the Web site).

A common mechanism of sexual selection through female choice is differential investment in reproduction by females mated to attractive males (Burley 1986). Such cryptic female choice after pair formation may dramatically affect the relative reproductive success of males. Fluctuating asymmetry has been associated with differential investment in domestic chickens *Gallus gallus* (Forkman and Corr 1996). Hens randomly assigned to roosters laid more eggs if housed with individuals with symmetrical wattles ($r_s = -0.50$, $P = 0.003$), while the symmetry of the wattles of the hens did not influence productivity. Additional examples of cryptic female choice in relation to asymmetry have been report by Otronen (1997) for the fly *Dryomyza anilis* and Thornhill *et al.* (1995) for copulatory orgasm and asymmetry in humans.

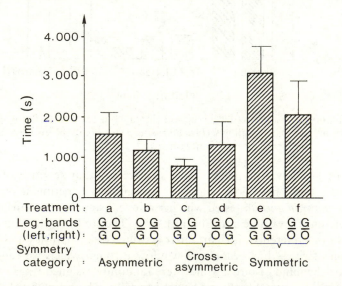

Fig. 8.5 Female preference for male zebra finches *Taeniopygia guttata* with symmetric colour rings as measured by the time females spent close to males. Values are means (+ s.e.). Adapted from Swaddle and Cuthill (1994a).

Success in male–male competition appears also to be influenced by morphological asymmetry. A study of contests between males of the damselfly species, *Hetaerina cruentata* and *H. americana* related the outcome of conflicts to wing asymmetry (Córdoba-Aguilar 1997). Both conflicts between residents and between residents and intruders were resolved to the advantage of the male with the more symmetric wings. Assessment of wing asymmetry may have taken place since the duration of the conflict was directly related to the difference in the degree of wing asymmetry.

If asymmetry reflects phenotypic condition we would predict that males with symmetric phenotypes should win in male–male competition more often than asymmetric males. This may be the mechanism behind the negative relationship between skeletal asymmetry and mating success in the lek-breeding black grouse *Lyrurus tetrix* (Rintamäki *et al.* 1997).

In accordance with this prediction, fights and verbal aggression in human males were inversely related to asymmetry in school children and college age men (Manning and Wood 1997; Furlow *et al.* 1997a). Furthermore, in species with sexual size dimorphism such as humans we would expect males with large body size to be more symmetric than males with small body size, while large female body size should be associated with a high degree of asymmetry, if large female body size is disadvantagous. Estimates of asymmetry in men and women are consistent with these predictions (Manning 1995).

Human facial attractiveness represents a particularly well-studied subject, but the role of facial symmetry has only recently received attention. Faces created by combining individual faces into composites have repeatedly been shown to be more attractive than the individual faces, and this has been interpreted as a preference for average facial features (Langlois and Roggman 1990; Symons 1979). This conclusion only holds to a certain extent because very attractive faces are not average (Alley and Cunningham 1991). Cultural conditioning may play an important role in human mate preferences, as already suggested by Charles Darwin in his book on sexual selection. However, this paradigm is not supported by recent studies emphasising that sexual preferences are cross-cultural with similar features being considered attractive (e. g. Perrett *et al.* 1994). The old paradigm that average faces are more attractive is problematic because composite faces are also more symmetrical since asymmetries are random with respect to side (Alley and Cunningham 1991; Benson and Perrett 1991). An apparent preference for averageness is thus potentially confounded by a preference for symmetry. Grammer and Thornhill (1994) were the first to investigate the relative importance of averageness and symmetry for facial attractiveness. Computer images of faces and of composites of faces revealed that symmetry rather than averageness accounted for independent measures of faces with respect to attractiveness, dominance, sexual appeal, and health. Two subsequent attempts to replicate these findings have failed, but apparently not because humans differ from other organisms in having explicit sexual preferences for asymmetry. Swaddle and Cuthill (1995) manipulated facial asymmetry without altering the mean size of facial features. Faces that were made more symmetrical were generally rated less attractive. Kowner (1996) in a study in Japan found a similar lack of preference for symmetrical mirror-image faces over natural, asymmetrical faces. A recent study by Perrett *et al.* (1997) suggested that the apparent preference for asymmetry in the studies by Langlois *et al.* (1994), Swaddle and Cuthill (1995), and Kowner (1996) arose because of the use of mirror-reflection or because of averaging a face and its mirror image. Such artificial faces are obviously perfectly symmetrical, but are also perceived as strange and unattractive because they have unnatural facial features of texture and facial shape. If asymmetry was removed from faces by means of computer techniques without mirror-imaging, there was indeed a strong preference for the more symmetrical faces (Perrett *et al.* 1997). This result emphasises the importance of experimental procedure for the outcome of mate preference tests, and it provides a cautionary note for future work in this very interesting area.

8.5.2 *Comparative studies of sexual selection*

Comparative studies of developmental instability of sexual characters based on measurements of museum specimens may just like field samples suffer from a number of potential biases (Swaddle *et al.* 1994, 1995; Simmons *et al.* 1995). The following studies should be viewed in the light of such potential biases.

8.5.2.1 *Characters involved in mate choice*

Comparative studies have investigated the level of developmental instability in secondary sexual characters and the relationship between developmental instability and the intensity of sexual selection. If directional sexual selection tends to reduce developmental stability, secondary sexual characters should demonstrate larger degrees of fluctuating asymmetry than the same trait in the non-ornamented sex, and than either sex of a closely related, non-ornamented species. In a pairwise comparative study of ornamented and non-ornamented closely related taxa of birds, Møller and Höglund (1991) found an increased relative level of fluctuating asymmetry in the secondary sexual feather characters of males of the ornamented species as compared with a non-sexual character (wing length), and as compared with the character homologous to the secondary sexual character of males in the females of the same species, and in both males and females of the closely related, non-ornamented species. There were also persistent negative relationships between ornament asymmetry and ornament size in males of the ornamented species, but not in females or in either sex of the non-ornamented species. These results may indicate that feather ornaments are reliable indicators of phenotypic quality, since only males able to produce large secondary sexual characters are able to produce symmetric traits. An alternative interpretation is that males with large secondary sexual traits are constrained into developing symmetry, if the ability to cope with a certain level of asymmetry depends on the size of a secondary sexual character (Evans 1993; see Chapter 7).

A subsequent study by Balmford *et al.* (1993) arrived at a conflicting conclusion. They found no tendency for relative asymmetry in tail length to be larger in the longer-tailed sex of a large number of evolutionarily independent cases of tail ornamentation. Furthermore, Balmford *et al.* (1993) found no evidence of a negative relationship between absolute fluctuating asymmetry and tail length. Balmford *et al.* (1993) investigated relative fluctuating asymmetry in tail length in relation to the shape of the tail. Relative asymmetry was higher in birds with graduated tails or pintails than in species with forked tails. However, this result is confounded by time spent flying, since an aerial lifestyle is likely to result in a more symmetric morphology, as found in their study. Since birds with forked tails are often aerial insectivores that spend a large proportion of their time flying, species with forked tails are predicted to have more symmetric tails than species with graduated tails or pintails. Balmford *et al.* (1993) also found that the outer tail feathers were more asymmetric in species with deep as compared with those with shallow forks. This result may be related to asymmetry being more costly for species with more deeply forked tails.

The conclusion of the study by Balmford *et al.* (1993) was that Møller and Höglund (1991) had used a biased sample. Balmford *et al.* used a larger sample of long-tailed species of birds, but several species were unlikely to have evolved long tails in response to sexual selection. Many species had sexually size-monomorphic tails, while that was not the case in the study by Møller and Höglund. A subsequent study on feather ornament asymmetry in an even larger sample of bird species revealed an enlarged degree of relative fluctuating asymmetry in the secondary sexual characters of males as compared with females (Cuervo and Møller 1997a). However, this study also found that the degree of relative fluctuating asymmetry initially increased, when a trait became the target of sexual selection, but subsequently decreased under intense sexual selection (Fig. 8.6). Lekking species with a considerable skew in mating success, and therefore presumably a higher intensity of sexual selection, were relatively less asymmetric in their secondary sexual characters than were monogamous species. Morphological variation must depend on mutation–selection balance, and if loci become fixed due to intense directional selection, the additive genetic variance must decrease unless the mutational input increases. The result of the comparative study suggests that as the intensity of sexual selection increases from socially monogamous mating systems to polygyny and lekking, and as the additive genetic variance for morphology therefore becomes depleted, the level of developmental instability is reduced.

The result of this comparative study is consistent with another study of developmental instability in relation to the intensity of sexual selection. Møller and Pomiankowski (1993c) made a comparative study of multiple sexual feather ornaments which are much more common in highly polygynous and lekking species than in socially monogamous birds. Individuals with larger mean ornaments were more symmetric in species with single ornaments, while that was generally not the case in species with multiple feather ornaments. Furthermore, when the most extravagantly ornamented individuals were most symmetric

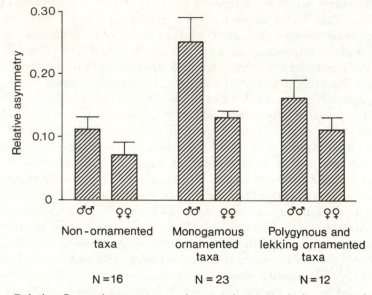

Fig. 8.6 Relative fluctuating asymmetry in secondary sexual characters of birds in relation to mating system. Values are means (+ s.e.), and numbers are number of evolutionarily independent species. Adapted from Cuervo and Møller (1997a).

among species with multiple ornaments, this was usually only the case for one of the ornaments. This result suggests that most multiple ornaments are not currently reliable indicators of phenotypic quality.

If symmetry of secondary sexual characters reliably reveals phenotypic quality, we might predict that the most extravagant traits had the least degree of asymmetry. Therefore, females should only pay attention to the expression of a secondary sexual character if males with the largest traits also generally are most symmetric. In a review of studies of female mate preferences for extravagant feather ornaments, Møller (1993c) found that some species of birds demonstrated current directional preferences for secondary sexual characters while a similar number did not. Species with a current directional preference generally had a negative relationship between asymmetry and ornament size, while there was no such general trend in species without a preference. This result suggests that females only show a preference for male traits if they reliably reveal the quality of males as demonstrated by their low level of symmetry.

8.5.2.2 *Characters involved in intrasexual competition*

The level of fluctuating asymmetry has been studied in relation to intrasexual selection in a number of comparative studies. Møller (1992a) investigated fluctuating asymmetry in two kinds of weapons used in male–male interactions, spurs of birds and horns of beetles. Beetle horns and spurs of birds had higher relative fluctuating symmetry than elytra and wings, respectively.

Individuals with the longest horns and spurs generally also had the most symmetric weapons, while this pattern was not found for the two non-sexual control characters. This suggests that the size of a weapon is a reliable indicator of phenotypic quality, since only individuals able to produce large weapons are also able to produce a symmetric character.

A comparative study of forceps asymmetry in earwigs (*Dermaptera*) revealed that forceps asymmetry was no larger than asymmetry in elytra (Tomkins and Simmons 1995). Forceps asymmetry in different species increased with increasing forceps exaggeration, suggesting that an enlargement of forceps size due to sexual selection also resulted in a larger degree of fluctuating asymmetry. There was no general tendency for males with larger forceps to be more symmetric, although this relationship was found for the single species *Forficula auricularia* in which a female preference for symmetric males has been suggested (Radesäter and Halldórsdóttir 1993). This result may indicate that forceps size and symmetry do not reveal phenotypic quality in the majority of earwig species, and that secondary sexual characters in most species are arbitrary traits.

A final study considered canines in primates, which are exaggerated secondary sexual characters used in male–male interactions (Manning and Chamberlain 1993). The degree of fluctuating asymmetry in canines of different primate species was positively correlated with three other measures of the intensity of sexual selection, namely canine dimorphism, canine size, and sexual size dimorphism. Primate species with a high intensity of male–male competition had a negative correlation between asymmetry and canine height in both sexes, while there was no negative correlation between asymmetry and size in species with less intense competition (Fig. 8.7). Symmetry of canine length may not necessarily be directly related to symmetry of the actual display of canines, although it is difficult to imagine how such biases can have generated the patterns observed.

8.5.3 *Benefits of sexual selection with respect to developmental stability*

Sexual selection may give rise to direct or indirect fitness benefits for the choosy party, with direct benefits accruing to the individual in the current generation and indirect benefits in the subsequent generation (Andersson 1994; Møller 1994b). Direct fitness benefits of sexual selection include nuptial gifts, territories, parental care, and the absence of contagious parasites, while examples of indirect fitness benefits include attractiveness, general viability or parasite resistance genes. Although a fair amount of work has been done on developmental stability and sexual selection, there is relatively little information on the fitness benefits accruing to choosy individuals. Only three studies have addressed the question of females choosing symmetric males for parental care reasons. In an experimental study of barn swallows in which tail length and asymmetry were manipulated independently, Møller (1994b) found that symmetric males provided relatively less parental care compared to their mates than did asymmetric males. This was opposite to what was predicted if asymmetry

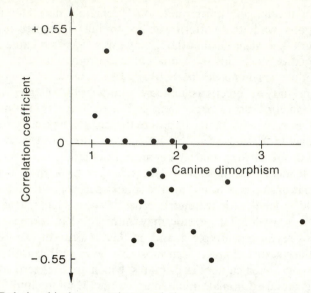

Fig. 8.7 Relationship between the correlation coefficient for the degree of fluctuating asymmetry in canine size and canine size in relation to sexual size dimorphism in canines of primates (a measure of the intensity of male–male competition). Adapted from Manning and Chamberlain (1993).

reflected direct fitness benefits, but is consistent with a hypothesis of differential female parental investment in the offspring of attractive males. A second study demonstrated that male sticklebacks with an asymmetric number of lateral plates more often had nests containing fry rather than eggs compared with symmetric fish (Moodie and Moodie 1996). A similar pattern was seen in humans with phenotypically symmetric men investing less in romantic relationships than asymmetric men (Gangestad and Thornhill 1997a). These results suggest that females do not benefit directly in terms of parental care from their choice of males with symmetric phenotypes.

If fluctuating asymmetry has an additive genetic component, as evidenced from a review of the literature (Møller and Thornhill 1997a; section 5.2), choosy individuals that mate with a symmetric partner will thus, on average, raise symmetric offspring. If developmental stability is expressed consistently at different stages of the life-cycle, a mate preference for symmetry will result in the production of zygotes with symmetric phenotypes (Møller 1996c). This mechanism will give rise to an indirect fitness benefit.

Obviously, detailed investigations have to be made of the fitness benefits accruing to individuals mated with symmetric partners.

8.5.4 *Plant sexual selection*

Sexual reproduction in plants depends on transfer of pollen by wind, insects, or other pollination agents. Flowers are usually assumed to have evolved as a way of attracting pollinators, and flowers generally show considerably more variation among species than vegetative parts of plants (Stebbins 1950). Møller and Eriksson (1994) studied fluctuating asymmetry in leaves and petals of a number of different flowering plant species. Although the level of fluctuating asymmetry did not differ between characters, the pattern of asymmetry in relation to character size differed for petals and leaves. While large leaves had a larger relative asymmetry, the pattern was the opposite for petals with the smallest petals generally being most asymmetric. Although these patterns of asymmetry have been confirmed in other studies (e.g. Eriksson 1996a, b, c), an extensive study of flowering plants from South Africa suggests that these patterns are not ubiquitous (Jennions 1996).

A subsequent study of a range of different plant species determined the relationship between floral size and asymmetry and visits by insect pollinators (Møller and Eriksson 1995). Flowers visited by pollinators were generally more symmetrical than the nearest neighbouring, non-visited flower, while larger flowers also tended to receive more visits by pollinators. Flower visits were assortative with respect to floral symmetry, since subsequently visited flowers were the most symmetric more often than expected by chance. The individuals with the most symmetric flowers may thus tend to reproduce with each other, and any genetic basis of developmental stability could result in an indirect fitness benefit of the pollinator preference, while an entirely environmental basis could result in a direct fitness benefit, if symmetric flowers provided larger pollinator rewards. Pollinators appeared to visit the largest, most symmetric flowers because nectar production generally was negatively related to floral asymmetry (Møller and Eriksson 1995).

The directional preference for the most symmetric flowers was investigated experimentally for the bilaterally symmetric fireweed *Epilobium angustifolium* (Møller 1995b). Flowers with the longest petals were more symmetric, and such flowers also produced more nectar than flowers with short and asymmetric petals. Experimental manipulation of the level of petal asymmetry in fireweed by means of clipping the lower petals revealed that pollinators preferred symmetric over asymmetric flowers and large over small flowers. Hence, pollinators demonstrated a clear preference with respect to floral symmetry, and this resulted in the elevated visitation rates of symmetric flowers. Another kind of asymmetry in pattern expressed as dots on petals of a species of Asteraceae was not affected by pollinator preferences for symmetric flowers (Midgley and Johnson 1996). The use of coloured discs mimicking flowers that differed in the position of the centre of the flower affected visitation rates by both Hymenoptera, Diptera, Coleoptera and Lepidoptera in two experiments in Spain and Denmark (Møller and Sorci 1997). Thus, the preference for symmetric flowers by these insects was present even in the complete absence of pollinator rewards.

8.6 Maintenance of developmental stability

Individuals with high degrees of developmental instability are often at a selective disadvantage both due to natural and sexual selection. It has previously been shown that there is a small, but statistically significant additive genetic component to variability in developmental stability (Møller and Thornhill 1996a; Chapter 5). Intense directional selection and high heritability should give rise to a strong micro-evolutionary response to selection. How can genotypic variation in developmental instability be maintained? At least six solutions may exist for this problem (see Møller and Thornhill 1996a). First, developmental instability may mainly, or entirely, be of environmental origin, and selection against asymmetric phenotypes will thus not result in an evolutionary response to selection. Since there is evidence for additive genetic variation in developmental instability (see Chapter 5), we will not consider this possibility further. Second, continuous directional selection may have displaced the phenotype of individuals from the developmental optimum. Third, stress may increase the frequency of mutations. Fourth, biased mutation on signals may disrupt developmental homeostasis and increase asymmetry. Fifth, fluctuating selection pressures will prevent many individuals from developing the optimal phenotype given the current environmental and genetic conditions. Sixth, a heterozygote advantage may maintain genetic variance in developmental instability when developmental instability is lowest in heterozygotes, as sometimes appears to be the case (see Mitton 1993; Chapter 5).

Mutation rates have been estimated in a large number of organisms for a number of characters, and they are usually in the range 10^{-5} to 10^{-6} per locus and generation (Dobzhansky 1970). Poor environmental conditions increase the frequency of mutations considerably, as suggested by the review in section 4.3.2. Environmental conditions known to elevate the level of mutations include irradiation, abnormal climatic conditions, abnormal composition of food, and levels of behavioural stress. Novel genetic variation on which selection may act will therefore be introduced. Mutations are generally detrimental since only a small fraction of all mutations ever results in an improvement in design. Mutations may often result in disease because mutations disrupt the pleiotropic effects of genes on physiological pathways (Evans 1988). This is particularly the case for characters with a recent evolutionary history of directional selection, since such traits particularly are subject to biased mutation (Pomiankowski *et al.* 1991; Mukai 1964). Induced mutations are likely to cause a reduction in the mean values of the characters themselves, particularly those most closely related to fitness, while the changes induced in other characters are less likely to be unidirectional (Sakai and Shimamoto 1965a). Since both signals and their developmental instabilities may be closely related to fitness, mutations can be considered to be biased with respect to signal value. Developmental stability can be considered to be subject to biased mutation, and mutation *per se* will result in disruption of the genome and hence generate novel fluctuating asymmetry.

An example may illustrate this point. Barn swallow males have considerably longer tails than females, and male tail length is subject to a directional female mate preference (Møller 1994b). A study of the phenotypes of barn swallows at Chernobyl, Ukraine, and a neighbouring area with no elevated level of irradiation was used to test indirectly for the presence and consequences of elevated mutation rates (Møller 1993b). Male barn swallows had considerably elevated levels of fluctuating asymmetry in the length of their outermost tail feathers in Chernobyl after the 1986 radiation event, but not before 1986, or in a control area during either period (Møller 1993b). The level of fluctuating asymmetry for males was also considerably larger than expected from the latitudinal cline in tail asymmetry in barn swallows, while that was not the case for females (Møller 1995c). The post-1986 sample from Chernobyl was also characterised by an elevated frequency of deformed tail feathers among males. Both tail asymmetry and deformity resulted in a delay in reproduction because of a later start of egg laying (Fig. 8.8; Møller 1993b), and delayed reproduction is known to result in reduced fitness in another barn swallow population (Møller 1994b). Alternative explanations such as a reduced food supply or the like after radioactive contamination are unlikely since breeding success was not reduced after the 1986 radiation event, and since feathers are moulted annually in the African winter quarters of the migratory barn swallows. These results are consistent with the assumption that secondary sexual characters are particularly susceptible to the environmental and genetic effects of irradiation, and that disruption of the preferred phenotype results in a disadvantage in terms of sexual selection.

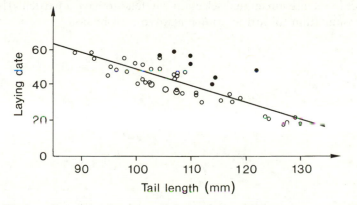

Fig. 8.8 Timing of reproduction by male barn swallows *Hirundo rustica* at Chernobyl, Ukraine, during 1991 in relation to tail feather deformations. Individuals with deformed tails (filled circles) reproduced significantly later than other males. Adapted from Møller (1993b).

In conclusion, there is only little evidence of mutations affecting the level of developmental stability of secondary sexual characters. Future studies should address this and other solutions to the question of maintenance of developmental instability.

8.7 Summary

- Developmental instability may play an important role in many signalling systems, and developmental instability may be an integral measure of all signals independently of whether they are reliable signals, conventional signals, or sexual signals in particular.

- Initial selection for visually symmetrical signals may be ascribed to sensory biases in the perceptive machinery of receivers, as supported by computer models of neural networks.
- The breakdown of symmetric colour patterns under domestication suggests that symmetry disappears in the absence of natural selection, probably due to a lack of predation pressure.
- Experimental studies of sexual signals have demonstrated unequivocal preferences for symmetric signals, and observational studies suggest that developmental stability plays an important but variable role in sexual selection.
- Among birds there was an initial increase in developmental instability, when morphological characters became the target of sexual selection. However, secondary sexual characters in species with extreme skew in male mating success were developmentally relatively stable.
- Directional selection as under sexual selection will tend to increase the mutational input that disrupts developmental stability, and characters subject to intense directional selection will tend to have a particularly strong bias in mutation toward inferior phenotypic expressions.

9

Developmental stability and fitness

9.1 Introduction

Developmental instability *per se* and its causes are often assumed to result in reduced performance in domains of fitness other than sexual competition (sexual selection was discussed in Chapter 8). The consequences of such reductions in fitness have often been presumed rather than estimated. Here we present a large body of evidence for a negative association between fitness and individual asymmetry. Significant correlations between morphological asymmetry and various fitness components do not demonstrate a causal relationship, and many potentially confounding variables may obscure attempts to unravel real relationships. Individuals with asymmetric phenotypes may be characterised by poor overall quality. For example, they may have experienced poor environmental conditions during their development, or they may have genotypes that make them less efficient in terms of growth, survival, or reproduction. Random experimental assignment of individuals to treatments will unravel the direct relationships between developmental instability and fitness components. The use of clones or split brood designs will facilitate investigation of the susceptibility of different genotypes to environmental and genetic stresses and the relationship between asymmetry and fitness components (section 2.5). This chapter provides a review of the fitness correlates of asymmetry (see Møller (1997a) for a more extensive review).

If developmentally stable individuals have higher metabolic efficiency, this should result in excess energy resources being available for maintenance, growth, and reproduction. The evidence for a relationship between developmental stability, metabolism, and growth is presented (section 9.2).

The consequences of individual asymmetry for various components of fitness are reviewed in sections 9.3–5. Various measures of fecundity such as clutch size and number of clutches and their relationship with asymmetry are treated (section 9.3).

Developmental selection was originally defined as being the process of plants (or other organisms) preferentially rearing outcrossed offspring in favour of self-fertilised offspring of poorer quality (Buchholz 1922; Kozlowski and

Stearns 1989). However, developmental selection can be extended to include selection against gametes and offspring of inferior quality (Møller 1997d). Gamete redundancy and abortion at various stages of the developmental process are very common phenomena in nature, and so is the culling of poor-quality offspring at later stages of the developmental process. Interestingly, developmental selection often appears to act against asymmetric offspring. The evidence for developmental selection against asymmetric gametes and offspring is treated (section 9.4).

A number of different mechanisms may generate relationships between individual asymmetry and fitness components. If asymmetric individuals have less efficient metabolism, and therefore allocate more energy to maintenance, there will be less energy available for other processes such as growth and reproduction. Superior developmental stability may be related to the competitive ability of individuals. If particular individuals demonstrate superior growth abilities associated with developmental stability, then such individuals will reach adult body size earlier than others, or achieve an absolutely larger size if the organism has indeterminate growth. Timing of maturation or size *per se* are both known to affect competitive ability (section 9.5).

A second mechanism that may generate a general relationship between asymmetry and fitness components arises from the negative effects of parasites on their hosts and the association between asymmetry and general health (Thornhill and Møller 1997). Parasites are ubiquitous, and most individuals are affected by parasites at one or more times during their lifetime. Individual asymmetry may reliably reflect the ability of individuals to cope with parasite attacks or their susceptibility to parasitism. A range of studies has demonstrated that asymmetric individuals are more often infected with parasites than more symmetric individuals, and some of these studies have demonstrated that this relationship arises because asymmetric individuals are more susceptible to parasitism. Elevated susceptibility to parasitism associated with an asymmetric phenotype may be due to overall poor phenotypic quality of asymmetric individuals, or morphological symmetry may reflect genetically based resistance to parasites. The role of parasites as a determinant of fitness components and the association with developmental stability is described (section 9.6).

If developmentally unstable individuals suffer from access to insufficient resources and elevated levels of parasitism, then such individuals may be more susceptible to predation. In a similar vein, developmental instability may reliably reflect the ability of individuals to succeed in intraspecific and interspecific competition. The intensity of competition may be measured directly from the effects of competitive interactions on the development of phenotypes. The relationship between asymmetry and risk of predation and intensity of competition, respectively, are treated (section 9.7).

Survival and longevity reflect the ability to acquire sufficient resources for growth, maintenance, and reproduction and an ability to avoid parasitism and predation. The relationships between developmental instability, survival, and longevity are described (section 9.8).

Studies of selection in the wild are usually based on comparisons of the phenotypes of individuals in a sample before and after selection has taken place. It is very important in such comparisons that the samples are unbiased with respect to the variables under investigation. This assumption is particularly pertinent for studies of relationships between developmental instability and fitness. Individuals with asymmetric phenotypes may be distributed differently from symmetric individuals in space and time if they are of inferior competitive ability. Furthermore, if there is viability selection against developmentally unstable phenotypes during embryonic or juvenile life-stages, this may obscure attempts to find correlations between fecundity or longevity and developmental instability. The sections of this chapter should be read with this caveat in mind.

9.2 Developmental stability, metabolism, and growth

Metabolic efficiency, growth rate, and developmental stability have been assumed to be directly related to heterozygosity due to the effects of heterosis (Mitton and Grant 1984). Sometimes there appears to be a relationship between heterozygosity and developmental stability in comparisons among individuals within populations and more often a relationship in comparisons among populations (Clarke 1993b; Vøllestad *et al.* 1997; Chapter 5).

The relationship between metabolism and developmental stability has only been investigated in humans. A study of adults demonstrated a positive association between skeletal asymmetry and resting metabolism in men (even when controlling for the effects of body size), but a non-significant negative relationship in women (Manning *et al.* 1997). Sexual size dimorphism in humans has been attributed to sexual selection with larger men that are less asymmetric having an advantage in male–male competition and female choice (Manning 1996b). Large men are skeletally less asymmetric than small men. Interestingly, the relationship between body size and asymmetry is positive in women. The higher metabolic rate of more asymmetric men may arise because less energy can be devoted to the control of developmental processes that ensure the ontogeny and the maintenance of a symmetric phenotype.

A second study of fluctuating asymmetry in children showed that both absolute and relative asymmetry decreased steadily from early childhood to at least 18 years of age with the exception of a slight increase during adolescence (Wilson and Manning 1996). This pattern is closely paralleled by a steady decrease in metabolism throughout childhood, and the decrease in asymmetry with age may thus be attributed to more energy being allocated to the control of growth processes. These studies of metabolism in humans are suggestive for a direct role of energetics in the maintenance of morphological symmetry. More studies are obviously needed in other organisms.

Lower growth rates of asymmetrical individuals have been reported for a range of different organisms (Table 9.1). A total of eleven studies have been reported, all of which showed a negative relationship between growth rate and asymmetry. For example, leaf asymmetry in teak *Tectonia grandis* is a predictor of the growth performance of individual trees (Bagchi *et al.* 1989). Symmetry was estimated for a number of different characters, and there was generally a positive relationship among estimates of stability for different characters. Pesticide resistance in the Australian sheep blowfly causes an increase in developmental instability of abdominal bristles until a genetic modifier re-establishes developmental homeostasis (Clarke and McKenzie 1987; see Chapter 2). Different genotypes of flies for the resistance and the modifier gene vary in their level of fluctuating asymmetry in number of bristles, and mean asymmetry is positively correlated with the duration of the developmental period (McKenzie and O'Farrell 1993). A final example concerns growth rates and asymmetry in the muskrat *Ondatra zibethicus* (Pankakoski 1985). A highly significant negative correlation was found between asymmetry in skeletal characters and growth indices among populations (Fig. 9.1). A weaker but similar relationship was also found within populations. Indirect evidence is found in humans where men with low levels of asymmetry start copulation activity earlier in life than men with high levels of fluctuating asymmetry (Thornhill and Gangestad 1994).

Since poor environmental conditions increase asymmetry (Chapter 6), a negative relationship between developmental instability and growth performance may arise simply because runts grow poorly and become very asymmetric. However, selection tends to act severely against developmentally unstable offspring (see section 9.4), since offspring of poor phenotypic quality will often be the first individuals to suffer mortality during brood reduction (e.g. Møller 1994d). The relationship between growth performance of offspring and asymmetry may therefore be weakened by intense natural selection against asymmetric offspring.

In conclusion, studies of growth rates demonstrate that slow growth is generally associated with asymmetry, or that there may be a trade-off between developmental instability and growth rate, and that is the case in both poikilothermic and homeothermic organisms. These preliminary data support the prediction of Mitton and Grant (1984) that developmental instability and performance in terms of growth are negatively associated due to the effects of heterosis. An alternative mechanism leading to the superior growth performance of symmetric individuals is that individuals in good environments with plenty of resources grow fast and become symmetric.

9.3 Developmental stability and fecundity

If asymmetric individuals use a larger proportion of their resources for maintenance, we might predict lower reproductive rates for asymmetric individuals. A total of fifteen studies have determined the relationship between developmental instability and fecundity (Table 9.2). Fourteen of these studies found a reduction in clutch or litter size, the number of clutches, or the quality of

Table 9.1 Growth rate of organisms in relation to developmental instability. 'Experiment' refers to whether asymmetry was experimentally induced, or whether survival prospects were estimated under standardized environmental conditions. Studies are based on comparisons among individuals within populations. Effects are either no relationship between asymmetry and growth rate ('0'), or negative relationships ('−'). Adapted from Møller (1997a).

Species	Character	Effect	Experiment	Reference
PLANTS				
Tectonia grandis	leaves	−	yes	Bagchi *et al.* (1989)
Ulmus glabra	leaves	−	no	Møller (1997c)
ANIMALS				
Lucilia cuprina	bristles	−	yes	McKenzie and O'Farrell (1993)
Various fish species	phenodeviants	−	no	Kirpichnikov (1981)
Thamnophis elegans	ventral scales	−	yes	Arnold (1988)
Gallus gallus	skeletal traits	−	yes	Møller *et al.* (1995b)
Hirundo rustica	tail length	−	yes	Møller *et al.* (1995a)
Parus caeruleus	wing length	0	no	Björklund (1996)
Sturnus vulgaris	primary feathers	0	yes	Swaddle and Witter unpubl.
Ondatra zibethicus	skeletal traits	−	no	Pankakoski (1985)
Rattus rattus	femora	−	yes	Gest *et al.* (1986)

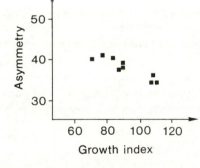

Fig. 9.1 Growth rate of muskrats *Ondatra zibethicus* in relation to developmental instability of morphological characters among populations. Adapted from Pankakoski (1985).

offspring with increasing asymmetry. Most of these studies are observational, but two are based on an experimental approach. Fecundity was negatively related to asymmetry in tail length of the barn swallow (Fig. 9.2; Møller 1992b, 1993e). Males with experimentally manipulated symmetric outermost tail feathers acquired females that raised a larger number of offspring. This effect was not due to asymmetric males being less able to provide parental care for their offspring. On the contrary, males with experimentally asymmetric tails provided more parental care relative to their mates than males with symmetric tails (Møller 1994d). Symmetric, attractive males thus appeared to acquire mates of higher phenotypic quality that were able to reproduce at a higher rate than the mates of male barn swallows with asymmetric tail feathers. An even more compelling experimental result arose from the leg-band experiments on male zebra finches (Swaddle 1996). Males with leg-bands of attractive (symmetric) arrangement had considerably larger reproductive success than male zebra finches with less attractive (asymmetric) bands.

The negative relationships between asymmetry and fecundity may arise for a number of different reasons. Individual asymmetry may directly affect the ability of individuals to acquire or utilise resources necessary for reproduction. Alternatively, the effect may arise because developmental instability is related to dominance rank and developmentally unstable individuals are subdominant and therefore have reduced access to essential resources. This is likely to be the case for gemsbok *Oryx gazella* in which females with asymmetric horns were subdominant during interactions at waterholes (Møller *et al.* 1996). Finally, asymmetry of a particular character may directly reflect reproductive potential because hormonal mechanisms involved in reproduction are also a causative agent of asymmetry. This is, for example, likely to be the case for fecundity and breast asymmetry in humans (Møller *et al.* 1995c).

Clutch size and the number of clutches are two important determinants of total fecundity. Other components include the size of offspring and the timing of reproduction. Breeding date was directly related to asymmetry in tail length in female barn swallows (Møller 1994c). Individual females with more asymmetric tails started laying later than symmetric females. Rate of recruitment drops rapidly as the season progresses (Møller 1994b), and early reproducing, symmetrical females thus will contribute disproportionately to the next generation.

A number of variables other than clutch size and reproductive rate affect reproductive success. In the following section we will describe how selection by parents of their developmentally most stable offspring may affect the average phenotypes of offspring produced.

9.4 Developmental stability and developmental selection

Gametes and offspring are frequently produced in excessive numbers, and developmental selection during early developmental stages can reduce the number of offspring to what can be safely reared. Developmental selection was originally proposed by Buchholz (1922) to occur when plants (or other organisms) preferentially reared outcrossed offspring in favour of self-fertilised offspring of poorer quality (see also Kozlowski and Stearns 1989). However, the concept of developmental selection can be applied to a greater range of

Table 9.2 Fecundity in relation to developmental instability. 'Experiment' refers to whether asymmetry was experimentally induced, or whether fecundity was estimated under standardized environmental conditions. Adapted from Møller (1997a).

Species	Character	Effect	Experiment	Reference
PLANTS				
Epilobium angustifolium	petal length	reduced seed set	yes	Møller (1996c)
Lychnis viscaria	petals	increased abortion	yes	Eriksson (1996c)
ANIMALS				
Drosophila melanogaster	bristles	reduced egg hatchability	no	Parsons (1962)
Cottus bairdi	otolith	egg number, egg mass	no	Downhower *et al.* (1990)
Gasterosteus aculeatus	lateral plates	number of eggs	no	Moodie and Moodie (1996)
Lacerta vivipara	ventral scales	smaller broods	no	Chenuil (1991)
Hirundo rustica	tail	reduced frequency of second clutches	yes	Møller (1992b)
Hirundo rustica	tail	reduced frequency of second clutches	yes	Møller (1993e)
Hirundo rustica	wing and tail	no effect	no	Møller (1994c)
Taeniopygia guttata	colour rings	fewer fledged offspring	yes	Swaddle (1996)
Sorex araneus	skeletal	smaller litters	no	Zakharov *et al.* (1991)
Homo sapiens	skeletal	pre-term infants	no	Livshits and Kobyliansky (1991)
Homo sapiens	breasts	reduced fecundity	no	Møller *et al* (1995c)
Homo sapiens	breasts	reduced and delayed fecundity	no	Manning *et al.* (1996)
Oryx gazella	horn	reduced fecundity	no	Møller *et al.* (1996)

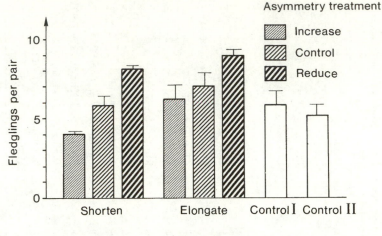

Fig. 9.2 Annual fecundity of male barn swallows *Hirundo rustica* in relation to experimental treatment of length and asymmetry of the outermost tail feathers. Values are mean (+ s.e.). Adapted from Møller (1992b).

phenomena related to parental selection of offspring of superior genetic or phenotypic quality. Cases of developmental selection disfavouring developmentally unstable offspring have been reviewed by Møller (1997d), who suggested that such selection may be a major component of parental selection for improved quality of offspring, if developmentally stable parents tend to produce developmentally stable offspring. If a common developmental program results in development of a stable phenotype at different stages of the life cycle, then sexual selection in favour of symmetric partners will concomitantly result in developmental selection against asymmetric offspring.

Gametes often appear to be produced in excess numbers relative to what is needed for reproduction. Excess production of sperm has been explained in terms of developmental selection at the gamete level (Møller 1997d). Rate of sperm production is negatively related to a range of environmental and genetic stress factors (Romanoff 1960; Sheldon 1994); exactly the same factors that are known to increase the level of developmental instability in overall morphology of organisms (Chapters 5 and 6) also decrease the quality of sperm (inbreeding (Wildt *et al.* 1987; O'Brien 1994), environmental factors such as malnutrition (Romanoff 1960), pesticides and other pollutants (Henderson *et al.* 1986; Harrison and Boer 1977), temperature (VanDemark and Free 1970), and parasites and diseases (Romanoff 1960; Harrison and Boer 1977)). It is tempting to speculate that malformed sperm that do not fertilise eggs have developmentally unstable phenotypes, this being the causal reason for their inferior success. A similar argument may be raised for excess production of eggs, for example by female mammals (Simpson *et al.* 1982; Birney and Baird 1985; Stearns 1987).

Developmental selection at the embryonic level and later developmental stages appears to be a very common phenomenon. Developmental selection is often associated with factors known to increase the level of developmental instability such as inferior genotypes and high levels of inbreeding (reviewed by Møller 1997d). Several studies have reported differential seed or fruit abortion of inferior phenotypes such as developmental abnormalities (Table 9.3; Stephenson 1981; Wiens *et al.* 1987).

When environmental conditions are poor, developmental selection is more intense in plants (reviewed by Møller 1997d). Poor environmental conditions may directly give rise to phenotypically abnormal pollen and hence result in a reduction in their fertilisation ability. Differences in fitness between selfing and out-crossing can increase under poor environmental conditions (Dudash 1990). Post-zygotic developmental selection is often more important under poor environmental conditions, and the frequency of malformations in zygotes may also increase under such conditions (reviewed in Møller 1997d).

The literature on developmental selection in animals is limited (Table 9.3). A clear example concerns the scorpion *Pandinus imperator* in which females provide extensive parental care, and newly born offspring must climb to the cephalothorax of the mother in order to obtain protection (Mahsberg 1996). Malformed offspring are unable to climb to the cephalothorax and are later eaten by the mother. Such malformed offspring are relatively common, but when reared in isolation together with offspring of normal phenotypes, they remain malformed and achieve poor reproductive and survival performance when becoming adults. There is an extensive literature on human abortion and infanticide related to developmental instability. For example, a large fraction of all spontaneous abortions have abnormal karyotypes, and individuals with such karyotypes have elevated morphological asymmetry (reviewed in Møller 1997d). Developmental selection also takes place by means of infanticide affecting offspring with physical and mental deformities, which is a widespread phenomenon in a range of different human cultures (reviewed by Møller 1997d).

Developmental selection may also be coupled directly to sexual selection (Møller 1997d). A female preference for males with developmentally stable phenotypes may be coupled with developmental selection, if a general developmental program affects the expression of phenotypes at different stages of the life cycle (Møller 1997c). Individuals that mate with symmetric partners may suffer from less intense developmental selection against asymmetric zygotes, if developmental stability of a genotype is expressed at both zygote and adult stages of the life cycle. This mechanism is illustrated by fireweed *Epilobium angustifolium*, in which there is a pollinator preference for the bilaterally most symmetrical flowers, probably due to the production of higher pollinator rewards (Møller and Eriksson 1995; Møller 1995b). Embryo abortion is extensive in this outcrossing plant, often reaching more than 70% (Wiens *et al.* 1987; Møller 1997c). A cross-pollination experiment revealed that the abortion rate was positively related to asymmetry in the flower of both the pollen donor and the pollen recipient (Fig. 9.3; Møller 1997c). A preference for

Table 9.3 Abortion and developmental selection in relation to developmental instability. 'Experiment' refers to whether asymmetry was experimentally induced, or whether abortion rates were estimated under standardized environmental conditions. Mainly adapted from Møller (1997d).

Species	Character	Effect	Experiment	Reference
PLANTS				
Prunus cerasus	phenodeviants	increased abortion	no	Bradbury (1929)
Malus silvestris	phenodeviant	increased abortion	no	Kraus (1915)
Pinus radiata	phenodeviant	increased abortion	no	Sweet (1973)
Pinus silvestris	phenodeviant	increased abortion	no	Sarvas (1962)
	phenodeviants	increased abortion	no	Bawa and Webb (1984)
Lychnis viscaria	petals	increased abortion	yes	Eriksson (1996c)
Epilobium angustifolium	phenodeviants	increased abortion	no	Wiens et al. (1987)
Epilobium angustifolium	petals	increased abortion	yes	Møller (1996c)
ANIMALS				
Lucilia cuprina	bristles	reduced hatch rate	yes	McKenzie and O'Farrell (1993)
Pandinus imperator	phenodeviants	infanticide	yes	Mahsberg (1996)
Pavo cristatus	train ocelli	reduced hatch rate	yes	M. Petrie pers. comm.
Turdus merula	wing and tail	increased mortality	no	Møller (1995d)
Mus musculus	phenodeviants	infanticide	no	Ehret (1975)
Macaca nemestrina	dermatoglyphics	increased perinatal mortality	no	Newell-Morris et al. (1989)
Homo sapiens	chromosomal abnormality	increased abortion	no	Boue et al. (1975)
Homo sapiens	chromosomal abnormality	increased abortion	no	Simpson et al. (1982)
Homo sapiens	chromosomal abnormality	increased abortion	no	Wolf et al. (1984)
Homo sapiens	phenodeviants	infanticide	no	Ford (1964)
Homo sapiens	phenodeviants	infanticide	no	Montag and Montag (1979)
Homo sapiens	phenodeviants	infanticide	no	Daly and Wilson (1984)

symmetrical flowers as mediated by pollinators therefore gave rise to the production of an enlarged number of developmentally stable embryos. A similar but less dramatic effect was found for the frequently selfing plant *Lychnis viscaria* (Eriksson 1996c), suggesting that this mechanism of developmental selection associated with sexual selection may be common.

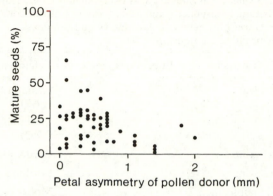

Fig. 9.3 Abortion rate of fireweed *Epilobium angustifolium* embryos (%) in relation to asymmetry in the length of the petals of the pollen donor. Adapted from Møller (1997c).

Developmental stability generally appears to have an additive genetic basis (Møller and Thornhill 1997a; Chapter 5). The ability to generate a symmetric phenotype at different stages of the life cycle may have a common genetic background. Individuals of the choosy sex that mate with a symmetric partner will tend to produce symmetric offspring. Preferences for symmetric mates appear to be common (Chapter 8), and production of symmetric offspring may prove to be a common benefit of mate choice.

9.5 Developmental stability and competitive ability

A superior environmental or genetic ability to cope with poor environmental conditions may provide individuals with a head-start in competition for social dominance. A number of studies have investigated the relationship between individual asymmetry and social dominance rank (Table 9.4). Interspecific dominance relationships may also be affected by developmental instability. In the Japanese scorpionflies *Panorpa ochraceopennis* and *P. nipponensis* asymmetry in wing length is associated with losing interspecific contests between the two species for dead insect food, which plays an important role as a nuptial gift. Male mating success, female fecundity, and survival prospects in both species are affected by the ability to acquire food (Thornhill 1992a).

These studies can basically be divided into those that concern characters directly involved in competition for social dominance and those concerning other morphological traits. For example, antlers, horns, and canines are weaponry directly involved in aggressive interactions, and it is not surprising that a structural asymmetry in the phenotypic expression of a weapon reduces the ability of an individual to win a fight. This was the case in fallow deer stags *Dama dama* and both sexes of gemsbok *Oryx gazella* (Malyon and Healy 1994; Møller *et al.* 1996).

Table 9.4 Studies of social dominance in relation to developmental instability. 'Experiment' refers to whether asymmetry was experimentally induced, or whether dominance rank was estimated under standardized environmental conditions. Effects are either positive relationships betweeen asymmetry and dominance ('+'), no relationship ('0'), or negative relationships ('–').

Species	Character	Effect	Experiment	Reference
Carcinus maenas	legs	–	yes	Sheddon and Swaddle (1997)
Panorpa vulgaris	wing	–	no	Thornhill and Sauer (1992)
Hetaerina cruentata	wing	–	no	Córdoba-Aguilar (1996)
Hetaerina americana	wing	–	no	Córdoba-Aguilar (1996)
Hirundo rustica	tail feathers	0	yes	Møller (1992b, 1993e)
Sturnus vulgaris	wing feathers	+	yes	Swaddle and Witter (1994)
Sturnus vulgaris	chest spots	0	yes	Swaddle and Witter (1995)
Taeniopygia guttata	leg bands	0	yes	Swaddle (1996)
Homo sapiens	skeletal	–	no	Gangestad and Thornhill (1997a)
Dama dama	antlers	–	no	Malyon and Healy (1994)
Oryx gazella	horns	–	no	Møller *et al.* (1996)

A series of experiments has determined (i) the relationship between dominance rank and asymmetry and (ii) the effect of social dominance on development of asymmetry in female European starlings (Swaddle and Witter 1994, 1995; Witter and Swaddle 1994). The relationship between dominance status and asymmetry was studied in two experiments. In the first, Swaddle and Witter (1994) investigated the relationship between nutritional stress, spotty chest feathers, which may be a sexual signal, and recently developed primary wing feathers. Nutritional stress caused enlarged asymmetries in wing feathers. Primary feather asymmetry was negatively correlated with fat stores, indicating that individuals in poor condition also developed large asymmetry. Primary feather asymmetry was positively correlated with social dominance rank (Fig. 9.4). Low levels of wing feather asymmetry were directly associated with chest spottiness, as expected if spottiness is a condition-dependent sexual signal. In other words, social dominance appeared to be positively associated with asymmetry. The most asymmetric individuals were the most dominant, and this study thus provides a counter-example to the bulk of the cases discussed in this chapter.

A second experiment investigated the relationship between asymmetry in

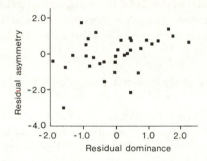

Fig. 9.4 Dominance rank of European starlings *Sturnus vulgaris* in relation to asymmetry in their wing feathers controlling for experimental treatments and fat stores. Adapted from Swaddle and Witter (1994).

chest feather spots and dominance rank among female starlings (Swaddle and Witter 1995). Social dominance rank was not affected by asymmetry levels, although the overall condition of the female chest plumage plays an important role in intra-sexual encounters. The reason for this lack of relationship may be that the chest feathers are exposed to extensive abrasion throughout the breeding season, and asymmetries due to damage may mask any quality-revealing asymmetry in the plumage trait.

Dominance rank may also affect levels of developmental instability. When competition for food was intense, starlings that differed in dominance rank did not develop a difference in asymmetry in wing feather length (Witter and Swaddle 1994). However, when competition was less intense, dominant individuals developed larger levels of asymmetry in the primary wing feathers than subdominants. Hence, the advantage of dominance in terms of easy access to food was only experienced under intense competition, while the costs of dominance resulted in an elevated asymmetry under less intense regimes of competition (Witter and Swaddle 1994).

Experimental manipulation of the degree of asymmetry in the outermost tail feathers of male barn swallows revealed no effect on rates of aggressive encounters or on the proportion of victorious fights (Møller 1992a, 1993e). A similar result was found in the leg-band experiments on male zebra finches (Swaddle 1996a). Hence, the experimental studies of asymmetry do not provide any supportive evidence for symmetry being a signal of dominance status in birds. This might have been expected because the traits under study were not used directly in dominance interactions. Further experimental studies of the relationship between developmental stability and dominance rank are needed.

A comparative study of primates has shown that canine fluctuating asymmetry was larger in species with a high intensity of male–male competition (Manning and Chamberlain 1993). Such species also generally demonstrated a negative correlation between canine asymmetry and canine height in both sexes. If individuals with large canines are the most dominant, canine symmetry will tend to be positively related to dominance rank in species with intense male–

male competition. A similar argument can be applied to patterns of fluctuating asymmetry in spurs of birds and horns of beetles, two other types of characters that are used as weapons in intrasexual competition (Møller 1992a).

9.6 Developmental stability and susceptibility to parasitism

Poor environmental conditions increase fluctuating asymmetry and other measures of developmental instability. Poor environmental conditions are also known to reduce the efficiency of the immune system (Chandra and Newberne 1977; Gershwin *et al.* 1985), and a direct coupling between developmental instability and parasitism is thus likely. Developmental stability may reflect a genetic ability to resist parasites, if genes for parasite resistance do not disrupt developmental stability. If point mutations confer resistance to parasites in the same way as pesticide resistance is achieved, then a novel mutant may result in disruption of developmental homeostasis until modifier alleles have countered this disruptive effect (Clarke and McKenzie 1987; Polak and Trivers 1994). In agreement with this idea, alleles that confer fruitflies with resistance towards nematodes give rise to increased asymmetry in sternopleural bristles (Polak 1997a), suggesting that resistance alleles result in disruption of a co-adapted genome.

We provide a review of cases on parasitism and developmental stability in Table W9.1 on the Web site. The literature on developmental stability and disease has already been reviewed by Møller (1996d) and Thornhill and Møller (1997), who presented detailed accounts of asymmetry as a reliable marker of susceptibility to parasitism and disease. A total of 37 studies have determined the relationship between asymmetry and parasitism and 32 of these studies reported higher levels of parasitism in asymmetric individuals, while four studies did not find any relationship, and one found disease to be more prevalent among symmetric individuals (Table W9.1). For example, wych elms *Ulmus glabra* with a high degree of leaf asymmetry suffered from higher frequencies of mines produced by the beetle *Rhynchaenus rufus* than trees with less leaf asymmetry (Fig. 9.5; Møller 1995a). Similarly, fruitflies of the species *Drosophila nigrospiracula* that had larger degrees of asymmetry in sternopleural bristles suffered from more severe infections with a nematode (Polak 1993, 1997b). However, another study of four different species of fruitflies reported no association between asymmetry and parasitism (Polak 1997a). The study reporting a negative association between asymmetry and disease is a revealing exception. Human males with symmetric phenotypes are more attractive than asymmetric individuals, and they appear for that reason to acquire more venereal diseases than asymmetric males (S. Gangestad and R. Thornhill unpublished data).

Less symmetric individuals may be more susceptible to parasitism, or more often encounter parasites (Møller 1996d). The latter explanation cannot account for the results of all the studies listed in Table W9.1. For example, studies of *Drosophila nigrospiracula*, *Musca domestica*, *Hirundo rustica*, and *Homo sapiens* have revealed higher susceptibilities to parasitism among asymmetric individuals.

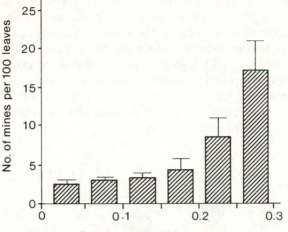

Fig. 9.5 Abundance of mines produced by the larvae of the beetle *Rhynchaenus rufus* in relation to leaf asymmetry of wych elm trees *Ulmus glabra*. Values are means (+ s.e.). Adapted from Møller (1995a).

Several studies also determined whether parasitised individuals become more asymmetric, as a response to parasitism. This was the case in *Drosophila nigrospiracula*, barn swallow, and reindeer *Rangifer tarandus*, and willow *Salix borealis* (Polak 1993c; Folstad *et al.* 1996; Zvereva *et al.* 1996). For example, experimental mite infections of the nests of barn swallows produced higher levels of asymmetry in a secondary sexual character after the annual moult, while that was not the case for two other, sexually monomorphic feather characters (Fig. 9.6; Møller 1992c). This study revealed that asymmetry of a character reliably reflected the infection history of an individual, and that a secondary sexual character was more susceptible to the stress caused by the infection than two ordinary morphological traits not currently subject to a directional mate preference. Levels of fluctuating asymmetry in secondary sexual characters thus may reveal the infection status and the susceptibility of males to parasitism for mate-prospecting females. This idea emphasises the role of developmental stability in parasite-mediated sexual selection (see Chapter 8). Herbivory also appears to give rise to increased foliar asymmetry in response to insect damage and experimental defoliation (Zvereva *et al.* 1997). Degree of asymmetry is directly related to parasite virulence (Agnew and Koella 1997).

The clear relationships between parasitism and disease and asymmetry may have environmental or genetic origins (Thornhill and Møller 1997). The genetics of disease susceptibility in relation to developmental instability have been investigated to some extent for humans by Russian geneticists. Botvinev *et al.* (1980) found that children with non-modal body mass suffered disproportionately from a range of diseases including infectious diseases. These groups of children were also characterised by deviant allele frequencies for blood group loci and lower levels of heterozygosity. Althukov *et al.* (1981)

noted that children with acute pneumonia have a marked disposition to viro-bacterial diseases, a high frequency of small developmental anomalies, and smaller body length and mass at birth. The study also demonstrated significant differences in four genetic systems between the two groups of children, a lower heterozygosity per locus, and a higher frequency of rare antigen combinations and rare electrophoretic protein variants. These two studies are suggestive of a relationship between the genetics of disease susceptibility and developmental instability measured in terms of phenodeviants, and this certainly calls for further studies on humans and other organisms.

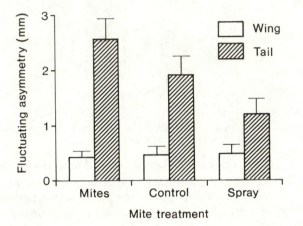

Fig. 9.6 Fluctuating asymmetry in wing length and length of the outermost tail feathers (only the latter is a secondary sexual character) of the barn swallow *Hirundo rustica* after experimental exposure to the tropical fowl mite *Ornithonyssus bursa* during the previous breeding season. Values are means (+ s.e.). Adapted from Møller (1992c).

Positive correlations between parasitism and asymmetry can arise if the level of immune defence is associated with asymmetry. This could be the case if environmental conditions simultaneously affected both developmental stability and the immune system (Chandra and Newberne 1977; Gershwin *et al.* 1985). Alternatively, genes for immune defence may have pleiotropic effects on, or be genetically linked to, asymmetry. Sexual selection may play an important role in the relationship between parasitism and developmental stability because secondary sexual characters may reliably reflect the immunocompetence of an individual (Folstad and Karter 1992). The basic idea is that testosterone or similar endocrine products are necessary for the development of secondary sexual characters, but these biochemicals simultaneously have a negative, immunosuppressive effect. Individual males thus have to optimise their level of sexual display without compromising their immunocompetence. Asymmetric secondary sexual characters may be considered immunocompetence handicaps, since they appear to develop in response to testosterone, they reflect the ability to raise an effective immune response, and they reliably reflect current levels of parasite infections (Thornhill and Gangestad 1993; Watson and Thornhill 1994).

Asymmetry of several secondary sexual characters is related to the level of parasitism (blue peafowl *Pavo cristatus* (Hasegawa 1995); barn swallow (Møller 1992c); reindeer (Folstad *et al.* 1996; Table W9.1). These studies suggest that secondary sexual characters sometimes reveal parasite infection status, and that secondary sexual characters are particularly susceptible to environmental perturbations caused by parasitism (Møller 1992c). Alternatively, the non-experimental study of blue peafowl may indicate that individuals in poor condition, or that experienced a bad time during ontogeny, are also not good at raising an effective immune defence.

The relationship between testosterone and developmental instability has been determined in two studies (Saino and Møller 1994; Jamison *et al.* 1993). Saino and Møller (1994) reported a weak, negative relationship between asymmetry in the length of the outermost tail feathers of male barn swallows and circulating testosterone. A second study of humans revealed an elevated level of asymmetry in fingerprints of males with high concentrations of testosterone (Jamison *et al.* 1993). Since dermatoglyphic asymmetries develop during the first three months of pregnancy, when testosterone production of the embryo is absent, the positive relationship between asymmetry and adult circulating testosterone titres must have developed in response to a prenatal common agent.

The relationship between developmental stability and defence systems against parasites such as the immune system has not been systematically investigated. Secondary chemical compounds of plants are used in defence against herbivorous parasites. Sakai and Shimamoto (1965b) found higher concentrations of nicotine (a chemical compound used in anti-herbivore defence) in strains of tobacco *Nicotiana tabacum* with lower levels of leaf asymmetry. A second example concerns reduced immunocompetence which is known to be associated with chromosomal abnormalities in humans (Todaro and Martin 1967; Lubiniecki *et al.* 1977). Individuals with chromosomal abnormalities have greatly increased asymmetry (Fraser 1994; Thornhill and Møller 1997). Further studies of the relationships between sexual selection, developmental stability, endocrinology, and immunology are clearly needed.

In conclusion, the clear associations between developmental instability and parasitism potentially have a number of evolutionary implications. Some of these have been spelt out in this section, while others are listed by Møller (1996d) and Thornhill and Møller (1997).

9.7 Developmental stability, predation, and competition

9.7.1 *Developmental stability and predation*

Developmentally unstable organisms may turn out to suffer from increased risks of predation either because morphological asymmetries directly affect their ability to evade predators, or because individual asymmetry reliably reflects a poor phenotypic or genetic constitution. Asymmetric individuals may therefore be more likely to suffer from predation.

Three studies have determined the relationship between developmental stability and risk of predation. Fluctuating asymmetry in wing and tibia length of the fly *Musca domestica* was studied in a free-living population. The relationship between morphology and risk of predation by an aerially insectivorous passerine bird, the barn swallow, was assessed by comparison of the morphology of individual flies captured by barn swallows and fed to their offspring and that of a random sample of flies from the same sites as where the birds were foraging. Fluctuating asymmetry of flies was indeed significantly larger in prey than among individuals not falling victim of barn swallows, even when controlling for character size (Fig. 9.7; Møller 1996a).

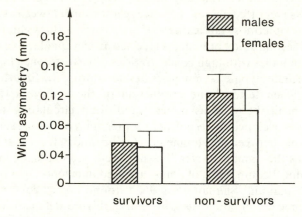

Fig. 9.7 Fluctuating asymmetry in wing length of *Musca domestica* captured by barn swallows and fed to their offspring and asymmetry of a random sample from the field. Values are means (+ s.e.). Adapted from Møller (1996a).

A second study of predation revealed a very similar result. Asymmetry in tail length of barn swallows was significantly higher among individuals captured and recovered from the nests of European sparrowhawks *Accipiter nisus* as compared with that of a field sample of live individuals (Møller and Nielsen 1997).

A third example investigated asymmetry in dung flies *Scatophaga stercoraria* and their house fly prey (Swaddle 1997b). Captured flies had more asymmetric tibia than flies that survived, and successful predators also had more symmetric tibia than unsuccessful dung flies.

These examples concern species in which the developmentally unstable morphology may directly affect the ability of individuals to evade predators (see Chapter 7). Thus, it remains unknown whether risk of predation also is related to developmental instability of characters not directly involved in locomotion.

A final study provided indirect evidence for an effect of predation on the frequency of developmentally unstable phenotypes. In a comparison of the level of fluctuating asymmetry in lateral plates of the three-spined stickleback

Gasterosteus aculeatus there was a significantly higher level of asymmetry in lakes without predatory fish when compared with lakes with predatory fish (Moodie and Reimchen 1976). A likely explanation for this result is intense directional natural selection against developmentally unstable individuals due to predation.

Predation risk may impose physiological stress on animals and therefore lead to increased morphological asymmetry, which then subsequently may lead to increased risks of predation. An experiment on European starlings exposed individual males to different risks of predation by varying the availability of cover and the position of food in relation to cover before and during the moulting period (Witter and Lee 1995). There was no effect of the two treatments on the timing or rate of moult, and the final size of the morphological characters was similarly unaffected. Male starlings from aviaries without cover had considerably higher degrees of asymmetry in their primary feathers than males with access to cover (Fig 9.8). Similarly, males with food presented near cover were more symmetric in their first primary feather than males with food in the open (Fig. 9.8). These results suggest that the perceived risk of predation is sufficient to increase the level of morphological asymmetry. Individuals living in habitats with little cover obviously may run higher risks of predation simply because they become more asymmetric.

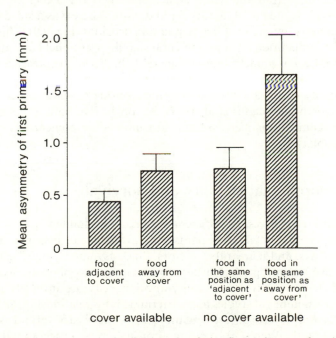

Fig. 9.8 Absolute asymmetry in the length of the first primary of male European starlings *Sturnus vulgaris* in relation to the availability of cover and the position of food in relation to cover during the moult period. Values are means (+ s.e.). Adapted from Witter and Lee (1995).

9.7.2 *Developmental stability and competition*

Intraspecific and interspecific competition often have differential effects on individuals of poor competitive ability. Measures of developmental stability of individuals such as individual asymmetry thus should correlate with intensity of exploitation or interference competition for access to limiting resources. This appeared to be the case in a study of European blackbirds *Turdus merula* breeding in a mosaic of habitat fragments (Møller 1995d). Nestlings produced in small habitat fragments were more asymmetric than nestlings from large fragments, and recruitment was differential with respect to morphological asymmetry. Recruitment to small habitat fragments was mainly by individuals of inferior competitive ability such as first-time breeders, and only during years of high population density. Individual blackbirds recruiting into small habitat patches mainly originated from large patches. These results suggest that competitive ability as reflected by recruitment was directly related to developmental stability of individuals.

Competition may also have direct effects on developmental stability of individuals experiencing different intensities of interactions. An experimental study of poplars *Populus euramericana* tested effects of both intraspecific and interspecific competition on asymmetry in leaves (Rettig *et al.* 1997). Individuals originating from the same clone, and therefore being of the same genotype, were planted in plots varying in density. There was a steady increase in leaf asymmetry among the three population densities (Fig. 9.9). This effect of intraspecific competition on developmental stability was upheld for interspecific competition. Leaf asymmetry was considerably higher in the presence of a herb layer than in its absence (Fig. 9.9). Grazing of competitors increased developmental stability in a grass species suffering from detrimental effects of interspecific competition (Alados *et* al. 1997). In conclusion, both intraspecific and interspecific competition increased one measure of developmental instability of leaves of poplars.

9.8 Developmental stability and survival

Survival depends on access to sufficient resources for maintenance, an ability to avoid predation and debilitating parasitism, avoidance of overly costly reproductive activities, and many other factors. Since asymmetry may be related to risk of parasitism and predation, as shown in the two previous sections, we could predict that survival probability and longevity are inversely related to asymmetry. A number of studies have determined this relationship, and 9 out of 21 studies have reported a negative relationship between survival and asymmetry (Table W9.2 on the Web site). A study of oribi *Ourebia ourebi* found no effect (Arcese 1994), a study of lion *Pathera leo* found a positive association in one sex and a negative one in the other (Packer and Pusey 1993), while a third study of brown hare *Lepus europaeus* found a positive relationship between skeletal

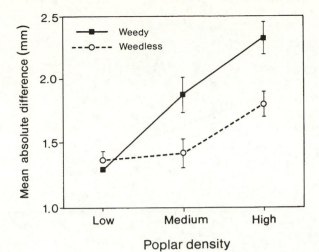

Fig. 9.9 Absolute leaf asymmetry of the two halves of poplar *Populus euramericana* leaves in weedy and weedless treatments for low, medium, and high tree densities. Values are means (s.e.). Adapted from Rettig *et al.* (1997).

asymmetry and age (Suchentrunk 1993). The relationship in the latter study was confounded by juveniles in populations with high levels of heterozygosity having a survival advantage, and, surprisingly, a positive association between asymmetry at the population level and heterozygosity.

Longevity has been directly related to developmental stability in three organisms. In the forest tent caterpillar moth *Malacosoma disstria*, longevity is inversely related to leg asymmetry (Fig. 9.10; Naugler and Leech 1994). A similar relationship has been found for the coccinellid beetle *Harmonia axyridis* (Ueno 1994) and the barn swallow (Møller 1994c).

Fecundity and other measures of reproductive success were higher among developmentally stable individuals. Life-history theory predicts that iteroparous organisms should balance investment in current reproduction to optimize survival until the next reproductive season (Roff 1992; Stearns 1992). Most observational studies of reproduction under field or laboratory conditions do not find a negative correlation between current reproductive investment and survival prospects (Roff 1992; Stearns 1992). Information on fecundity and survival in relation to developmental stability is consistent with this conclusion. Developmentally unstable individuals survive less well despite their low current fecundity (section 9.3).

The single most important determinant of lifetime reproductive success (at least when disregarding the effects of sperm competition) in a diverse assemblage of species is longevity (Clutton-Brock 1988; Newton 1989). If survival prospects and longevity in general will prove to be negatively associated with developmental instability, as suggested by the studies in Table W9.2, fluctuating asymmetry may indeed turn out to be an easily accessible predictor of fitness.

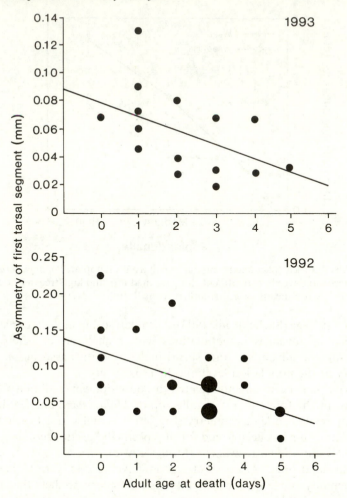

Fig. 9.10 Longevity of the forest tent caterpillar moth *Malacosoma disstria* in relation to fluctuating asymmetry in first tarsal segment. Adapted from Naugler and Leech (1994).

9.9 Summary

- Individuals with more efficient metabolisms should have excess energy available for maintenance, growth, and reproduction, and growth performance is often inversely associated with developmental instability.
- Several measures of fecundity such as clutch and litter size and number of clutches and litters were inversely associated with developmental instability in a range of different organisms. Relationships between developmental stability and reproductive success appear to arise because of developmental selection against developmentally abnormal gametes or offspring.
- Prevalence and intensity of parasite infection is often associated with

developmental instability. This relationship may arise because developmentally unstable individuals are more susceptible to parasite infections. Alternatively, developmental instability may reflect an inferior immune system, which is known to have important environmental components.

- Developmentally unstable individuals also appear to be more susceptible to predation and the negative effects of competition.
- Survival prospects and longevity are often inversely related to developmental instability.

References

Adams, M.S. and Niswander, J.D. 1967 Developmental 'noise' and a congenital malformation. *Genet. Res., Camb.* **10**, 313–317.

Agladze, K., Dulos, E. and De Kepper, P. 1992 Turing patterns in confined gel and gel-free media. *J. Phys. Chem.* **96**, 2400–2403.

Agnew, P. and Koella, J. 1997 Virulence, parasite mode of transmission, and host fluctuating asymmetry. *Proc. R. Soc. Lond. B* **264**, 9–15.

Aitken, M., Anderson, D., Francis, B. and Hinde, J. 1989 *Statistical Modelling in GLIM.* Oxford University Press, Oxford.

Akam, M. 1989 Hox and HOM: homologous gene clusters in insects and vertebrates. *Cell* **57**, 347–349.

Alados, C.L., Emlen, J.M., Wachocki, B. and Freeman, D.C. 1997 Developmental instability and the fractal architecture in desert plant species under grazing pressure. *Acta Oecol.* (in press).

Alados, C.L., Escós, J. and Emlen, J.M. 1993 Developmental instability as an indicator of environmental stress in the Pacific hake (*Merluccius productus*). *Fish. Bull.* **91**, 587–593.

Alados, C.L., Escós, J.M. and Emlen, J.M. 1994 Scale asymmetry: a tool to detect developmental instability under the fractal geometry scope. In *Fractals in the Natural and Applied Sciences.* M.M. Novak (Ed.), pp. 25–36. Elsevier, Amsterdam.

Alados, C.L., Escós, J. and Emlen, J.M. 1996 Fluctuating asymmetry and fractal dimension of the saggital suture as inbreeding depression indicators in dama and dorcas gazelle. *Can. J. Zool.* **73**, 1967–1974.

Alanis, L.B. 1966 Environmental and genotype-environmental components of variability. I. Inbred lines. *Heredity* **21**, 387–397.

Alatalo, R.V., Höglund, J. and Lundberg, A. 1988 Sexual selection models and patterns of variation in tail ornaments in birds. *Biol. J. Linn. Soc.* **34**, 363–374.

Alekseeva, T.A., Zinichev, V.V. and Zotin, A.I. 1992 Energy criteria of reliability and stability of development. *Acta Zool. Fenn.* **191**, 159–165.

Alerstam, T. 1990 *Bird Migration.* Cambridge University Press, Cambridge.

Alexander, R. McN., Brandwood, A., Currey, J.D. and Jayes, A.S. 1984 Symmetry and precision control of strength in limb bones of birds. *J. Zool.* **203**, 135–143.

Alibert, P., Renaud, S., Dod, B., Bonhomme, F. and Auffray, J.-C. 1994 Fluctuating asymmetry in the *Mus musculus* hybrid zone: a heterotic effect in disrupted co-adapted genomes. *Proc. R. Soc. Lond. B* **258**, 53–59.

Allard, R.W., Kahler, A.L. and Clegg, M.T. 1977 Estimation of mating cycle components of selection in plants. In *Measuring Selection in Natural Populations*. F.B. Christiansen and T.M. Fenchel (Eds.), pp. 1–19. Springer, Berlin.

Allen, F.L., Comstock, R.E. and Rasmusson, D.C. 1978 Optimal environments for yield testing. *Crop Sci.* **18**, 747–751.

Allendorf, F.W. and Leary, R.F. 1986 Heterozygosity and fitness in natural populations of animals. In *Conservation Biology: the Science of Scarcity and Diversity*. M.E. Soulé (Ed.), pp. 57–76. Sinauer, Sunderland.

Allendorf, F.W., Knudsen, K.L. and Leary, R.F. 1983 Adaptive significance in the tissue-specific expression of a phophoglucomutase gene in rainbow trout. *Proc. Natl. Acad. Sci. USA* **80**, 1397–1400.

Alley, T.R. and Cunningham, M.R. 1991 Averaged faces are attractive, but very attractive faces are not average. *Psychol. Sci.* **2**, 123–125.

Altukhov, Y.P., Kurbatova, O.L., Botviniev, O.K., Afanasiev, K.I., Malinina, T.V., Kholod, O.N., Strelkova, L.K. and Ivanova, N.S. 1981 Gene markers and disease: genetic anthropometric and clinical characteristics of children with acute pneumonia. *Genetika* **17**, 920–931.

Ames, L.J., Felley, J.D. and Smith, M.H. 1979 Amounts of asymmetry in Centrarchid fish inhabiting heated and nonheated reservoirs. *Trans. Am. Fish. Soc.* **108**, 489–495.

Ancel, P. 1950 *La Chimiotératogenése*. Doin, Paris.

Andersson, L. and Sandberg, K. 1984 Genetic linkage in the horse. II. Distribution of male recombination estimates and the influence of age, breed and sex on recombination frequency. *Genetics* **106**, 109–122.

Andersson, M. 1982 Female choice selects for extreme tail length in a widowbird. *Nature* **299**, 818–820.

Andersson, M. 1994 *Sexual Selection*. Princeton University Press, Princeton.

Aparicio, J.M. 1995 Patterns of growth and fluctuating asymmetry. Manuscript.

Arcese, P. 1994 Harem size and horn symmetry in oribi. *Anim. Behav.* **48**, 1485–1488.

Arnold, S.J. 1988 Quantitative genetics and selection in natural populations: microevolution of vertebral numbers in the garter snake *Thamnophis elegans*. In *Second International Conference on Quantitative Genetics*. B.S.Weir, M.M. Goodman, E.J. Eiser and G. Namtung (Eds.), pp. 619–636. Sinauer, Sunderland.

Arnqvist, G. and Wooster, D. 1995 Meta-analysis: synthesizing research findings in ecology and evolution. *Trends Ecol. Evol.* **10**, 236–240.

Astheimer, L.B., Buttemer, W.A. and Wingfield, J.C. 1992 Interactions of corticosterone with feeding, activity and metabolism in passerine birds. *Ornis Scand.* **23**, 355–365.

Atchley, W.R., Herring, S.W., Riska, B. and Plummer, A.A. 1984 Effects of the muscular dysgenesis gene on developmental stability in the mouse mandible. *J. Craniofac. Gen. Develop. Biol.* **4**, 179–189.

Attneave, F. 1954 Some informational aspects of visual perception. *Psychol. Rev.* **61**, 183–193.

Auerbach, C. 1976 *Mutation Research*. Chapman and Hall, London.

Bader, R.S. 1965 Fluctuating asymmetry in the dentition of the house mouse. *Growth* **29**, 291–200.

Bagchi, S.K., Sharma, V.P. and Gupta, P.K. 1989 Developmental instability in leaves of *Tectonia grandis*. *Silvae Genetica* **38**, 1–6.

Bailly, F., Gaill, F. and Mosseri, R.A. 1991 A dynamical system for biological development: the case of *Caenorhabditis elegans*. *Acta Biotheor.* **39**, 167–184.

Baker, C.M.A. and Manwell, C. 1977 Heterozygosity of the sheep: polymorphism of 'malic enzyme', isocitrate dehydrogenase (Nadp), catalase and esterase. *Aust. J. Biol. Sci.* **30**, 127–140.

Balmford, A., Jones, I.L. and Thomas, A.L.R. 1993 On avian asymmetry: evidence of natural selection for symmetrical tails and wings in birds. *Proc. R. Soc. Lond. B* **252**, 245–251.

Barden, H.S. 1980 Fluctuating dental asymmetry: a measure of developmental stability in Down syndrome. *Am. J. Phys. Anthropol.* **52**, 169–173.

Barlow, H.B. 1980 The absolute efficiency of perceptual decisions. *Phil. Trans. R. Soc. B* **290**, 71–82.

Barlow, H.B. and Reeves, B.C. 1979 The versatility and absolute efficiency of detecting mirror symmetry in random dot displays. *Vision Res.* **19**, 783–793.

Barraclough, T.G., Harvey, P.H. and Nee, S. 1995 Sexual selection and taxonomic diversity in passerine birds. *Proc. R. Soc. Lond. B* **259**, 211–215.

Barton, N.H. 1989 Founder effect speciation. In *Speciation and its Consequences*. D. Otte and J.A. Endler (Eds.), pp. 229–256. Sinauer, Sunderland.

Bateman, K.G. 1959 The genetic assimilation of four venation phenocopies. *J. Genet.* **56**, 443–447.

Bawa, K.S. and Webb, C.J. 1984 Flower, fruit and seed abortion in tropical forest trees: implications for the evolution of paternal and maternal reproductive patterns. *Am. J. Bot.* **7**, 736–751.

Beacham, T.D. and Withler, R.E. 1985 Heterozygosity and morphological variability of pink salmon (*Oncorhynchus gorbuscha*) from southern British Columbia and Puget Sound. *Can. J. Genet. Cytol.* **27**, 571–579.

Beardmore, J.A. 1960 Developmental stability in constant and fluctuating temperatures. *Heredity* **14**, 411–422.

Beardmore, J.A. 1965 A genetic basis for lateral bias. *Symposium on the Mutational Process. Mutation in Population, Prague, Czechoslovakia*, pp. 75–83. Publishing House of the Czechoslovak Academy of Sciences, Prague.

Beardmore, J.A. and Shami, S.A. 1976 Parental age, genetic variation and selection. In *Population Genetics and Ecology*. S. Karlin and E. Nevo (Eds.), pp. 3–22. Academic Press, New York.

Beardmore, J.A. and Shami, S.A 1979 Heterozygosity and the optimum phenotype under stabilizing selection. *Aquilo, Ser. Zool.* **20**, 100–110.

Belyaev, D.K. and Borodin, P.M. 1982 The influence of stress on variation and its role in evolution. *Biol. Zentralbl.* **100**, 705–714.

Bengtsson, B.-E., Bengtsson, Å. and Himberg, M. 1985 Fish deformities and pollution in some Swedish waters. *Ambio* 14, 32–35.

Benkman, C.W. and Lindholm, A.K. 1991 The advantages and evolution of a morphological novelty. *Nature* **349**, 519–520.

Bennett, A.T.D., Cuthill, I.C., Partridge, J.C. and Maier, E.J. 1996 Ultraviolet vision and mate choice in zebra finches. *Nature* **380**, 433–435.

Benson, P. and Perrett, D. 1991 Computer averaging and manipulations of faces. In *Photovideo: Photography in the Age of the Computer*. P. Wombell (Ed.), pp. 32–38. Rivers Oram Press, London.

Bergé, P., Pomeau, Y. and Vidal, C. 1984 *Order within Chaos: Towards a Deterministic Approach to Turbulence*. Wiley, New York.

Berger, E. 1976 Heterosis and the maintenance of enzyme polymorphism. *Am. Nat.* **110**, 823–829.

Bethe, A., Bergmann, G.V., Embden, G. and Ellinger, A. 1930 *Handbuch der normalen und pathologischen Physiologie. Bd XIII. Schutz- und Angriffseinrichtungen. Reaktion auf Schädigungen.* Springer, Berlin.

Bijlsma-Meeles, E. and Bijlsma, R. 1988 The alcohol dehydrogenase polymorphism in *Drosophila melanogaster*: fitness measurement and predictions under conditions with no alcohol stress. *Genetics* **120**, 743–753.

Birch, M.C., Poppy, G.M. and Baker, T.C. 1990 Scents and eversable scent structures of male moths. *Ann. Rev. Entomol.* **35**, 25–58.

Birney, E.C. and Baird, D.D. 1985 Why do some mammals polyovulate to produce a litter of two? *Am. Nat.* **126**, 136–140.

Björklund, M. 1996 The effect of male presence on nestling growth and fluctuating asymmetry in the blue tit (*Parus caeruleus*). *Condor* **98**, 172–175.

Blackman, R.L. 1979 Stability and variation in aphid clonal lineages. *Biol. J. Linn. Soc.* **11**, 259–277.

Blanco, G., Sanchez, J.A., Vazquez, E., Garcia, E. and Rubio, J. 1990 Superior developmental stability of heterozygotes at enzyme loci in *Salmo salar* L. *Aquaculture* **84**, 199–209.

Blum, A. 1988 *Plant Breeding for Stress Environments.* CRC Press, Boca Raton.

Bodmer, W.F. 1961 Viability effects and recombination differences in a linkage test with pallid and fidget in the house mouse. *Heredity* **16**, 485–495.

Bohr, V.A. and Wassermann, K. 1988 DNA repair at the level of the gene. *Trends Biochem. Sci.* **13**, 429–433.

Booth, C.L., Woodruff, D.S. and Gould, S.J. 1990 Lack of significant associations between allozyme heterozygosity and phenotypic traits in the land snail *Cerion*. *Evolution* **44**, 210–213.

Bornstein, M.H., Ferdinandsen, K. and Gross, C.G. 1981 Perception of symmetry in infancy. *Devl. Psychol.* **17**, 82–86.

Bottini, E., Gloria-Bottini, F., Lucarelli, P., Polzonetti, A., Santoro, F. and Varveri, A. 1979 Genetic polymorphisms and intrauterine development: evidence of decreased heterozygosity in light of dates for human newborn babies. *Experientia* **35**, 1565–1567.

Botvinev, O.K., Kurbatova, O.L. and Altukhov, Y.P. 1980 Population-genetic approach to the problem of non-specific biological resistance of human organism. II. Clinical characteristics, congenital malformations and genetic structure of diseased newborns and infants with respect to their weight and body length at birth. *Genetika* **16**, 1884–1894.

Boue, J., Boue, A. and Lazar, P. 1975 Retrospective and prospective epidemiological studies of 1500 karyotyped spontaneous human abortions. *Teratology* **12**, 11–25.

Bownes, M. 1990 Preferential insertion of P elements into genes expressed in the germline of *Drosophila melanogaster*. *Molec. Gen. Genet.* **222**, 457–460.

Bradbury, D. 1929 A comparative study of the developing and aborting fruits of *Prunus cerasus*. *Am. J. Bot.* **16**, 525–542.

Bradley, B.P. 1980 Developmental stability of *Drosophila melanogaster* under artificial and natural selection in constant and fluctuating environments. *Genetics* **95**, 1033–1042.

Brakefield, P.M. and Breuker, C.J. 1996 The genetical basis of fluctuating asymmetry for developmentally integrated traits in a butterfly eyespot pattern. *Proc. R. Soc. Lond. B.* **263**, 1557–1563.

Brakefield, P.M. and French, V. 1993 Butterfly wing patterns: developmental mechanisms and evolutionary change. *Acta Biotheor.* **41**, 447–468.

Brandt, M. and Siegel, M.I. 1978 The effects of stress on cortical bone thickness in rodents. *Am. J. Phys. Anthropol.* **49**, 3–34.

Bridges, C.B. 1927 The relation of the age of the female to crossing over in the third chromosome of *Drosophila melanogaster*. *J. Gen. Physiol.* **8**, 689–700.

Briggs, J.C. 1991 A Cretaceous-Tertiary mass extinction? *BioScience* **41**, 619–624.

Brightwell, L.R. 1951 Some experiments with the common herit crab (*Eupagurus bernhardus* Linn.) and transparent univalve shells. *Proc. Zool. Soc. Lond.* **121**, 279–283.

Brown, N.A. and Lander, A. 1993 On the other hand ... *Nature* 363, 303–304.

Brown, N.A. and Wolpert, L. 1990 The development of handedness in left/right asymmetry. *Development* **109**, 1–9.

Brown, N.A., McCarthy, A. and Wolpert, L. 1991 Development of handed body asymmetry in mammals. In *Biological Asymmetry and Handedness*. G.R. Bock and J. Marsh (Eds.), pp. 182–201. Wiley, Chichester.

Brueckner, M., McGrath, J., D'Eustachio, P. and Horwich, A.L. 1991 Establishment of left–right asymmetry in vertebrates: genetically distinct steps are involved. In *Biological Asymmetry and Handedness*. G.R. Bock and J. Marsh (Eds.), pp. 202–218. Wiley, Chichester.

Buchholz, J.T. 1922 Developmental selection in vascular plants. *Bot. Gaz.* **73**, 249–286.

Bullock, S. and Cliff, D. 1997 The role of 'hidden preferences' in the artificial ω-evolution of symmetrical signals. *Proc. R. Soc. Lond.* B **264**, 505–511.

Bulmer, M.G. 1985 *The Mathematical Theory of Quantitative Genetics*. Clarendon, Oxford.

Burley, N. 1981 Sex-ratio manipulation and selection for attractiveness. *Science* 211, 721–722.

Burley, N. 1986 Sexual selection for aesthetic traits in species with biparental care. *Am. Nat.* **127**, 415–445.

Burnaby, P. 1966 Growth-invariant discriminant functions and generalized distances. *Biometrics* **22**, 96–110.

Burton, R.S. 1990 Hybrid breakdown in physiological response: a mechanistic approach. *Evolution* **44**, 1806–1813.

Buss, L.W. 1987 *The Evolution of Individuality*. Princeton University Press, Princeton.

Cairns, J., Overbaugh, J. and Miller, S. 1988 The origin of mutants. *Nature* **335**, 142–145.

Carmeliet, P., Ferreira, V., Breier, G., Pollefeyt, S., Kieckens, L., Gertsenstein, M., Fahrig, M., Vandenhoeck, A., Harpal, K., Eberhardt, C., Declercq, C., Pawling, J., Moons, L., Collen, D., Risau, W. and Nagy, A. 1996. Abnormal blood vessel development and letality in embryos lacking single VEGT allele. *Nature* **380**, 435–439.

Caro, T.M. 1986 The function of stotting: a review of the hypotheses. *Anim. Behav.* **34**, 649–662.

Carson, H.L. and Lande, R. 1984 Inheritance of a secondary sexual character in *Drosophila silvestris*. *Proc. Natl. Acad. Sci. USA* **81**, 6904–6907.

Castets, V., Dulos, E., Boissonade, J. and De Kepper, P. 1990 Experimental evidence of a sustained standing Turing-type nonequilibrium chemical pattern. *Phys. Rev. Lett.* **64**, 2953–2956.

Chakraborty, R. 1981 The distribution of the number of heterozygous loci in an individual in natural populations. *Genetics* **98**, 461–466.

Chakraborty, R. 1987 Biochemical heterozygosity and phenotypic variability of poly-genic traits. *Heredity* **59**, 19–28.

Chakraborty, R. and Ryman, N. 1983 Relationship of mean and variance of genotypic values with heterozygosity per individual in a natural population. *Genetics* **103**, 149–152.

Chandley, A.C. 1968 The effect of X-rays on female germ cells of *Drosophila melanogaster*. III. A comparison with heat-treatment on crossing-over in the X-chromosome. *Mutation Res.* **5**, 93–107.

Chandra, R.K. and Newberne, P.M. 1977 *Nutrition, Immunity, and Infection*. Plenum Press, New York.

Charlesworth, B. 1984 The evolutionary genetics of life histories. *Evolutionary Ecology*. B. Shorrocks (Ed.), pp. 117–133. Blackwell, Oxford.

Charlesworth, B. 1987 The heritability of fitness. In *Sexual Selection: Testing the Alternatives*. J.W. Bradbury and M.B. Andersson (Eds.), pp. 21–40. Wiley, Chichester.

Charlesworth, D. and Charlesworth, B. 1987 Inbreeding depression and its evolutionary consequences. *Ann. Rev. Ecol. Syst.* **18**, 237–268.

Chenuil, C. 1991 *Asymetrie du Nombre d'Ecailles Ventrales chez le Lézard Vivipare*. MSc thesis, Université de Paris-Sud, Paris.

Chippindale, A.K. and Palmer, A.R. 1993 Persistence of subtle departures from symmetry over multiple molts in individual brachyuran crabs: relevance to developmental stability. *Genetica* **89**, 185–199.

Clarke, G.M. 1992 Fluctuating asymmetry: a technique for measuring developmental stress of genetic and environmental origin. *Acta Zool. Fenn.* **191**, 31–35.

Clarke, G.M. 1993a Fluctuating asymmetry of invertebrate populations as a biological indicator of environmental quality. *Environ. Poll.* **82**, 207–211.

Clarke, G.M. 1993b The genetic basis of developmental stability. I. Relationships between stability, heterozygosity and genomic coadaptation. *Genetica* **89**, 15–23.

Clarke, G.M. 1995 The genetic basis of developmental stability. II. Asymmetry of extreme phenotypes revisited. *Am. Nat.* **146**, 708–725.

Clarke, G.M. and McKenzie, J.A. 1987 Developmental stability of insecticide resistant phenotypes in blowfly; a result of canalizing natural selection. *Nature* **325**, 345–346.

Clarke, G.M. and McKenzie J.A. 1992 Fluctuating asymmetry as a quality control indicator for insect mass rearing projects. *J. Econ. Entomol.* **85**, 2045–2050.

Clarke, G.M. and Oldroyd, B.P. 1996 The genetic basis of developmental stability in *Apis mellifera*. II. Relationships between character size, asymmetry and single-locus heterozygosity. *Genetica* **97**, 211–224.

Clarke, G.M. and Ridsdill-Smith, T.J. 1990 The effect of Avermectin B$_1$ on developmental stability in the bush fly, *Musca vetustissima*, as measured by fluctuating asymmetry. *Entomol. Exp. Appl.* **54**, 265–269.

Clarke, G.M., Brand, G.W. and Whitten, M.J. 1986 Fluctuating asymmetry: a technique for measuring developmental stress caused by inbreeding. *J. Aust. Biol. Sci.* **39**, 145–153.

Clarke, G.M., Oldroyd, B.P. and Hunt, P. 1992 The genetic basis of developmental stability in *Apis mellifera*: heterozygosity versus genic balance. *Evolution* **46**, 753–762.

Clutton-Brock, T.H. (Ed.) 1988 *Reproductive Success: Studies of Individual Variation in Contrasting Breeding Systems*. University of Chicago Press, Chicago.

Coelho, J.R. and Mitton, J.B. 1988 Oxygen consumption during hovering is associated with genetic variation of enzymes in honey-bees. *Funct. Ecol.* **2**, 141–146.

Conway, C.J., Eddleman, W.R. and Simpson, K.L. 1994 Evaluation of lipid indices of the wood thrush. *Condor* **96**, 783–790.

Cook, N.D. 1995 Artefact or network evolution? *Nature* **374**, 313.

Corballis, M.C. and Roldán, C.E. 1975 Detection of symmetry as a function of angular orientation. *J. exp. Psychol.* **1**, 221–230.

Córdoba-Aguilar, A. 1997 Male contests in damselflies (Odonata: Calopterygidae): winners are more symmetrical. *J. Insect Behav.* (in press).

Cornuet, J., Oldroyd, B.P. and Crozier, R.H. 1994 Unequal thermostability of allelic forms of malate dehydrogenase in honey bees. *J. Apic. Res.* **34**, 45–47.

Cothran, E.G., Chesser, R.K., Smith, M.H. and Johns, P.E. 1983 Influences of genetic variation and maternal factors on fetal growth rate in white-tailed deer. *Evolution* **37**, 282–291.

Cott, H.B. 1940 *Adaptive Coloration in Animals*. Methuen, London.

Coxeter, H.S.M. 1961 *Introduction to Geometry*. Wiley, London.

Coyne, J.A. 1987 Lack of respsonse to selection for directional asymmetry in *Drosophila melanogaster*. *J. Hered.* **78**, 119.

Coyne, J.A. 1992 Genetics and speciation. *Nature* **355**, 511–515.

Crescitelli, F. 1935 The respiratory metabolism of *Galleria mellonella* (bee moth) during pupal development at different constant temperatures. *J. Cell. Comp. Physiol.* **6**, 351–358.

Cross, S.S., Start, R.D., Silcocks, P.B., Bull, A.D., Cotton, D.W.K. and Underwood, J.C.E. 1993 Quantitation of the renal arteri analysis. *J. Pathol.* **170**, 479–484.

Crow, J.F. and Kimura, M. 1970 *An Introduction to Population Genetics Theory*. Burgess Publishing, Minneapolis.

Crusio, W.E. and van Abeelen, J.H.F. 1993 Canalization of behavioural development and heterozygosity in mice. *Behav. Genet.* **23**, 550.

Cuervo, J.J. and Møller, A.P. 1997a Phenotypic variation and fluctuating asymmetry in avian feather ornaments. MS.

Cuervo, J.J. and Møller, A.P. 1997b Divergence in secondary sexual characters and non-sexual characters among bird species. MS.

Cuervo, J.J. and Møller, A.P. 1997c Fluctuating asymmetry in migratory and resident passerine bird populations. MS.

Cullis, C.A. 1984 Environmentally induced DNA changes. In *Evolutionary Theory: Paths into the Future*. J.W. Pollard (Ed.), pp. 203–216. Wiley, Chichester.

Cuthill, I. and Guilford, T. 1990 Perceived risk and obstacle avoidance in flying birds. *Anim. Behav.* **40**, 188–190.

Cuthill, I.C., Swaddle, J.P. and Witter, M.S. 1993 Fluctuating asymmetry. *Nature* **363**, 217–218.

Daly, M. and Wilson, M. 1984 A sociobiological analysis of human infanticide. In *Infanticide: Comparative and Evolutionary Perspectives*. G. Hausfater and S.B. Hrdy (Eds.), pp. 487–502. Aldine, New York.

Danzmann, R.G., Ferguson, M.M., Allendorf, F.W. and Knudsen, K.L. 1986 Heterozygosity and developmental rate in a strain of rainbow trout, *Salmo gairdneri*. *Evolution* **40**, 86–93.

Darwin, C. 1868 *The Variation of Animals and Plants under Domestication*. John Murray, London.

Darwin, C. 1871 *The Descent of Man, and Selection in Relation to Sex*. John Murray, London.

Davies, A.G., Game, A.Y., Chen, Z., Williams, T.J., Goodall, S., Yen, J.L., McKenzie, J.A. and Batterham, P. 1996 *Scalloped wings* is the *Lucilia cuprina Notch* homologue and a candidate for the *modifier* of fitness and asymmetry of diazinon resistance. *Genetics* **143**, 1321–1337.

Davis, A. P., Witte, D. P., Hsieh-Li, H., Potter, S. S. and Capecchi, M. R. 1995 Absence of radius and ulna in mice lacking *hoxa-11* and *hoxd-11*. *Nature* **375**, 791–795.

Davis, T.A. 1962 The non-inheritance of asymmetry in *Cocos nucifera*. *J. Genet.* **58**, 42–50.

Davis, T.A. 1963 The dependence of yield on asymmetry in coconut palms. *J. Genet.* **58**, 186–215.

Dawkins, M.S. and Guilford, T. 1995 An exaggerated preference for simple neural network models of signal evolution? *Proc. R. Soc. Lond.* B **261**, 357–360.

De Marinis, F. 1959 The nature of asymmetry and variability in the double bar-eyeless *Drosophila*. *Genetics* **44**, 1101–1111.

De Pomerai, D. 1991 *From Gene to Animal*. 2nd edn. Cambridge University Press, Cambridge.

Delius, J.D. and Habers, G. 1978 Symmetry: can pigeons conceptualize it? *Behav. Biol.* **22**, 336–342.

Delius, J.D. and Nowak, B. 1982 Visual symmetry recognition by pigeons. *Psychol. Res.* **44**, 199–212.

Dennis, R. 1993 *Butterflies and Climatic Change*. Manchester University Press, Manchester.

Derr, J.A. 1980 The nature of variation in life history characters of *Dysdercus bimaculatus* (Heteroptera: Pyrrhocoridae), a colonizing species. *Evolution* **34**, 548–557.

Deutsch, J. 1994 Horn asymmetry and mating success on an antelope lek. MS.

Devroetes, P. 1989 *Dictyostelium discoideum*: a model system for cell-cell interactions in development. *Science* **245**, 1054–1058.

Dobzhansky, T. 1951 *Genetics and the Origin of Species*. Columbia University Press, New York.

Dobzhansky, T. 1955 A review of some fundamental concepts and problems of populations genetics. *Cold Spring Harbor Symp. Quant. Biol.* **20**, 1–15.

Dobzhansky, T. 1970 *Genetics and the Evolutionary Process*. Columbia University Press, New York.

Dobzhansky, T. and Wallace, B. 1953 The genetics of homeostasis in *Drosophila*. *Proc. Natl. Acad. Sci. USA* **39**, 162–170.

Downhower, J.F., Blumer, L.S., Lejeune, P., Gaudin, P., Marconato, A. and Bisazza, A. 1990 Otolith asymmetry in *Cottus bairdi* and *C. gobio*. *Polsk. Arch. Hydrobiol.* **37**, 209–220.

Doyle, W.J., Kelley, C. and Siegel, M.I. 1977 The effects of audiogenic stress on the growth of long bones in the laboratory rat (*Rattus norvegicus*). *Growth* **41**, 183–189.

Drake, J.W. 1970 *The Molecular Basis of Mutation*. Holden-Day, San Francisco.

Dubinin, N.P. and Romashov, D.D. 1932 Die genetische Struktur der Art und ihre Evolution. *Biol. Zh.* **1**, 52–95.

Dubrova, Y.E., Dambueva, I.K., Kholod, O.N., Prokhorovskaya, V.D., Pushkina, E.I. and Blank, M.L. 1991 Influence of mother's heterozygosity on the variation of anthropometric traits in newborns. *Genetika* **27**, 2168–2176.

Dudash, M.R. 1990 Relative fitness of selfed and outcrossed progeny in a self-compatible, protandrous species, *Sabatia angularis* L. (Gentianaceae): a comparison in three environments. *Evolution* **44**, 1129–1149.

Dufour, K. and Weatherhead, P. J. 1996 Estimation of organism-wide asymmetry in red-winged blackbirds and its relation to studies of mate selection. *Proc. R. Soc. Lond.* B **263**, 769–775.

Durrant, A. 1971 Induction and growth of flax genotrophs. *Heredity* **27**, 277–298.

Eanes, W.F. 1978 Morphological variance and enzyme heterozygosity in the monarch butterfly. *Nature* **276**, 263–264.

Efimov, V.M., Galaktionov, Y.K., Akimov, I.A. and Zaloznaya, L.M. 1987 Fluctuating asymmetry and its variability (an ontogenic aspect). *Dopovidi Akademii Nauk Ukrainskoi Rsr Seriya B—Geologichni Khimickni Ta Biologichni Nauki* **8**, 62–65.

Ehret, G. 1975 Schallsignale der Hausmaus (*Mus musculus*). *Behaviour* **52**, 38–55.

Eilbeck, J.C. 1989 Pattern formation and pattern selection in reaction-diffusion systems. In *Theoretical Biology: Epigenetic and Evolutionary Order from Complex Systems*. B. Goodwin and P. Saunders (Eds.), pp. 31–41. Edinburgh University Press, Edinburgh.

Eisenman, R. and Rappaport, J. 1967 Complexity preference and semantic differential ratings of complexity-simplicity and symmetry-asymmetry. *Psychonom. Sci.* **7**, 147–148.

Eisner, F.F. and Reznichenko, L.P. 1977 Heritability and repeatability of some characters of functional activity of adrenal cortex in cattle. *Genetica*, **13**, 430–438.

Elston, R.C., Lange, K. and Namboodiri, K.K. 1976 Age trends in human chiasmata frequencies and recombination fractions. II. Method for analysing recombination fractions and application to the ABO: nail-patella linkage. *Am. J. Human Genet.* **28**, 69–76.

Emlen, J.M., Freeman, C.D. and Graham, J.D. 1993 Nonlinear growth dynamics and the origin of fluctuating asymmetry. *Genetica* **89**, 77–96.

Endler, J.A. 1977 *Geographic Variation, Speciation, and Clines*. Princeton University Press, Princeton.

Endler, J.A. 1986 *Natural Selection in the Wild*. Princeton University Press, Princeton.

Enquist, M. and Arak, A. 1994 Symmetry, beauty and evolution. *Nature* **372**, 169–172.

Eriksson, M. 1996a Fluctuating asymmetry in *Lychnis viscaria*: measuring methods and patterns. In *Consequences for Plant Reproduction of Pollinator Preference for Symmetric Flowers*. M. Eriksson. Ph.D. thesis, Department of Zoology, Uppsala University, Sweden.

Eriksson, M. 1996b Bumblebee preference for symmetric flowers in *Lychnis viscaria* (Caryophyllaceae) and *Epilobium hirsutum* (Onagraceae). In *Consequences for Plant Reproduction of Pollinator Preference for Symmetric Flowers*. M. Eriksson. Ph.D. thesis, Department of Zoology, Uppsala University, Sweden.

Eriksson, M. 1996c Floral asymmetry in pollen donors affects female fitness in *Lychnis viscaria*. In *Consequences for Plant Reproduction of Pollinator Preference for Symmetric Flowers*. M. Eriksson. Ph.D. thesis, Department of Zoology, Uppsala University, Sweden.

Erway, L., Hurley, L.S. and Fraser, A.S. 1970 Congenital ataxia and otolith defects due to manganese deficiency in mice. *J. Nutr.* **100**, 643–654.

Escós, J.M., Alados, C.L. and Emlen, J.M. 1995 Fractal structures and fractal functions as disease indicators. *Oikos* **74**, 310–314.

Evans, H.J. 1988 Mutation as a cause of genetic disease. *Phil. Trans. R. Soc. Lond. B* **319**, 325–340.

Evans, M.R. 1993 Fluctuating asymmetry and long tails: the mechanical effects of asymmetry may act to enforce honest advertisement. *Proc. R. Soc. Lond. B* **253**, 205–209.

Evans, M.R. and Hatchwell, B.J. 1993 New slants on ornament asymmetry. *Proc. R. Soc. Lond. B* **251**, 171–177.

Evans, M.R., Martins, T.L.F. and Haley, M. 1994 The asymmetrical cost of tail elongation in red-billed streamertails. *Proc. R. Soc. Lond. B* **256**, 97–103.

Ewald, P. 1994 *Evolution of Infectious Disease*. Oxford University Press, Oxford.

Fahn, A. 1974 *Plant Anatomy*. Pergamon, New York.

Falconer, D.S. 1960 Selection of mice for growth on high and low planes of nutrition. *Genet. Res., Camb.* **1**, 91–113.

Falconer, D.S. 1981 *Introduction to Quantitative Genetics*. 2nd edn. Longman, New York.

Falconer, D.S. 1989 *Introduction to Quantitative Genetics*. 3rd edn. Longman, New York.

Falconer, D.S. and Latyszewski, M. 1952 The environment in relation to selection for size in mice. *J. Genet.* **51**, 67–80.

Farris, M.A. and Mitton, J.B. 1984 Population density, outcrossing rate, and heterozygote superiority in ponderosa pine. *Evolution* **38**, 1151–1154.

Feder, J. 1988 *Fractals*. Plenum, New York.

Fedoroff, N.V. 1989 About maize transposable elements and development. *Cell* **56**, 181–191.

Felley, J. 1980 Analysis of morphology and asymmetry in bluegill sunfish (*Lepomis macrochirus*) in the southeastern United States. *Copeia* **1980**, 18–29.

Ferguson, M.M. 1986 Developmental stability of rainbow trout hybrids: genomic coadaptation or heterozygosity? *Evolution* **40**, 323–330.

Ferguson, M.M. 1992 Enzyme heterozygosity and growth rate in rainbow trout—genetic and physiological explanations. *Heredity* **68**, 115–122.

Ferrara, N., Carver-Moore, K., Chen, H., Dowd, M., Lu, L., O'Shea, K.S., Powell-Braxton, L., Hillan, K.J. and Moore, M.W. 1996 Heterozygous embryonic lethality induced by targeted inactivation of the VEGF gene. *Nature* **380**, 439–442.

Fields, S.J., Spiers, M., Hershokovitz, I. and Livshits, G. 1995 Reliability of reliability coefficients in the estimation of asymmetry. *Am. J. Phys. Anthropol.* **96**, 83–87.

Fisher, R.A. 1930 *The Genetical Theory of Natural Selection*. Clarendon Press, Oxford.

Fisher, R.A. 1949 A preliminary linkage test with agouti and undulated mice. *Heredity* **3**, 299–341.

Fitter, A.H., Fitter, R.S.R., Harris, I.T.B. and Williamson, M.H. 1995 Relationships between first flowering date and temperature in the flora of a locality in central England. *Funct. Ecol.* **9**, 55–60.

Fleischer, R.C., Johnston, R.F. and Kiltz, W.J. 1983 Allozymic heterozygosity and morphological variation in house sparrows. *Nature* **304**, 628–629.

Folstad, I. and Karter, A.J. 1992 Parasites, bright males, and the immunocompetence handicap. *Am. Nat.* **139**, 603–622.

Folstad, I., Arneberg, P. and Karter, A.J. 1996 Antlers and parasites. *Oecologia* **105**, 556–558.

Forbes, M., Leung, B. and Schalk, G. 1997 Fluctuating asymmetry in *Coenagrion resolutum* damselflies in relation to size, age, and parasitism. *Odonatologica* **26**, 9–16.

Ford, C.S. 1964 *A Comparative Study of Human Reproduction*. Yale University Press, New Haven.

Forkman, B. and Corr, S. 1996 Influence of size and asymmetry of sexual characteristics in the rooster and hen on the number of eggs laid. *Appl. Anim. Behav. Sci.* **49**, 285–291.

Fox, G.A., Collins, B., Hayakawa, E., Weseloh, D.V., Ludwig, J.P., Kubiak, T.J. and Erdman, T.C. 1991 Reproductive outcomes in colonial fish-eating birds: a biomarker for developmental toxicants in Great Lakes food chains. II. Spatial variation in the occurrence and prevalence of bill defects in young double-crested cormorants in the Great Lakes, 1979–1987. *J. Great Lakes Res.* **17**, 158–167.

Fox, W., Gordon, C. and Fox, M.H. 1961 Morphological effects of low temperatures during the embryonic development of the garter snake. *Zoologica* **46**, 57–71.

Fraser, A.S. 1962 Survival of the mediocre. In *The Evolution of Living Organisms*. G.W. Leeper (Ed.), pp. 74–80. Melbourne University Press, Melbourne.

Fraser, A.S. and Kindred, B.M. 1960 Selection for an invariant character, vibrissa number, in the house mouse. II. Limits to variability. *Aust. J. Biol. Sci.* **13**, 48–58.

Fraser, F.C. 1994 Developmental instability and fluctuating asymmetry in man. In *Developmental Instability: Its Origins and Evolutionary Implications*. T.A. Markow (Ed.), pp. 319–334. Kluwer, Dordrecht.

Freeman, D.C., Graham, J.H. and Emlen, J.M. 1993 Developmental stability in plants: symmetries, stress and epigenesis. *Genetica* **89**, 97–119.

Freeman, D.C., Graham, J.H., Byrd, D.W., McArthur, E.D. and Turner, W.A. 1995 Narrow hybrid zone between two subspecies of big sagebrush, *Artemisia tridentata* (Asteraceae). III. Developmental instability. *Am. J. Bot.* **82**, 1144–1152.

French, V. and Brakefield, P.M. 1992 The development of eyespot patterns on butterfly wings: morphogen sources or sinks? *Development* **116**, 103–109.

Freyd, J. and Tversky, B. 1984 Force of symmetry in form perception. *Am. J. Psychol.* **97**, 109–126.

Fujio, Y. 1982 A correlation of heterozygosity with growth rate in the Pacific oyster, *Crassostrea gigas. Tohoku J. Agric. Res.* **33**, 66–75.

Furlow, F. B., Armijo-Prewitt, T. and Gangestad, S. 1997a Fluctuating asymmetry and aggression in college age men. MS.

Furlow, F. B., Marshall, M.C. and Kimball, R. 1997b Crow bioacoustics and phenotypic quality in red junglefowl. *Anim. Behav.* (in press).

Furlow, F. B., Armijo-Prewitt, T. Gangestad, S. and Thornhill, R. 1997c Fluctuating asymmetry and psychometric intelligence. *Proc. R. Soc. Lond. B* **264**, 823–830.

Gabriel, M.L. 1945 Factors affecting the number and form of vertebrae in *Fundulus heteroclitus. J. exp. Zool.* **95**, 105–147.

Gajardo, G.M. and Beardmore, J.A. 1989 Ability to switch reproductive mode in *Artemia* is related to maternal heterozygosity. *Mar. Ecol. Prog. Ser.* **55**, 191–195.

Gangestad, S.W. and Thornhill, R. 1997a An evolutionary psychological analysis of human sexual selection: developmental stability, male sexual behavior, and mediating features. In *Evolutionary Social Psychology*. J.A. Simpson and D.T. Kenrick (Eds.), pp. 169–195. Lawrence Erlbaum, Hillsdale.

Gangestad, S. W. and Thornhill, R. 1997b The analysis of fluctuating asymmetry redux: the robustness of parametric tests. *Anim. Behav.* (in press).

García-Vázquez, E. and Rubio, J. 1988 Canalization and phenotypic variation caused by changes in the temperature of development. *J. Hered.* **79**, 448–452.

Garland, T., Jr., and Carter, P.A. 1994 Evolutionary physiology. *Ann. Rev. Ecol. Syst.* **56**, 579–621.

Garner, W.R. and Clement, D.E. 1963 Goodness of pattern and pattern uncertainty. *Percept. Psychophys.* **2**, 446–452.

Garton, D.W. 1984 Relationship between multiple locus heterozygosity and physiological energetics of growth in the estuarine gastropod, *Thais haemastoma. Physiol. Zool.* **57**, 520–543.

Garton, D.W., Koehn, R.K. and Scott, T.M. 1984 Multiple locus heterozygosity and the physiological energetics of growth in the coot clam, *Mulina lateralis*, from a natural population. *Genetics* **108**, 445–455.

Gavrilets, S. and Hastings, A. 1994 A quantitative-genetic model for selection on developmental noise. *Evolution* **48**, 1478–1486.

Gebhardt-Heinrich, S.G. and van Noordwijk, A. 1991 The genetical ecology of nestling growth in the great tit. I. Heritability estimates under different environmental conditions. *J. evol. Biol.* **3**, 341–362.

Gershwin, M.E., Beach, R.S. and Hurley, L. S. 1985 *Nutrition and Immunity.* Academic Press, Orlando.

Gest, T.R., Siegel, M.I. and Anistranski, J. 1983 Increased fluctuating asymmetry in the long bones of neonatal rats stressed in cold, heat, and noise. *Am. J. Phys. Anthropol.* **60**, 196–197.

Gest, T.R., Siegel, M.I. and Anistranski, J. 1986 The long bones of neonatal rats stressed by cold, heat, and noise exhibit increased fluctuating asymmetry. *Growth* **50**, 385–389.

Gibson, J.B. and Bradley, B.P. 1974 Stabilizing selection in constant and fluctuating environments. *Heredity* **33**, 293–302.

Gilbert, S.F. 1991 *Developmental Biology.* Sinauer, Sunderland.

Gileva, É.A. and Kosareva, N.L. 1994 Decrease in fluctuating asymmetry among house mice in territories polluted with chemical and radioactive mutagens. *Russian J. Ecol.* **25**, 225–228.

Gilliard, E.T. 1969 *Birds of Paradise and Bowerbirds.* Natural History Press, Garden City.

Giurfa, M., Eichmann, B. and Menzel, R. 1996 Symmetry perception in an insect. *Nature* **382**, 458–461.

Glass, L. (Ed.) 1991 Nonlinear dynamics of physiological function and control. *Chaos* **1**, 1–132.

Goater, C.P., Raymond, R.D. and Bernasconi, M.V. 1993 Effects of body size and parasite infection on the locomotory performance of juvenile toads, *Bufo bufo.* *Oikos* **66**, 129–136.

Goldberger, A.L., Rigney, D.R. and West, B.J. 1990 Chaos and fractals in human physiology. *Sci. Am.* **262**, 43–49.

Goldschmidt, R.B. 1920 Die quantitativen Grundlagen von Vererbung und Artbildung. *Roux's Vorträge und Aufsätze* **24**, 1–163.

Goldschmidt, R.B. 1940 *The Material Basis of Evolution.* Yale University Press, New Haven.

Goldschmidt, R.B. 1955 *Theoretical Genetics.* Cambridge University Press, Berkeley.

Goodwin, B. 1971 A model of early amphibian development. In *British Society of Experimental Biology Symposium 25.* D. D. Davies and M. Balls (Eds.), pp. 417–425. Cambridge University Press, Cambridge.

Gould, S.J. and Eldredge, N. 1993 Punctuated equilibria: tempo and mode of evolution reconsidered. *Paleobiology* **3**, 115–151.

Gould, S.J. and Eldredge, N. 1977 Punctuated equilibrium comes of age. *Nature* **366**, 223–227.

Govind, C.K. and Pearce, J. 1986 Differential reflex activity determines claw and closer muscle asymmetry in developing lobsters. *Science* **233**, 354–356.

Govind, C.K. and Pearce, J. 1992 Mechanoreceptors and minimal reflex activity determining claw laterality in developing lobsters. *J. exp. Biol.* **171**, 149–162.

Govindaraju, D.R. and Dancik, B.P. 1987 Allozyme heterozyosity and homeostasis in germinating seeds of jack pine. *Heredity* **59**, 279–283.

Grafen, A. 1990 Biological signals as handicaps. *J. theor. Biol.* **144**, 517–546.

Graham, J.H. 1992 Genomic coadaptation and developmental stability in hybrid zones. *Acta Zool. Fenn.* **191**, 121–131.

Graham, J.H. and Felley, J.D. 1985 Genomic coadaptation and developmental stability within introgressed populations of *Enneacanthus gloriosus* and *E. obesus* (Pisces: Centrarchidae). *Evolution* **39**, 104–114.

Graham, J.H., Freeman, D.C. and Emlen, J.M. 1993a Developmental stability: a sensitive indicator of populations under stress. In *Environmental Toxicology and Risk Assessment*, ASTM STP vol. 1179. W.G. Landis, J.S. Hughes and M.A. Lewis (Eds.), pp. 136–158. American Society for Testing and Materials, Philadelphia.

Graham, J.H., Roe, K.E. and West, T.B. 1993b Effects of lead and benzene on the developmental stability of *Drosophila melanogaster*. *Ecotoxicology* 2, 185–195.

Graham, J.H., Freeman, D.C. and Emlen, J.M. 1993c Antisymmetry, directional asymmetry, and chaotic morphogenesis. *Genetica* **89**, 121–137.

Grammer, K. and Thornhill, R. 1994 Human (*Homo sapiens*) facial attractiveness and sexual selection: the role of symmetry and averageness. *J. Comp. Psychol.* **108**, 233–242.

Graubard, M.A. 1932 Inversion in *Drosophila melanogaster*. *Genetics* **17**, 81–105.

Greene, D.L. 1984 Fluctuating dental asymmetry and measurement error. *Am. J. Phys. Anthropol.* **65**, 283–289.

Grell, R.F. 1978a A comparison of heat and interchromosomal effects on recombination and interference in *Drosophila melanogaster*. *Genetics* **89**, 65–77.

Grell, R.F. 1978b High frequency recombination in centromeric and histone regions of *Drosophila* genomes. *Nature* **272**, 78–80.

Griffing, B. and Langridge, J. 1963 Factors affecting crossing over in the tomato. *Aust. J. Biol. Sci.* **16**, 826–837.

Gualtieri C.T., Adams, A., Shen, C.D. and Loiselle, D. 1982 Minor physical anomalies in alcoholic and schizophrenic adults and hyperactive and autistic children. *Am. J. Psychiatr.* **139**, 640–643.

Guilford, T. and Dawkins, M.S. 1995 What are conventional signals? *Anim. Behav.* **49**, 1689–1695.

Gustafsson, L., Møller, A.P. and Merilä, J. 1997 Genotype-by-environment interactions and environmental gradients in the barn swallow and the collared flycatcher. I. Morphological traits. *J. evol. Biol.* (in press).

Guthrie, R.D. 1965 Variability in characters undergoing rapid evolution: an analysis of *Microtus* molars. *Evolution* **19**, 214–233.

Gvozdev, V.A. 1986 Mobile genetic elements in *Drosophila melanogaster*: a study of distribution and saltatory transpositions coupled with fitness changes. *Sov. Sci. Rev. D. Physiochem. Biol.* **6**, 107–138.

Hademenos, G.J., Massoud, T., Valentino, D.J., Duckwiler, G. and Ninuela, F. 1994 A nonlinear mathematical model for the development and rupture of intracranial saccular aneurysms. *Neurol. Res.* **16**, 376–384.

Hagen, D.W. 1973 Inheritance of numbers of lateral plates and gillrakers in *Gasterosteus aculeatus*. *Heredity* **30**, 303–312.

Hall, B.G. 1988 Adaptive evolution that requires multiple spontaneous mutations. I. Mutations involving an insertion sequence. *Genetics* **120**, 887–897.

Hall, B.G. 1990 Spontaneous point mutations that occur more often when advantageous than when neutral. *Genetics* **126**, 5–16.

Hall, B.G. 1991 Adaptive evolution that requires multiple spontaneous mutations: mutations involving base substitutions. *Proc. Natl. Acad. Sci. USA* **88**, 5882–5886.

Hallgrímsson, B. 1993 Fluctuating asymmetry in *Macaca fascicularis*: a study of the etiology of developmental noise. *Int. J. Primatol.* **14**, 421–443.

Hamilton, W.D. 1982 Pathogens as causes of genetic diversity in their host populations. In *Population Biology of Infectious Disease Agents*. R.M. Anderson and R.M. May (Eds.), pp. 269–296. Verlag-Chemie, Weinheim.

Hamilton, W.D. 1986 Instability and cycling of two competing hosts with two parasites. In *Evolutionary Processes and Theory*. S. Karlin and A. Nevo (Eds.), pp. 645–668. Academic Press, New York.

Hanawalt, P.C. 1987 Preferential repair of damage in actively transcribed DNA sequences in vivo. *Genome* **31**, 605–611.

Handford, P. 1980 Heterozygosity at enzyme loci and morphological variation. *Nature* **286**, 261–262.

Hardin, R.T. and Bell, A.E. 1967 Two-way selection for body weight in *Tribolium* on two levels of nutrition. *Genet. Res., Camb.* **9**, 309–330.

Harrison, R.G. and Boer, C.H. 1977 *Sex and Infertility*. Academic Press, London.

Hartl, G.B., Lang, G., Klein, F. and Willing, R. 1991 Relationship between allozymes, heterozygosity, and morphological characters in red deer (*Cervus elaphus*) and the influence of selective hunting on allele frequency distributions. *Heredity* **66**, 343–350.

Hartl, G.B., Suchentrunk, F., Willing, R. and Petznek, R. 1993 Allozyme heterozygosity and fluctuating asymmetry in the brown hare (*Lepus europaeus*): a test of the developmental homeostasis hypothesis. *Phil. Trans. R. Soc. Lond. B* **350**, 313–323.

Harvey, G.A. and Semlitsch, R.D. 1988 Effects of temperature on growth, development, and color polymorphism in the ornate chorus frog, *Pseudacris ornata. Copeia* **1988**, 1001–1007.

Harvey, S., Phillips, J.G., Rees, A. and Hall, T.R. 1984 Stress and adrenal function. *J. exp. Zool.* **232**, 633–645.

Hasegawa, M. 1995 Sexual selection among peafowl: fluctuating asymmetry and parasite resistance. *Bull. Ass. Nat. Sci. Senshu Univ.* **26**, 19–25.

Hasson, O. 1994 Cheating signals. *J. theor. Biol.* **167**, 223–238.

Hastings, H.M. and Sugihara, G. 1993 *Fractals: A User's Guide for the Natural Sciences*. Oxford University Press, Oxford.

Hawkins, A.J.S., Bayne, B.L., Day, A.J., Rusing, J. and Worrall, C.M. 1989 Genotype-dependent interrelations between energy metabolism, protein metabolism and fitness. In *Reproduction, Genetics and Distributions of Marine Organisms*. J. S. Ryland and P.A. Taylor (Eds.), pp. 283–292. Olsen and Olsen, Fredensborg.

Hayman, D.L. and Parsons, P.A. 1960 The effect of temperature, age and an inversion on recombination values and interference in the X-chromosome of *Drosophila melanogaster. Genetica* **32**, 74–88.

Heiskanen, A., Jokela, P., Laitinen, M., Savontaus, M.-L. and Portin, P. 1984 Effect of temperature on the intraindividual variation of sternopleural bristles in *D. melanogaster. Drosophila Inf. Serv.* **60**, 123–124.

Held, L.I., Jr. 1992 Models of embryonic periodicity. *Monogr. Develop. Biol.* **24.**

Henderson, J., Baker, H.W.G. and Hannah, P.J. 1986 Occupationally related male infertility: a review. *Clin. Reprod. Fertil.* **4**, 87–106.

Hershkovitz, I., Livshits, G., Moskona, D., Arensburg, B. and Kobyliansky, E. 1993 Variables affecting dental fluctuating asymmetry in human isolates. *Am. J. Phys. Anthropol.* **91**, 349–365.

Herzog, A., Volmer, A. and Döll, G. 1983 Die sogenannte 'Haarseuche' beim Reh (*Capreolus capreolus* L.), eine Haarparakeratose. *Berl. Münch. Tierärztl. Wschr.* **96**, 17–23.

Hewitt, J.K., Hahn, M.E. and Karkowski, L.M. 1987 Genetic selection disrupts stability of mouse brain weight development. *Brain Res.* **417**, 225–231.

Hilbish, T.J. and Koehn, R.K. 1985 The physiological basis of natural selection at the Lap locus. *Evolution* **85**, 1302–1317.

Ho, L., Bohr, V.A. and Hanawalt, P.C. 1989 Demethylation enhances removal of pyrimidine dimers from the overall genome and from specific DNA sequences in Chinese hamster ovary cells. *Molec. Cell. Biol.* **9**, 1594–1603.

Hoffmann, A.A. 1983 Paleobiology at the crossroads: a critique of some paleobiological research programs. In *Dimensions of Darwinism*. M. Grene (Ed.), pp. 241–272. Cambridge University Press, Cambridge.

Hoffmann, A.A. 1995 Acclimation: increasing survival at a cost. *Trends Ecol. Evol.* **10**, 1–2.

Hoffmann, A.A. and Parsons, P.A. 1989 Selection for dessication resistance in *Drosophila melanogaster*: additive genetic control and correlated responses for other stresses. *Genetics* **122**, 837–845.

Hoffmann, A.A. and Parsons, P.A. 1993 Direct and correlated responses to selection for desiccation resistance: a comparison of *Drosophila melanogaster* and *D. simulans*. *J. evol. Biol.* **6**, 643–657.

Holloway, G.J., Povey, S.R. and Sibly, R.M. 1990 The effect of new environment on adapted genetic architecture. *Heredity* **64**, 323–330.

Holt, S.B. 1968 *The Genetics of Dermal Ridges*. Thomas, Springfield.

Horridge, G.A. 1996 The Honeybee (*Apis mellifera*) detects bilateral symmetry and discriminates its axis. *J. Insect Physiol.* **42**, 755–764.

Horridge, G.A. and Zhang, S.W. 1995 Pattern vision in honeybees (*Apis mellifera*): flower-like patterns with no predominant orientation. *J. Insect Physiol.* **41**, 681–688. **42**, 755–764.

Houle, D. 1989 Allozyme-associated heterosis in *Drosophila melanogaster*. *Genetics* **123**, 789–801.

Houle, D. 1992 Comparing evolvability and variability of quantitative traits. *Genetics* **130**, 195–204.

Houle, D. 1994 Adaptive distance and the genetic basis of heterosis. *Evolution* **48**, 1410–1417.

Houle, D. 1997 Comment on 'A meta-analysis of the heritability of developmental stability' by Møller and Thornhill. *J. evol. Biol.* **10**, 17–20.

Hsieh, C.L. and Lieber, M.R. 1992 CpG methylated minichromosmes become inaccessible for V(D)J recombination after undergoing replication. *EMBO J.* **11**, 315–325.

Hubbs, C.L. 1922 Variations in the number of vertebrae and other meristic characters of fishes correlated with the temperature of water during development. *Am. Nat.* **56**, 360–372.

Hubbs, C.L. and Hubbs, L.C. 1945 Bilateral asymmetry and bilateral variation in fishes. *Pap. Mich. Acad. Sci., Arts Lett.* **30**, 229–310.

Hubert, W.A. and Alexander, C.B. 1995 Observer variation in counts of meristic traits affects fluctuating asymmetry. *N. Am. J. Fish. Mgmt.* **15**, 156–158.

Huether, C. A., Jr. 1969 Constancy of the pentamerous corolla phenotype in natural populations of *Linanthus*. *Evolution* **23**, 572–588.

Huntley, H.E. 1970 *The Divine Proportion, a Study in Mathematical Beauty*. Dover, New York.

Huxley, T.H. 1860 The origin of species. *West. Rev.* **17**, 541–570.

Hwu, T.-K. and Thseng, F.-S. 1982 Environmental effects of agronomic characters and their developmental stabilities in rice (*Oryza sativa* L.) *J. Agric. Ass. China* **118**, 48–61.

Hyde, J.E. 1990 *Molecular Parasitology*. Open University Press, Milton Keynes.

Ivanovics, A.M. and Wiebe, W. J. 1981 Toward a working definition of stress: a review and critique. In *Stress Effects on Natural Ecosystems*. G.W. Barrett and R. Rosenberg (Eds.), pp. 13–27. Wiley, New York.

Jablonka, E. and Lamb, M.J. 1995 *Epigenetic Inheritance and Evolution: The Lamarckian Dimension*. Oxford University Press, Oxford.

Jackson, J.F. 1973a A search for the population asymmetry parameter. *Syst. Zool.* **22**, 166–170.

Jackson, J.F. 1973b The phenetics and ecology of a narrow hybrid zone. *Evolution* **27**, 58–68.

Jaenisch, R. 1988 Transgenic animals. *Science* **240**, 1468–1474.

Jagoe, C.H. and Haines, T.A. 1985 Fluctuating asymmetry in fishes inhabiting acidified and unacidified lakes. *Can. J. Zool.* **63**, 130–138.

Jamison, C.S., Meier, R.J. and Campbell, B.C. 1993 Dermatoglyphic asymmetry and testosterone levels in normal males. *Am. J. Phys. Anthropol.* **90**, 185–198.

Jennions, M.D. 1996 The allometry of fluctuating asymmetry in Southern African plants: flowers and leaves. *Biol. J. Linn. Soc.* **59**, 127–142.

Johns, P.E., Baccus, R., Manlove, M.N., Pinder, J.E., III and Smith, J.H. 1977 Reproductive patterns, productivity, and genetic variability in adjacent white-tailed deer populations. *Proc. Ann. Southeast Assoc. Game Fish. Comm.* **31**, 167–172.

Johnson, G.R. and Frey, K.J. 1967 Heritabilities of quantitative attributes of oats (*Avena* sp.) at varying levels of environmental stress. *Crop Sci.* **7**, 43–46.

Johnstone, R.A. 1994 Female preference for symmetrical males as a by-product of selection for mate recognition. *Nature* **372**, 172–175.

Johnstone, R.A. 1995 Artefact or network evolution? *Nature* **374**, 314.

Johnstone, R.A. and Grafen, A. 1992 Error-prone signalling. *Proc. R. Soc. Lond. B* **248**, 229–233.

Jokela, P. and Portin, P. 1991 Effect of extra Y chromosome on number and fluctuating asymmetry of sternopleural bristles in *Drosophila melanogaster*. *Hereditas* **114**, 177–187.

Jones, K.L. 1988 *Smith's Recognizable Patterns of Human Malformation*. 4th edn. Saunders, Philadelphia.

Jones, R.T. 1946 Properties of low aspect ratio pointed wings above and below the speed of sound. NACA TR 835.

Jones, R.T. 1990 *Wing Theory*. Princeton University Press, Princeton.

Joswiak, G.R., Smith, C.R. and Moore, W.S. 1985 Allozyme markers in the lizard hybrids *Sceloporus undulatus × S. woodi*. *Isozyme Bull.* **18**, 81.

Julesz, B. 1971 *Foundations of Cyclopean Perception*. University of Chicago Press, Chicago.

Kacser, H. and Burns, J.A. 1981 The molecular basis of dominance. *Genetics* **97**, 639–666.

Kartavstev, Y.F. 1990 Allozyme heterozygosity and morphological homeostasis in pink salmon *Oncorhynchus gorbuscha* (Pisces: Salmonidae). *Genetika* **8**, 1399–1407.

Kat, P.W. 1982 The relationship between heterozygosity for enzyme loci and developmental homeostasis in peripheral populations of aquatic bivalves (Unionidae). *Am. Nat.* **119**, 824–832.

Kazazian, H.H., Jr. and Scott, A.F. 1993 'Copy and paste' transposable elements in the human genome. *J. Clin. Invest.* **91**, 1859–1860.

Kerkis, J. J. 1940 Physiological changes in a cell as a cause of mutation pressure. *Adv. Curr. Biol.* **12**, 143–159.

Kerkis, J. J. 1975 Some problems of spontaneous and induced mutagenesis in mammals and man. *Mutation Res.* **29**, 271–277.

Kerszberg, M. 1989 Developmental canalization can enhance species survival. *J. theor. Biol.* **139**, 287–309.

Kieser, J.A. 1987 Epigenetic canalization and phenotypic change: a minimax model. *Med. Hypoth.* **22**, 105–110.

Kieser, J.A. 1992 Fluctuating odontometric asymmetry and maternal alcohol consumption. *Ann. Human Biol.* **19**, 513–520.

Kieser, J.A. 1993 Evolution, developmental instability, and the theory of acquisition. *Genetica* **89**, 219–225.

Kieser, J.A. and Groeneveld, H. 1991 Fluctuating odontometric asymmetry, morphological variability, and genetic monomorphism in the cheetah *Acinonyx jubatus*. *Evolution* **45**, 1175–1183.

Kieser, J.A. and Groeneveld, H. 1994 Effects of prenatal exposure to tobacco smoke on developmental stability of children. *J. Craniofacial Genetics Dev. Biol.* **14**, 43–47.

Kilias, G., Alahiotis, S.N. and Onoufriou, A. 1979 The alcohol dehydrogenase locus affects meoitic crossing-over in *Drosophila melanogaster*. *Genetica* **50**, 173–177.

Kindred, B. 1963 Selection for an invariant character, vibrissa number, in the house mouse. IV. Probit analysis. *Genetics* **48**, 621–632.

Kindred, B. 1965 Selection for temperature sensitivity in scute *Drosophila*. *Genetics* **52**, 232–238.

Kindred, B. 1967 Selection for canalisation in mice. *Genetics* **55**, 635–644.

King, D.P.F. 1985 Enzyme heterozygosity associated with anatomical character variance and growth in the herring (*Clupea harengus* L.). *Heredity* **54**, 289–296.

Kirpichnikov, V.S. 1981 *Genetic Bases of Fish Selection*. Springer, Berlin.

Kitchell, J.A. 1990 Biological selectivity of extinction. In *Extinction Events in Earth History*. E.G. Kauffman and O.H. Walliser (Eds.), pp. 31–43. Springer, Berlin.

Klein, J. 1986 *Natural History of the Major Histocompatibility Complex*. Wiley, New York.

Koehn, R.K. 1991 The cost of enzyme synthesis in the genetics of energy balance and physiological performance. *Biol. J. Linn. Soc.* **44**, 231–247.

Koehn, R.K. and Shumway, S.R. 1982 A genetic/physiological explanation for differential growth rate among individuals of the American oyster *Crassostrea virginica* (Gmelin). *Mar. Biol. Lett.* **3**, 35–42.

Koehn, R.K., Milkman, R. and Mitton, J.B. 1976 Population genetics of marine pelecypods. IV. Selection, migration and genetic differentiation in the blue mussel *Mytilus edulis*. *Evolution* **30**, 2–32.

Koji, T., Nakane, P.K., Murakoski, M., Watanabe, K. and Terayama, H. 1988 Cell density dependent morphological changes in adult rat hepatocytes during primary culture. *Cell Biochem. Funct.* **6**, 237–243.

Kondo, S. and Asai, R. 1995 A reaction-diffusion wave on the skin of the marine angelfish *Pomacanthus*. *Nature* **376**, 765–768.

Kovach, J.K. 1993 Sources of behavioral deviation modelled by early color preferences in quail. III. Developmental stability and normative canalization. *Behav. Genet.* **23**, 369–377.

Kowner, R. 1996 Facial asymmetry and attractiveness judgment in developmental perspective. *Human Perception and Performance.* **22**, 662–675.

Kozlov, M.V., Wilsey, B.J., Koricheva, J. and Haukioja, E. 1996 Fluctuating asymmetry of birch leaves increases under pollution impact. *J. Appl. Ecol.* **33**, 1489–1495.

Kozlowski, J. and Stearns, S.C. 1989 Hypotheses for the production of excess zygotes: models of bet-hedging and selective abortion. *Evolution* **43**, 1369–1377.

Krakauer, D.C. and Pagel, M. 1996 Selection by somatic signals: the advertisement of phenotypic state through costly intercellular signals. *Phil. Trans. R. Soc. Lond. B* *351,* 647–658.

Kraus, E.J. 1915 The self-sterility problem. *J. Hered.* **6**, 549–557.

Krebs, R.A. and Loeschcke, V. 1994a Costs and benefits of activation of the heat-shock response in *Drosophila melanogaster. Funct. Ecol.* **8**, 730–737.

Krebs, R.A. and Loeschcke, V. 1994b Effects of exposure to short-term heat stress on fitness components in *Drosophila melanogaster. J. evol. Biol.* **7**, 39–49.

Kushev, V.A. 1971 *Mechanisms of Genetic Recombination.* Nauka, Leningrad.

Kuzin, A.M. 1993 The key mechanisms of hormesis. *Izv. Akad. Nauk SSR Biol.* **6**, 824–832.

Lamb, B.C. 1969 Evidence from *Sordaria* that recombination and conversion frequencies are partly determined before meiosis, and for a general model of the control of recombination frequencies. *Genetics* **63**, 807–820.

Lamb, T., Novak, J.M. and Mahoney, D.L. 1990 Morphological asymmetry and interspecific hybridization: a case study using hylid frogs. *J. evol. Biol.* **3**, 295–309.

Lambert, D.M. and Spencer, H.G. (Eds.) 1995 *Speciation and the Recognition Concept.* Johns Hopkins University Press, Baltimore and London.

Lambrecht, M. and Dhondt, A.A. 1988a Male quality, reproduction, and survival in the great tit (*Parus major*). *Behav. Ecol. Sociobiol.* **19**, 57–64.

Lambrecht, M. and Dhondt, A.A. 1988b The anti-exhaustion hypothesis: a new hypothesis to explain song performance and song switching in the great tit. *Anim. Behav.* **36**, 327–334.

Langlois, J.H. and Roggman, L.A. 1990 Attractive faces are only average. *Psychol. Sci.* **1**, 115–121.

Langlois, J.H., Roggman, L.A. and Musselman, L. 1994 What is average and what is not average about attractive faces. *Psychol. Sci.* **5**, 214–220.

Langridge, J. and Griffing, B. 1959 A study of high temperature lesions in *Arabidopsis thaliana. Aust. J. Biol. Sci.* **12**, 117–135.

Larsson, K. 1993 Inheritance of body size in the barnacle goose under different environmental conditions. *J. evol. Biol.* **6**, 195–208.

Lauder, J.M. 1993 Neurotransmitters as growth regulatory signals: role of receptors and second messengers. *Trends Neurosci.* **16**, 233–240.

Lawrence, M.J. 1963 The control of crossing-over in the X-chromosome of *Drosophila melanogaster. Heredity* **18**, 27–46.

Leamy, L. 1984 Morphometric studies in inbred and hybrid house mice. V. Directional and fluctuating asymmetry. *Am. Nat.* **123**, 579–593.

Leamy, L. 1997 Is developmental stability heritable? *J. evol. Biol.* **10,** 21–29.

Leamy, L. and Atchley, W. 1985 Directional selection and developmental stability: evidence from fluctuating asymmetry of morphometric characters in rats. *Growth* **49**, 8–18.

Leary, R.F. and Allendorf, F.W. 1989 Fluctuating asymmetry as an indicator of stress: implications for conservation biology. *Trends Ecol. Evol.* **4**, 214–217.

Leary, R.F., Allendorf, F.W. and Knudsen, R.L. 1983 Developmental stability and enzyme heterozygosity in rainbow trout. *Nature* **301**, 71–72.

Leary, R.F., Allendorf, F.W. and Knudsen, K.L. 1984 Superior developmental stability of heterozygotes at enzyme loci in salmonid fishes. *Am. Nat.* **124**, 540–551.

Leary, R.F., Allendorf, F.W. and Knudsen, R.L. 1985a Developmental instability as an indicator of reduced genetic variation in hatchery trout. *Trans. Am. Fish. Soc.* **114**, 230–235.

Leary, R.F., Allendorf, F.W. and Knudsen, K.L. 1985b Developmental instability and high meristic counts in interspecific hybrids of salmonid fishes. *Evolution* **39**, 1318–1326.

Leary, R.F., Allendorf, F.W. and Knudsen, R.L. 1985c Inheritance of meristic variation and the evolution of developmental stability in rainbow trout. *Evolution* **39**, 308–314.

Leary, R.F., Allendorf, F.W. and Knudsen, K.L. 1991 Effects of rearing density on meristics and developmental stability of rainbow trout. *Copeia* **1991**, 44–49.

Leary, R.F., Allendorf, F. and Knudsen, K.L. 1992 Genetic, environmental, and developmental causes of meristic variation in rainbow trout. *Acta Zool. Fenn.* **191**, 79–95.

Leary, R.F., Allendorf, F. and Knudsen, K.L. 1993 Null alleles at two lactate dehydrogenase loci in rainbow trout are associated with decreased developmental stability. *Genetica* **89**, 3–13.

Lee, Y.-H. and Vaquier, V.D. 1992 The divergence of species-specific abalone lysins is promoted by positive Darwinian selection. *Biol. Bull.* **182**, 97–104.

Lenski, R.E. and Mittler, J.E. 1993 The directed mutation controversy and neo-Darwinism. *Science* **259**, 188–194.

Lerner, I.M. 1954 *Genetic Homeostasis*. Oliver and Boyd, London.

Leroi, A.M., Bennett, A.F. and Lenski, R.E. 1994 Temperature acclimation and competitive fitness: an experimental test of the beneficial acclimation assumption. *Proc. Natl. Acad. Sci. USA* **91**, 1917–1921.

Leung, B. and Forbes, M.R. 1996 Fluctuating asymmetry in relation to stress and fitness: effects of trait type as revealed by meta-analysis. *Ecoscience* **3**, 400–413.

Leung, B. and Forbes, M.R. 1997a Modelling fluctuating asymmetry in relation to stress and fitness. *Oikos* **78**, 397–405.

Leung, B. and Forbes, M. 1997b Fluctuating asymmetry in relation to indices of quality and fitness in the damselfly, *Enallagma ebrium* (Hagen). *Oecologia.* **110**, 472–477.

Leung, B. and Forbes, M.R. 1997c Composite measures of fluctuating asymmetry: A critical appraisal of the options for increasing power. *Oikos* (in press).

Levin, D.A. 1970a Developmental instability and evolution in peripheral isolates. *Am. Nat.* **104**, 343–353.

Levin, D.A. 1970b Developmental instability in species and hybrids of *Liatris*. *Evolution* **24**, 613–624.

Levitan, D.R. and Petersen, C. 1995 Sperm limitation in the sea. *Trends Ecol. Evol.* **10**, 228–231.

Lewontin, R.C. 1974 *The Genetic Basis of Evolutionary Change*. Columbia University Press, New York.

Lewontin, R.C. 1983 Gene, organism and environment. In *Evolution from Molecules to Men*. D.S. Bendall (Ed.), pp. 273–285. Cambridge University Press, Cambridge.

Lindén, M., Gustafsson, L. and Pärt, T. 1992 Selection on fledgling mass in the collared flycatcher and the great tit. *Ecology* **73**, 336–343.

Lindgren, D. 1972 The temperature influence on the spontaneous mutation rate. *Hereditas* **70**, 165–178.

Lindsey, C.C. and Harrington, R.W., Jr. 1972 Extreme vertebral variation induced by temperature in a homozygous clone of the self-fertilizing cyprinodontid fish *Rivulus marmoratus*. *Can. J. Zool.* **50**, 733–744.

Livshits, G. and Kobyliansky, E. 1984 Biochemical heterozygosity as a predictor of developmental homeostasis in man. *Ann. Human Genet.* **48**, 173–184.

Livshits, G. and Kobyliansky, E. 1985 Lerner's concept of developmental homeostasis and the problem of heterozygosity level in natural populations. *Heredity* **55**, 341–353.

Livshits, G. and Kobyliansky, E. 1989 Study of genetic variance in the fluctuating asymmetry of anthropometrical traits. *Ann. Human Biol.* **116**, 121–129.

Livshits, G. and Kobyliansky, E. 1991 Fluctuating asymmetry as a possible measure of developmental homeostasis in humans: a review. *Human Biol.* **63**, 441–466.

Livshits, G. and Smouse, P.E. 1993a Multivariate bilateral asymmetry in human adults. *Human Biol.* **65**, 547–578.

Livshits, G. and Smouse, P.E. 1993b Relationship between fluctuating asymmetry, morphological modality and heterozygosity in an elderly Israeli population. *Genetica* **89**, 155–166.

Livshits, G., Otremski, I. and Kobyliansky, E. 1987 Longitudinal growth of infants in families of 'mixed' and 'non-mixed' ethnic origin in Israel. *Human. Biol.* **59**, 933–949.

Lobashov, M.E. 1976 Physiological hypothesis of mutagenesis. *Genet. Res., Camb.* **6**, 3–5.

Locke, J., Kotarski, M.A. and Tartof, K.D. 1988 Dosage-dependent modifiers of position effect variegation in *Drosophila* and a mass action model that explains their effect. *Genetics* **120**, 181–198.

Lockwood, R., Swaddle, J.P. and Rayner, J.M.V. 1997 Avian wingtip shape reconsidered: wingtip shape indices and morphological adaptations to migration. *J. Avian Biol.* (in press)

Loidl, J. 1969 Effects of elevated temperature on meiotic chromosome synapsis in *Allium ursinum. Chromosoma* **97**, 449–458.

Loveless, A. 1966 *Genetic and Allied Effects of Alkylating Agents.* Butterworths, London.

Lu, B.C. 1969 Genetic recombination in *Coprinus.* I. Its precise timing as revealed by temperature treatment experiments. *Can. J. Genet. Cytol.* **11**, 834–847.

Lu, B.C. 1974 Genetic recombination in *Coprinus.* IV. A kinetic study of the temperature effect on recombination frequency. *Genetics* **78**, 661–677.

Lubiniecki, A.S., Blattner, W.A. and Fraumeni, J.F., Jr. 1977 Elevated expression of T-antigen in simian papovavirus 40-infected skin fibroblasts from individuals with cytogenetic defects. *Cancer Res.* **37**, 1580–1583.

Lucas, A.M. and Stettenheim, P.R. 1972 *Avian Anatomy: Integument.* Agricultural Handbook 825. Superintendent of Documents, Washington.

Lucchesi, J.C. 1976 Interchromosomal effects. In *The Genetics and Biology of Drosophila.* Vol. 1A. M. Ashburner and E. Novitski (Ed.), pp. 315–329. Academic Press, London.

Luckey, T. 1980 *Hormesis with Ionizing Radiation.* CRC Press, Boca Raton.

Ludwig, W. 1932 *Das Rechts-Links Problem im Tierreich und beim Menschen.* Springer, Berlin.

Lundström, A. 1960 Asymmetries in the number and size of the teeth and their aetiological significance. *Trans. Europ. Orthodont. Soc.* **36**, 167–185.

Mackay, T.F.C. 1980 Genetic variance, fitness, and homeostasis in varying environments: an experimental check of the theory. *Evolution* **34**, 1219–1222.

Mahsberg, D. 1997 Brood care and sociality. In *Scorpions and Research.* P.H. Brownell and G.A. Polis (Eds.), (in press) Oxford University Press, Oxford.

Makaveev, T., Venev, I. and Baulov, M. 1978 Investigations on activity level and polymorphism of some blood enzymes in farm animals with different growth energy. II Correlations between homo- and heterozygosity of some protein and enzyme phenotypes and fattening ability and slaughter indices in various breeds of fattened pigs. *Genetic Selection* **10**, 229–236.

Malina, R.M. and Buschang, P.H. 1984 Anthropometric asymmetry in normal and mentally retarded males. *Ann. Human Biol.* **11**, 515–531.

Malyon, C. and Healy, S. 1994 Fluctuating asymmetry in antlers of fallow deer, *Dama dama*, indicates dominance. *Anim. Behav.* **48**, 248–250.

Mandelbrot, B.B. 1977 *Fractals: Form, Chance and Dimension.* W.H. Freeman, San Francisco

Mandelbrot, B.B. 1982 *The Fractal Geometry of Nature.* W.H. Freeman, New York.

Manning, J.T. 1995 Fluctuating asymmetry and body weight in men and women: implications for sexual selection, *L. Ethol. Sociobiol.* **16**, 145–153.

Manning, J.T. and Chamberlain, A.T. 1993 Fluctuating asymmetry in gorilla canines: a sensitive indicator of environmental stress. *Proc. R. Soc. Lond. B* **255**, 189–193.

Manning, J.T. and Chamberlain, A.T. 1994 Fluctuating asymmetry, sexual selection and canine teeth in primates. *Proc. R. Soc. Lond. B* **251**, 83–87.

Manning, J.T. and Ockenden, L. 1994 Fluctuating asymmetry in racehorses. *Nature* **370**, 185–186.

Manning, J.T. and Wood, D. 1997 Fluctuating asymmetry and aggression in boys. MS.

Manning, J.T., Scutt, D., Whitehouse, G.H. and Leinster, S.J. 1996 Breast asymmetry and phenotypic quality in women. *Ethol. Sociobiol.* (in press).

Manning, J.T., Koukourakis, K., Brodie, D.A. 1997b Fluctuating asymmetry, metabolic rate and sexual selection in human males. *Ethol. Solciobiol.* **18**, 15–21.

Markel, A.L. and Borodin, P.M. 1980 The stress phenomenon: genetic and evolutionary aspects. In *Problems of General Genetics.* Vol. 2.2, pp. 51–65. Mir, Moscow.

Markow, T. 1992 Human handedness and the concept of developmental stability. *Genetica* **87**, 87–94.

Markow, T.A. 1992 Genetics and developmental stability: an intergrative conjecture on etiology and neurobiology of schizophrenia. *Psychol. Med.* **22**, 295–305.

Markow, T.A. and Clarke, G.M. 1997 Meta-analysis of the heritability of developmental stability: a giant step backward. *J. evol. Biol.* **10**, 31–37.

Markow, T.A. and Gottesman, I.I. 1989 Dermatoglyphic fluctuating asymmetry in psychotic twins. *Psychiatr. Res.* **29**, 37–43.

Markow, T.A. and Gottesman, I.I. 1993 Behavioral phenodeviance: 7a Lerneresque approach. *Genetica* **89**, 297–305.

Markow, T.A. and Wandler, K. 1986 Fluctuating dermatoglyphic asymmetry and the genetics of liability to schizophrenia. *Psychiatr. Res.* **19**, 323–328.

Markowski, J. 1993 Fluctuating asymmetry as an indicator for differentiation among roe deer *Capreolus capreolus* populations. *Acta Theoriol.* **38**(Suppl. 2), 19–31.

Mason, L.G., Ehrlich, P.R. and Emmel, T.C. 1976 The population biology of the butterfly, *Euphydryas editha*. V. Character clusters and asymmetry. *Evolution* **21**, 85–91.

Mason, S.F. 1991 Origins of the handedness of biological molecules. In *Biological Asymmetry and Handedness.* G.R. Bock and J. Marsh (Eds.), pp. 3–15. Wiley, Chichester.

Mather, K. 1953 Genetic control of stability in development. *Heredity* **7**, 297–336.

Mather, K. 1973 *Genetical Structure of Populations.* Chapman and Hall, London.

Maynard Smith, J. 1960 Continuous, quantized and modal variation. *Proc. R. Soc. Lond. B* **152**, 397–409.

Maynard Smith, J. 1975 *The Theory of Evolution*. Penguin, Harmondsworth.

Maynard Smith, J. 1991 Theories of sexual selection. *Trends Ecol. Evol.* **6**, 146–151.

Maynard Smith, J. and Harper, D. 1988 The evolution of aggression: can selection generate variability? *Phil. Trans. R. Soc. Lond. B* **319**, 557–570.

Maynard Smith, J. and Sondhi, K.C. 1960 The genetics of a pattern. *Genetics* **45**, 1039–1050.

Maynard Smith, J. and Szathmáry, E. 1995 *The Major Transitions in Evolution*. W. H. Freeman, San Francisco.

Mayr, E. 1942 *Systematics and the Origin of Species*. Columbia University Press, New York.

Mayr, E. 1963 *Animal Species and Evolution*. Harvard University Press, Cambridge, Mass.

Mayr, E. 1982 *The Growth of Biological Thought*. Belknap Press, Cambridge, Mass.

McAndrew, B.J., Ward, R.D. and Beardmore, J.A. 1986 Growth rate and heterozygosity in the plaice, *Pleuronectes platessa*. *Heredity* **57**, 171–180.

McClintock, B. 1984 The significance of responses of the genome to challenge. *Science* **226**, 792–801.

McDonald, J.F. 1995 Transposable elements: possible catalysts of organismic evolution. *Trends Ecol. Evol.* **10**, 123–126.

McKenzie, J.A. and Batterham, P. 1994 The genetic, molecular and phenotypic consequences of selection for insecticide resistance. *Trends Ecol. Evol.* **9**, 166–169.

McKenzie, J.A. and Clarke, G.M. 1988 Diazinon resistance, fluctuating asymmetry and fitness in the Australian sheep blowfly. *Genetics* **120**, 213–220.

McKenzie, J.A. and McKechnie, S.W. 1978 Ethanol tolerance and the ADH polymorphism in a natural population of *Drosophila melanogaster*. *Nature* **272**, 75–76.

McKenzie, J.A. and O'Farrell, K. 1993 Modification of developmental instability and fitness: malathion-resistance in the Australian sheep blowfly, *Lucilia cuprina*. *Genetica* **89**, 67–76.

McKenzie, J.A. and Parsons, P. A. 1974 Microdifferentiation in a natural population of *Drosophila melanogaster* to alcohol in the environment. *Genetics* **77**, 385–394.

McKenzie, J.A. and Yen, J.L. 1995 Genotype, environment and the asymmetry phenotype. Dieldrin-resistance in *Lucilia cuprina* (the Australian sheep blowfly). *Heredity* **75**, 181–187.

McKenzie, J.A., Batterham, P. and Baker, L. 1990 Fitness and asymmetry modification as an evolutionary process. A study in the Autralian sheep blowfly, *Lucilia cuprina* and *Drosophila melanogaster*. In *Ecological and Evolutionary Genetics of Drosophila*, J.S.F. Barker, W.T. Starmer and R.J. MacIntyre (Eds.), pp. 57–73. Plenum, New York.

McManus, I.C. 1991 The inheritance of left-handedness. In *Biological Asymmetry and Handedness*. G.R. Bock and J. Marsh (Eds.), pp. 251–281. Wiley, Chichester.

McManus, I.C. 1992 Are paw preference differences in Hi and Low mice the result of specific genes or of heterosis and fluctuating asymmetry. *Behav. Genet.* **22**, 435–451.

McNab, B.K. 1973 Energetics and the distribution of vampires. *J. Mammal.* **54**, 131–144.

McNary, H.W. and Bell, A.E. 1962 The effect of environment on response to selection for body weight in *Tribolium castaneum*. *Genetics* **47**, 969–970.

McNelly-Ingle, C.A., Lamb, B. and Frost, L.C. 1966 The effect of temperature on recombination frequency in *Neurospora crassa*. *Genet. Res., Camb.* **7**, 169–183.

Meinhardt, H. 1982 *Models of Biological Pattern Formation*. Academic Press, New York.

Menne, M. and Curio, E. 1978 Untersuchungen zum Symmetriekonzept bei Kohlmeisen (*Parus major*). *Z. Tierpsychol.* **47**, 299–322.

Merilä, J. 1996 *Genetic and Quantitative Trait Variation in Natural Bird Populations*. PhD thesis, Uppsala University, Uppsala.

Merilä, J. and Björklund, M. 1995 Fluctuating asymmetry and measurement error. *Syst. Biol.* **44**, 97–101.

Michod, R.E. and Levin, B.R. (Eds.) 1988 *The Evolution of Sex*. Sinauer, Sunderland.

Midgley, J.J., and Johnson, S.D. 1997 Some pollinators do not prefer symmetrically marked or shaped daisy (Asteraceae) flowers. *Evol. Ecol.* (in press).

Mitton, J.B. 1978 Relationship between heterozygosity for enzyme loci and variation of morphological characters in natural populations. *Nature* **273**, 661–662.

Mitton, J.B. 1993 Enzyme heterozygosity, metabolism, and developmental stability. *Genetica* **89**, 47–65.

Mitton, J.B. 1995 Enzyme heterozygosity and developmental stability. *Acta Theriol., Suppl.* **3**, 33–54.

Mitton, J.B. and Grant, M.C. 1984 Associations among protein heterozygosity, growth rate, and developmental homeostasis. *Ann. Rev. Ecol. Syst.* **15**, 479–499.

Mitton, J.B. and Koehn, R.K. 1975 Genetic organization and adaptive response of allozymes to ecological variables in *Fundulus heteroclitus*. *Genetics* **79**, 97–111.

Mitton, J.B. and Koehn, R.K. 1985 Shell shape variation in the blue mussel, *Mytilus edulis* L., and its association with enzyme heterozygosity. *J. Exp. Mar. Biol. Ecol.* **90**, 73–80.

Mitton, J.B., Zelenka, D.J. and Carter, P.A. 1994 Selection of breeding stock in pigs favours 6PGD heterozygotes. *Heredity* **73**, 177–184.

Møller, A.P. 1990a Fluctuating asymmetry in male sexual ornaments may reliably reveal male quality. *Anim. Behav.* **40**, 1185–1187.

Møller, A.P. 1990b Effects of parasitism by the haematophagous mite *Ornithonyssus bursa* on reproduction in the barn swallow *Hirundo rustica*. *Ecology* **71**, 2345–2357.

Møller, A.P. 1991 Sexual ornament size and the cost of fluctuating asymmetry. *Proc. R. Soc. Lond.* B **243**, 59–62.

Møller, A.P. 1992a Patterns of fluctuating asymmetry in weapons: evidence for reliable signalling of quality in beetle horns and bird spurs. *Proc. R. Soc. Lond.* B **248**, 199–206.

Møller, A.P. 1992b Female swallow preference for symmetrical male sexual ornaments. *Nature* **357**, 238–240.

Møller, A.P. 1992c Parasites differentially increase the degree of fluctuating asymmetry in secondary sexual characters. *J. evol. Biol.* **5**, 691–699.

Møller, A.P. 1992d Fluctuating asymmetry and the evolution of signals. In *Workshop on Behavioural Mechanisms in Evolutionary Perspective*. P. Bateson and M. Gomendio (Eds.), pp. 96–98. Instituto Juan March de Estudios e Investigaciones, Madrid.

Møller, A.P. 1993a Fluctuating asymmetry. *Nature* **363**, 217.

Møller, A.P. 1993b Morphology and sexual selection in the barn swallow *Hirundo rustica* in Chernobyl, Ukraine. *Proc. R. Soc. Lond.* B **252**: 51–57.

Møller, A.P. 1993c Patterns of fluctuating asymmetry in sexual ornaments predict female choice. *J. evol. Biol.* **6**, 481–491.

Møller, A.P. 1993d Developmental stability, sexual selection and speciation. *J. evol. Biol.* **6**, 493–509.

Møller, A.P. 1993e Female preference for apparently symmetrical male sexual ornaments in the barn swallow *Hirundo rustica*. *Behav. Ecol. Sociobiol.* **32**, 371–376.

Møller, A.P. 1994a Directional selection on directional asymmetry: testes size and secondary sexual characters in birds. *Proc. R. Soc. Lond. B* **258**, 147–151.

Møller, A.P. 1994b *Sexual Selection and the Barn Swallow*. Oxford University Press, Oxford.

Møller, A.P. 1994c Sexual selection in the barn swallow (*Hirundo rustica*). IV. Patterns of fluctuating asymmetry and selection against asymmetry. *Evolution* **48**, 658–670.

Møller, A.P. 1994d Symmetrical male sexual ornaments, paternal care, and offspring quality. *Behav. Ecol.* **5**, 188–194.

Møller, A.P. 1995a Leaf mining insects and developmental stability in leaves of broad-leaved elm (*Ulmus glabra* L.) *J. Anim. Ecol.* **64**, 697–707.

Møller, A.P. 1995b Bumblebee preference for symmetrical flowers. *Proc. Natl. Acad. Sci. USA* **92**, 2288–2292.

Møller, A.P. 1995c Sexual selection in the barn swallow (*Hirundo rustica*). V. Geographic variation in ornament size. *J. evol. Biol.* **8**, 3–19.

Møller, A.P. 1995d Developmental stability and ideal despotic distribution of blackbirds in a patchy environment. *Oikos* **72**, 228–232.

Møller, A.P. 1995e Patterns of fluctuating asymmetry in sexual ornaments of birds from marginal and central populations. *Am. Nat.* **145**, 316–327.

Møller, A.P. 1996a Sexual selection, viability selection, and developmental stability in the domestic fly *Musca domestica*. *Evolution* **50**, 746–752.

Møller, A.P. 1996b Development of fluctuating asymmetry in tail feathers of the barn swallow *Hirundo rustica*. *J. evol. Biol.* **9**, 677–694.

Møller, A.P. 1996c Floral asymmetry, embryo abortion, and developmental selection in plants. *Proc. R. Soc. Lond. B* **263**, 53–56.

Møller, A.P. 1996d Parasitism and developmental instability of hosts: a review. *Oikos* **77**, 189–196

Møller, A.P. 1997a Developmental stability and fitness: a review. *Am. Nat.* **149**, 916–932.

Møller, A.P. 1997b Female preference for symmetrical calls in a grasshopper. MS.

Møller, A.P. 1997c Developmental stability and reliable signalling in plants: development of asymmetry in elm *Ulmus glabra* under different environmental conditions. MS.

Møller, A.P. 1997d Developmental selection against developmentally unstable offspring and sexual selection. *J. theor. Biol.* **185**, 415–422.

Møller, A.P. and Eriksson, M. 1994 Patterns of fluctuating asymmetry in flowers: implications of honest signalling for pollinators. *J. evol. Biol.* **7**, 97–113.

Møller, A.P. and Eriksson, M. 1995 Pollinator preference for symmetrical flowers and sexual selection in plants. *Oikos* **73**, 15–22.

Møller, A.P. and Höglund, J. 1991 Patterns of fluctuating asymmetry in avian feather ornaments: implications for models for sexual selection. *Proc. R. Soc. Lond. B* **245**, 1–5.

Møller, A.P. and Moreno, E. 1997 Muscular adaptations to tail length and asymmetry in the barn swallow *Hirundo rustica*. MS.

Møller, A.P. and Nielsen, J.T. 1997 Differential predation cost of a secondary sexual character: sparrowhawk predation on barn swallows. *Anim. Behav.* (in press).

Møller, A.P. and Pagel, M. 1997 Developmental stability and signalling among cells. MS.

Møller, A.P. and Pomiankowksi, A. 1993a Fluctuating asymmetry and sexual selection. *Genetica* **89**, 267–279.

Møller, A.P. and Pomiankowksi, A. 1993b Punctuated equilibria or gradual evolution: fluctuating asymmetry and variation in the rate of evolution. *J. theor. Biol.* **161**, 359–367.

Møller, A.P. and Pomiankowksi, A. 1993c Why have birds got multiple sexual ornaments? *Behav. Ecol. Sociobiol.* **32**, 167–176.

Møller, A.P. and Sorci G. 1997 Insect preference for symmetrical flower models. MS.

Møller, A.P. and Thornhill, R. 1997a A meta-analysis of the heritability of developmental stability. *J. evol. Biol.* **10**, 1–16.

Møller, A.P. and Thornhill, R. 1997b Developmental instability is heritable. *J. evol. Biol.* **10**, 69–76.

Møller, A.P. and Thornhill, R. 1997c Developmental stability and sexual selection: a meta-analysis. *Am. Nat.* (submitted).

Møller, A.P., Magnhagen, K., Ulfstrand, A. and Ulfstrand, S. 1995a Phenotypic quality and moult in the barn swallow *Hirundo rustica. Behav. Ecol.* **6**, 242–249.

Møller, A.P., Sanotra, G.S. and Vestergaard, K.S. 1995b Developmental stability in relation to population density and breed of chickens *Gallus gallus. Poultry Sci.* **74**, 1761–1771.

Møller, A.P., Soler, M. and Thornhill, R. 1995c Breast asymmetry, sexual selection, and human reproductive success. *Ethol. Sociobiol.* **16**, 207–219.

Møller, A.P., Cuervo, J.J., Soler, J.J. and Zamora-Muñoz, C. 1996 Horn asymmetry and fitness in gemsbok *Oryx g. gazella. Behav. Ecol.* **7**, 247–253.

Møller, A.P., Sanotra, G.S. and Vestergaard, K.S. 1997 Developmental instability and light regime in chickens *Gallus gallus.* MS.

Molnar, V. and Molnar, F. 1986 Symmetry-making and -breaking in visual art. *Comp. Math. Appls.* **12B**, 291–301.

Montag, B.A. and Montag, T.W. 1979 Infanticide: a historical perspective. *Minnesota Med.* May, 368–372.

Moodie, G.E.E. and Moodie, P.F. 1996 Do asymmetric sticklebacks make better fathers? *Proc. R. Soc. Lond. B* **263**, 535–539.

Moodie, G.E.E. and Reimchen, T.E. 1976 Phenetic variation and habitat differences in *Gasterosteus* populations of the Queen Charlotte Islands. *Syst. Zool.* **25**, 49–61.

Mooney, M.P., Siegel, M.I. and Gest, T.R. 1985 Prenatal stress and increased fluctuating asymmetry in the parietal bones of neonatal rats. *Am. J. Phys. Anthropol.* **68**, 131–134.

Moreno, G. 1994 Genetic architecture, genetic behavior, and character evolution. *Ann. Rev. Ecol. Syst.* **25**, 31–44.

Morgan, M.J. 1991 The asymmetrical genetic determination of laterality: flatfish, frogs and human handedness. In *Biological Asymmetry and Handedness.* G.R.Bock and J. Marsh (Eds.), pp. 234–250. Wiley, Chichester.

Moss, M.L. and Young, R.W. 1960 A functional approach to craniology. *Am. J. Phys. Anthrop.* **18**, 281–292.

Mukai, T. 1964 Polygenic mutation affecting quantitative character of *Drosophila melanogaster. Proc. Gamma Field Symp.* **3**, 13–29.

Munn, C. A. 1986 Birds that cry 'wolf'. *Nature* **319**, 143–145.

Murphy, P.A., Giesel, J.T. and Manlove, M.N. 1983 Temperature effects on life-history variation in *Drosophila simulans. Evolution* **37**, 1181–1192.

Murray, J.D. 1990 A pre-pattern formation mechanism for animal coat markings. *J. theor. Biol.* **88**, 161–199.

Nagorcka, B.N. 1989 Wavelike isomorphic prepatterns in development. *J. theor. Biol.* **137**, 127–162.

Nahon, E., Atzmony, D., Zahavis, A. and Granot, D. 1995 Mate selection in yeast: a reconsideration of the signals and the messages encoded by them. *J. theor. Biol.* **172**, 315–322.

256 · *References*

Naugler, C.T. and Leech, S.M. 1994 Fluctuating asymmetry and survival ability in the forest tent caterpillar moth *Malacosoma disstria*: implications for pest management. *Entomol. exp. appl.* **70**, 295–298.

Neel, J.V. 1941 A relation between larval nutrition and the frequency of crossing over in the third chromosome of *Drosophila melanogaster*. *Genetics* **26**, 506–516.

Nei, M. 1987 *Molecular Evolutionary Genetics*. Columbia University Press, New York.

Neville, A.C. 1976 *Animal Asymmetry*. Edward Arnold, London.

Nevo, E., Kim, Y.J., Shaw, C.R. and Thaeler Jr., C.S. 1974 Genetic variation, selection and speciation in *Thomomys talpoides* pocket gophers. *Evolution* **28**, 1–23.

Newell-Morris, L.L., Fahrenbruch, C.E. and Sackett, G.P. 1989 Prenatal psychological stress, dermatoglyphic asymmetry and pregnancy outcome in the pigtailed macaque (*Macaca nemestrina*). *Biol. Neonate* **56**, 61–75.

Newton, I. (Ed.) 1989 *Lifetime Reproduction in Birds*. Academic Press, London.

Nijhout, H.F. 1991 *The Development and Evolution of Butterfly Wing Patterns*. Smithsonian Institution Press, Washington.

Nilsson, J.-Å. 1994 Energetic stress and the degree of fluctuating asymmetry: implications for a long-lasting, honest signal. *Evol. Ecol.* **8**, 248–255.

Norberg, R.Å. 1978 Skull asymmetry, ear structure and function, and auditory localization in Tengmalm's Owl, *Aegolius funereus* (Linné). *Phil. Trans. R. Soc. Lond. B* **282**, 325–410.

Norberg, R.Å. 1994 Swallow tail streamer is a mechanical device for self-deflection of tail leading edge, enhancing aerodynamic efficiency and flight manoeuvrability. *Proc. R. Soc. Lond. B* **257**, 227–233.

Norberg, U.M. 1990 *Vertebrate Flight*. Springer, Berlin.

Novak, J.M., Rhodes, Jr., O.E., Smith, M.H. and Chesser, R.K. 1993 Morphological asymmetry in mammals: genetics and homeostasis reconsidered. *Acta Theriol.* **38**, Suppl. **2**, 7–18.

Numerical Algorithms Group. 1985 *The GLIM System Release 3.77 Manual*. Numerical Algorithms Group Ltd., Oxford.

O'Brien, S.J. 1994 A role for molecular genetics in biological conservation. *Proc. Natl. Acad. Sci. USA* **91**, 5748–5755.

Olsen, L.F. and Degn, H. 1977 Chaos in an enzymatic reaction. *Nature* **267**, 177–178.

Osborne, R. 1954 Sexual maturity in brown leghorns, and the relationship of genetic variance to differences in environment. *Proc. R. Soc. Edinb. B* **65**, 285–298.

Osgood, W. 1978 Effects of temperature on the development of meristic characters in *Natrix fasciata*. *Copeia* **1978**, 33–47.

Osorio, D. 1994 Symmetry versus crypsis. *Trends Ecol. Evol.* **9**, 346.

Osorio, D. 1996 Symmetry detection by categorization of spatial phase, a model. *Proc. R. Soc. Lond. B* **263**, 105–110.

Otronen, M. 1997 Cryptic female choice and asymmetry in the fly *Dryomyza anilis*. MS.

Otte, D. and Endler, J.A. (Eds.) 1988 *Speciation and its Consequences*. Sinauer, Sunderland.

Owen, R.D. and McBee, K. 1990 Analysis of asymmetry and morphometric variation in natural populations of chromosome-damaged mice. *Texas J. Sci.* **42**, 319–332.

Ozernyuk, N.D. 1989 The principle of minimum of energy in ontogenesis and canalization of developmental processes. *Ontogenesis* **20**, 117–127.

Ozernyuk, N.D., Dyomin, V.I., Prokofyev, E.A. and Androsova, I.M. 1992 Energy homeostasis and developmental stability. *Acta Zool. Fenn.* **191**, 167–175.

Packard, G.C. and Boardman, T.J. 1987 The misuse of ratios to scale physiological data that vary allometrically with body size. In *New Directions in Ecological Physiology*. M.E. Feder, A.F. Bennet, W.W. Burgren and R.B. Huey (Eds.), pp. 216–239. Cambridge University Press, Cambridge.

Packer, C. and Pusey, A.E. 1993 Should a lion change its spots? *Nature* **362**, 595.

Pagel, M. 1993 Honest signalling among gametes. *Nature* **363**, 539–541.

Palmer, A.R. 1994 Fluctuating asymmetry analyses: a primer. In *Developmental Instability: Its Origins and Evolutionary Implications*. T.A. Markow (Ed.), pp. 335–364. Kluwer, Dordrecht.

Palmer, A.R. 1996a Waltzing with asymmetry. *BioScience* **46**, 518–532.

Palmer, A.R. 1996b From symmetry to asymmetry: phylogenetic patterns of asymmetry variation in animals and their evolutionary significance. *Proc. Natl. Acad. Sci. USA* **93**, 14279–14286.

Palmer, A.R. and Strobeck, C. 1986 Fluctuating asymmetry: measurement, analysis, patterns. *Ann. Rev. Ecol. Syst.* **17**, 391–421.

Palmer, A.R. and Strobeck, C. 1992 Fluctuating asymmetry as a measure of developmental stability: implications of non-normal distributions and power of statistical tests. *Acta Zool. Fenn.* **191**, 57–72.

Palmer, A.R. and Strobeck, C. 1997 Fluctuating asymmetry and developmental stability: heritability of observable variation vs. heritability of inferred cause. *J. evol. Biol.* **10**, 39–49

Pankakoski, E. 1985 Epigenetic asymmetry as an ecological indicator in muskrats. *J. Mammal.* **66**, 52–57.

Pankakoski, E., Koivisto, I. and Hyvärinen, H. 1992 Reduced developmental stability as an indicator of heavy metal pollution in the common shrew *Sorex araneus*. *Acta Zool. Fenn.* **191**, 137–144.

Parisi, J., Röhricht, B., Peinke, J. and Rössler, O.E. 1987 Turbulent morphogenesis of a prototype model reaction-diffusion system. In *Chaos in Biological Systems*. H. Degn, A.V. Holden and L. F. Olsen (Eds.), pp. 91–95. Plenum Press, New York.

Parker, H.R., Philipp, D.P. and Whitt, G.S. 1985 Gene regulatory divergence among species estimated by altered developmental patterns in interspecific hybrids. *Mol. Biol. Evol.* **2**, 217–250.

Parker, L.T. and Leamy, L. 1991 Fluctuating asymmetry of morphometric characters in house mice: the effects of age, sex, and phenotypical extremeness in a randombred population. *J. Hered.* **82**, 145–150.

Parsons, P.A. 1961 Fly size, emergence time and sternopleural chaeta number in *Drosophila*. *Heredity* **16**, 455–473.

Parsons, P.A. 1962 Maternal age and developmental variability. *J. exp. Biol.* **39**, 251–260.

Parsons, P.A. 1964 Parental age and the offspring. *Q. Rev. Biol.* **39**, 258–275.

Parsons, P.A. 1974 Genetics of resistance to environmental stresses in *Drosophila* populations. *Ann. Rev. Genet.* **7**, 239–265.

Parsons, P.A. 1983 *The Evolutionary Biology of Colonizing Species*. Cambridge University Press, New York.

Parsons, P.A. 1988 Evolutionary rates: effects of stress upon recombination. *Biol. J. Linn. Soc.* **35**, 49–68.

Parsons, P.A. 1990a Fluctuating asymmetry: an epigenetic measure of stress. *Biol. Rev.* **65**, 131–145.

Parsons, P.A. 1990b Fluctuating asymmetry and stress intensity. *Trends Ecol. Evol.* **5**, 97–98.

Parsons, P.A. 1990c Extreme environmental stress: asymmetry, metabolic cost and conservation. In *Evolutionary Genetics of Drosophila*. J. S. F. Barker (Ed.), pp. 75–86. Plenum, New York.

Parsons, P.A. 1991a Can atmospheric pollution be monitored from the longevity of stress sensitive behavioural mutants in *Drosophila*? *Funct. Ecol.* **5**, 713–715.

Parsons, P.A. 1991b Evolutionary rates: stress and species boundaries. *Ann. Rev. Ecol. Syst.* **22**, 1–18.

Parsons, P.A. 1992 Fluctuating asymmetry: a biological monitor of environmental and genomic stress. *Heredity* **68**, 361–364.

Parsons, P.A. 1993a The importance and consequences of stress in living and fossil populations: from life-history variation to evolutionary change. *Am. Nat.* **142**(Suppl.), 5–20.

Parsons, P.A. 1993b Evolutionary adaptation and stress: energy budgets and habitats preferred. *Behav. Genet* **23**, 231–238.

Parsons, P.A. 1993c Stress, metabolic cost and evolutionary change: from living organisms to fossils. In *Evolutionary Patterns and Processes*, Linnean Society Symposium Vol. 14. D.R. Lees and D. Edwards (Eds.), pp. 139–156.

Parsons, P.A. 1993d Habitat preference: an interaction between genetic variability and the costs of stress. *Etología* **3**, 1–9.

Parsons, P.A. 1993e Stress, extinctions and evolutionary change: from living organisms to fossils. *Biol. Rev.* **68**, 313–333.

Parsons, P.A. 1993f Developmental variability and the limits of adaptation: interactions with stress. *Genetica* **89**, 245–253.

Parsons, P.A. 1994a Morphological stasis: an energetic and ecological perspective incorporating stress. *J. theor. Biol.* **171**, 409–414.

Parsons, P.A. 1994b Habitats, stress, and evolutionary rates. *J. evol. Biol.* **7**, 387–397.

Parsons, P.A. 1994c The energetic cost of stress. Can biodiversity be preserved? *Biodiv. Lett.* **2**, 11–15.

Parsons, P.A. 1995 Stress and limits to adaptation: sexual selection. *J. evol. Biol.* **8**, 445–461.

Pashler, H. 1990 Coordinate frame for symmetry detection and object recognition. *J. exp. Psychol.: Human Percept. Perform.* **16**, 150–163.

Pasteur, L. 1860 *Researches on Molecular Asymmetry*. (Alembic Club Reprint 14, Edinburgh, 1948).

Patterson, B.D. and Patton, J.L. 1990 Fluctuating asymmetry and allozymic heterozygosity among natural populations of pocket gophers (*Thomomys bottae*). *Biol. J. Linn. Soc.* **40**, 21–36.

Peitgen, H., Jurgens, H. and Saupe, D. 1992 *Chaos and Fractals*. Springer, New York.

Pennycuick, C.J. 1975 Mechanics of flight. In *Avian Biology*, Vol. 5. D.S. Farner and J.R. King (Eds.), pp. 1–75. Academic Press, London.

Pennycuick, C.J. 1989 *Bird Flight Performance: A Practical Calculation Manual*. Oxford University Press, Oxford.

Perrett, D.I., May, K.A. and Yoshikawa, S. 1994 Facial shape and judgements of female attractiveness. *Nature* **368**, 239–242.

Perrett, D.I., Burt, D.M., Lee, K.J. and Rowland, D.A. 1997 Fluctuating asymmetry in human faces: symmetry is beautiful. MS.

Petrie, M., Halliday, T. and Sanders, C. 1991 Peahens prefer peacocks with elaborate trains. *Anim. Behav.* **41**, 323–331.

Phelan, J.P. and Austad, S.N. 1994 Selecting animal models of human aging: inbred strains often exhibit less biological uniformity than F1 hybrids. *J. Gerontol.* **49**, B1-B11.

Picton, H.D., Palmisciano, D. and Nelson, G. 1990 Fluctuating asymmetry and testing isolation of Montana grizzly bear populations. *Int. Conf. Bear Res. Manag.* **8**, 421–424.

Pierce, B.A. and Mitton, J.B. 1982 Allozyme heterozygosity and growth in the tiger salamander, *Ambystoma tigrinum*. *J. Hered.* **73**, 250–253.

Piñeiro, R. 1992 Selection for canalization at extra dorsocentral and scutellar bristles in *Drosophila melanogaster*. *J. Hered.* **83**, 445–448.

Plough, H.H. 1917 The effect of temperature on crossing over in *Drosophila*. *J. exp. Zool.* **24**, 148–209.

Plough, H.H. 1921 Further studies on the effect of temperature on crossing over. *J. exp. Zool.* **32**, 187–202.

Polak, M. 1993 Parasites increase fluctuating asymmetry of male *Drosophila nigrospiracula*: implications for sexual selection. *Genetica* **89**, 255–265.

Polak, M. 1997a Parasites, fluctuating asymmetry and sexual selection. In *Parasites: Effects on Host Hormones and Behavior*. N. Beckage (Ed.), pp. 246–276. Chapman and Hall, New York.

Polak, M. 1997b Ectoparasitism in mothers causes higher fluctuating asymmetry in their sons: implications for sexual selection. *Am. Nat.* **149**, 955–974.

Polak, M. and Trivers, R. 1994 The science of symmetry in biology. *Trends Ecol. Evol.* **9**, 122–124.

Pomiankowski, A. 1997 Genetic variation in fluctuating asymmetry. *J. evol. Biol.* **10**, 51–55.

Pomiankowski, A. and Møller, A.P. 1995 A resolution of the lek paradox. *Proc. R. Soc. Lond. B.* **260**, 21–29.

Pomiankowski, A., Iwasa, Y. and Nee, S. 1991 The evolution of costly mate preferences. I. Fisher and biased mutation. *Evolution* **45**, 1422–1430.

Pomiankowski, A.N. 1987 The costs of choice in sexual selection. *J. theor. Biol.* **128**, 195–218.

Powers, D.A., Greaney, G.S. and Place, A.R. 1979 Physiological correlation between lactate dehydrogenase genotype and haemoglobin function in killifish. *Nature* **277**, 240–241.

Price, T. and Schluter, D. 1991 On the low heritability of life-history traits. *Evolution* **45**, 853–861.

Price, T., Chi, E., Pavelka, M. and Hack, M. 1991 Population and developmental variation in the feather tip. *Evolution* **45**, 518–533.

Price, T.D. 1985 Reproductive responses to varying food supply in a population of Darwin's finches: clutch size, growth rates and hatching asynchrony. *Oecologia* **66**, 411–416.

Purnell, D.J. and Thompson, J.N., Jr. 1973 Selection for asymmetrical bias in a behavioural character of *Drosophila melanogaster*. *Heredity* **31**, 401–405.

Qazi, Q.H., Masakawa, A., McGann, B. and Woods, J. 1980 Dermatoglyphic abnormalities in the fetal alcohol syndrome. *Teratology* **21**, 157–160.

Radesäter, T. and Halldórsdóttir, H. 1993 Fluctuating asymmetry and forceps size in earwigs, *Forficula auricularia*. *Anim. Behav.* **45**, 626–628.

Rasmuson, M. 1960 Frequency of morphological deviations as a criterion of developmental stability. *Hereditas* **46**, 511–536.

Ratner, V.A., Zabanov, S.A., Kolesnikova, O.V. and Vasilyeva, L.A. 1992 Induction of the mobile genetic element DM-412 transpositions in the *Drosophila* genome by heat shock treatment. *Proc. Natl. Acad. Sci. USA* **89**, 5650–5654.

Reeve, E.C.R. 1960 Some genetic tests on asymmetry of sternopleural chaeta number in *Drosophila. Genet. Res., Camb.* **1**, 151–172.

Reid, D.H. and Parsons, P.A. 1963 Sex of parent and variation of recombination with age in the mouse. *Heredity* **18**, 107–108.

Reid, J. and Inbau, F. 1966 *Truth and Deception: The Polygraph ('Lie-Detector') Technique*. Williams and Wilkins, Baltimore.

Rendel, J.M. 1967 *Canalization and Gene Control*. Academic Press, London.

Rendel, J.M. and Sheldon, B.L. 1959 Selection for canalization of the scute phenotype in *Drosophila melanogaster. Aust. J. Biol. Sci.* **13**, 36–47.

Rensch, B. 1957 Ästetische Faktoren bei Farb- und Formbevorzugungen von Affen. *Z. Tierpsychol.* **14**, 71–99.

Rensch, B. 1958 Die Wirksamkeit ästhetischer Faktoren bei Wirbeltieren. *Z. Tierpsychol.* **15**, 447–461.

Rettig, J.E., Fuller, R.C., Corbett, A.L. and Getty, T. 1997 Fluctuating asymmetry is an indicator of ecological stress. *Oikos* (in press).

Rhodes, F.H.T. 1983 Gradualism, punctuated equilibrium and the Origin of Species. *Nature* **305**, 269–270.

Rice, W.R. 1989 Analyzing tables of statistical tests. *Evolution* **43**, 223–225.

Rice, W.R. and Hostert, E.E. 1993 Laboratory experiments on speciation: what have we learned in 40 years? *Evolution* **47**, 1637–1653.

Richards, R.A. 1978 Genetic analysis of drought stress response in rapeseed (*Brassica campestris* and *B. napus*). I. Assessment of environments for maximum selection response in grain yield. *Euphytica* **27**, 609–615.

Ricker, J.P., Boring, P.H., Harris, A.C. and Markow, T. 1989 Fluctuating dermatoglyphic asymmetry and its association with positive and negative symptoms in schizophrenia. *Schizophr. Res.* **2**, 73.

Riesenfeld, A. 1966 The effects of experimental bipedalism and upright posture in the rat and their significance for the study of human evolution. *Acta Anat.* **65**, 449–521.

Riesenfeld, A. 1973 The effects of extreme temperatures and starvation on the body proportions of rats. *Am. J. Phys. Anthropol.* **39**, 426–460.

Rifaat, O.M. 1959 Effect of temperature on crossing-over in *Neurospora crassa. Genetics* **30**, 312–323.

Rintamäki, P., Alatalo, R.V., Höglund, J. and Lundberg, A. 1997 Fluctuating asymmetry in ornamental and non-ornamental traits in relation to copulation success in black grouse. *Anim. Behav.* (in press).

Rodhouse, P.G., McDonald, J.H., Newell, R.I.E. and Koehn, R.K. 1986 Gamete production, somatic growth and multiple locus heterozygosity in *Mytilus edulis* L. *Mar. Biol.* **90**, 209–214.

Roelofs, W.L. and Brown, R.L. 1982 Pheromones and evolutionary relationships of Tortricidae. *Ann. Rev. Ecol. Syst.* **13**, 395–422.

Roff, D.A. 1992 *The Evolution of Life Histories*. Chapman and Hall, New York.

Romanoff, A. L. 1960 *The Avian Embryo: Structural and Functional Development*. Macmillan, New York.

Root, M. 1990 Biological monitors of pollution. *BioScience* **40**, 83–86.

Root, T. 1988a Environmental factors associated with avian distributional limits. *J. Biogeogr.* **15**, 489–505.

Root, T. 1988b Energy constraints on avian ranges. *Ecology* **69**, 330–339.

Rose, A.M. and Baillie, D.I. 1979 The effect of temperature and parental age on recombination and nondisjunction in *Caenorhabditis elegans*. *Genetics* **92**, 409–418.

Ross, K.G. and Robertson, J.L. 1990 Developmental stability, heterozygosity, and fitness in two introduced fire ants (*Solenopsis invicta* and *S. richteri*) and their hybrid. *Heredity* **64**, 93–103.

Ruban, G.I. 1992 Plasticity of development in natural and experimental populations of Siberian sturgeon *Acipenser baeri* Brandt. *Acta Zool. Fenn.* **191**, 43–46.

Rutherford, T.A. and Webster, J.M. 1978 Some effects of *Mermis nigrescens* on the haemolymph of *Schistocerca gregaria*. *Can. J. Zool.* **56**, 339–347.

Ryabova, G.D., Ofitserov, M.V. and Shishanova, E.I. 1995 Investigation of the relationship between allozyme variation and some fitness components in stellate sturgeons *Acipenser stellatus* (Pallas). *Genetika* **31**, 1679–1692.

Ryan, B.F, Joiner, B.I. and Ryan, T.A. 1985 *MINITAB Handbook*. 2nd edn. PWS-Kent, Boston, Mass.

Ryan, M.J. 1990 Sexual selection, sensory systems, and sensory exploitation. *Oxford Surv. Evol. Biol.* **7**, 157–195.

Sachs, T. 1994 Variable development as a basis for robust pattern formation. *J. theor. Biol.* **170**, 423–425.

Sachs, T., Novoplansky, A. and Cohen, D. 1993 Plants as competing populations of redundant organs. *Plant, Cell and Environment* **16**, 765–770.

Sagan, L.A. 1989 On radiation, paradigms, and hormesis. *Science* **245**, 574 + 621.

Saino, N. and Møller, A.P. 1994 Secondary sexual characters, parasites and testosterone in the barn swallow, *Hirundo rustica*. *Anim. Behav.* **48**, 1325–1333.

Sakai, K.-I. and Shimamoto, Y. 1965a A developmental-genetic study on panicle characters in rice, *Oryza sativa* L. *Genet. Res., Camb.* **6**, 93–103.

Sakai, K.-I. and Shimamoto, Y. 1965b Developmental instability in leaves and flowers of *Nicotiana tabacum*. *Genetics* **51**, 801–813.

Sankaranarayanan, K. 1982 *Genetic Effects of Ionizing Radiation in Multi-Cellular Eukaryotes and the Assessment of Genetic Radiation Hazards in Man*. Elsevier, Amsterdam.

Sargent, T.D. 1976 *Legion of Night: The Underwing Moths*. University of Massachusetts Press, Amherst.

Sarre, S. and Dearn, J.M. 1991 Morphological variation and fluctuating asymmetry among insular populations of the sleepy lizard, *Trachydosaurus rugosus* Gray (Squamata: Scincidae). *Aust. J. Zool.* **39**, 91–104.

Sarre, S., Dearn, J.M. and Georges, A. 1994 The application of fluctuating asymmetry in the monitoring of animal populations. *Pacific Conservation Biol.* **1**, 118–122.

Sarvas, R. 1962 Investigations on the flowering and seed crop of *Pinus sylvestris*. *Commun. Inst. For. Fenn.* **53**, 1–198.

Saunders, S.R. and Mayhall, J.T. 1982 Fluctuating asymmetry of dental morphological trairs: new interpretations. *Human Biol.* **54**, 789–799.

Schaeffer, A.A. 1928 Spiral movement in man. *J. Morph. Physiol.* **45**, 293–398.

Scharloo, W. 1991 Canalization: genetic and developmental aspects. *Ann. Rev. Ecol. Syst.* **22**, 65–93.

Scharloo, W., Zweep, A., Schuitema, K.A. and Wijnstra, J.G. 1972 Stabilizing and disruptive selection on a mutant character in *Drosophila*. IV. Selection on sensitivity to temperature. *Genetics* **71**, 551–566.

Scheiner, S.M., Caplan, R.L. and Lyman, R.F. 1991 The genetics of phenotypic plasticity. III. Genetic correlations and fluctuating asymmetry. *J. evol. Biol.* **4**, 51–68.

Schmalhausen, I.I. 1940 Variability and the change of adaptive norms in the course of evolution. *Zh. Obshch. Biol.* **1**, 509–528.

Schmalhausen, I.I. 1949 *Factors of Evolution*. Blakiston, Philadelphia.

Schroeder, M. 1991 *Fractals, Chaos, Power Laws: Minutes from an Infinite Paradise.* Freeman, New York.

Schwabl, U, and Delius, J.D. 1984 Visual bar length discrimination threshold in the pigeon. *Bird Behav.* **5**, 118–121.

Sciulli, P.W., Doyle, W.J., Kelley, C., Siegel, P. and Siegel, M.I. 1979 The interaction of stressors in the induction of increased levels of fluctuating asymmetry in the laboratory rat. *Am. J. Phys. Anthropol.* **50**, 279–284.

Seavey, S.R. and Bawa, K.S. 1986 Late-acting self-incompatibility in angiosperms. *Bot. Rev.* **52**, 195–219.

Selker, E.U. 1990 Premeiotic instability of repeated sequences in *Neurospora crassa. Ann. Rev. Genet.* **24**, 579–613.

Selye, H. 1974 *Stress without Distress*. Lippincott, Philadelphia.

Seradilla, J.M. and Ayala, F.J. 1983 Alloprocoptic selection: a mode of natural selection promoting polymorphism. *Proc. Natl. Acad. Sci. USA* **80**, 2022–2025.

Seredenin, S.B., Durnev, A.V. and Vedernikov, A.A. 1980 Effect of emotional stress on chromosomal aberration frequency in bone marrow cells in mice. *Bull. Exp. Biol. Med.* **10**, 91–92.

Serova, I.A. and Kerkis, J.J. 1974 Cytogenetical effect of some steroid hormones and change in activity of lysosomal enzymes in vitro. *Genetika* **10**, 142–149.

Shackell, N.L. and Doyle, R.W. 1991 Scale morphology as an index of developmental stability and stress resistance of tilapia (*Oreochronis niloticus*). *Can. J. Fish Aquat. Sci.* **48**, 1662–1669.

Shapiro, B.L. 1970 Prenatal dental anomalies in mogolism: comment on the basis and implications of variability. *Ann. N.Y. Acad. Sci.* **171**, 562–577.

Shapiro, B.L. 1983 Down syndrome—a disruption of developmental homeostasis. *Am. J. Med. Genet.* **14**, 241–269.

Shapiro, J.A. 1984 Observations on the formation of clones containing araB-lacZ cistron fusions. *Molec. Gen. Genet.* **194**, 79–90.

Shapiro, J.A. 1992 Natural genetic engineering in evolution. *Genetica* **86**, 99–111.

Shaposhnikov, G.K. 1965 Morphological divergence and convergence in an experiment with aphids (Homoptera, Aphidinea). *Entomol. Rev. (USSR)* **44**, 3–25.

Shaposhnikov, G.K. 1966 Origin and breakdown of reproductive isolation and the criterion of the species. *Entomol. Rev. (USSR)* **45**, 1–18.

Shaposhnikov, G.K. 1987a Evolution of aphids in relation to evolution of plants. In *Aphids: Their Biology, Natural Enemies and Control*. A.K Minks and P. Harrewijn (Eds.), pp. 409–414. Elsevier, Amsterdam.

Shaposhnikov, G.K. 1987b Organization (structure) of populations and species, and speciation. In *Aphids: Their Biology, Natural Enemies and Control*. A.K Minks and P. Harrewijn (Eds.), pp. 415–430. Elsevier, Amsterdam.

Sheldon, B.C. 1994 Male phenotype, fertility, and the pursuit of extra-pair copulations by female birds. *Proc. R. Soc. Lond. B* **257**, 25–30.

Sheppard, P.M. 1959 The evolution of mimicry; a problem in ecology and genetics. *Cold Spring Harbor Symp. Quant. Biol.* **24**, 131–140.

Shorey, H.H. 1976 *Animal Communication by Pheromones*. Academic Press, New York.

Shykoff, J.A. and Moller, A.P. 1997 Developmental stability, asymmetry, and genotypic quality. MS.

Siegel, M.I. and Doyle, W.J. 1975a The differential effects of prenatal and postnatal stress on fluctuating dental asymmetry. *J. exp. Zool.* **191**, 211–214.

Siegel, M.I. and Doyle, W.J. 1975b Stress and fluctuating limb asymmetry in various species of rodents. *Growth* **39**, 363–369.

Siegel, M.I. and Doyle, W.J. 1975c The effects of cold stress on fluctuating asymmetry in the dentition of the mouse. *J. exp. Zool.* **193**, 385–391.

Siegel, M.I. and Mooney, M.P. 1987 Perinatal stress and increased fluctuating asymmetry of dental calcium in the laboratory rat. *Am. J. Phys. Anthropol.* **41**, 331–339.

Siegel, M.I. and Smookler, H.H. 1973 Fluctuating dental asymmetry and augiogenic stress. *Growth* **37**, 35–39.

Siegel, P., Siegel, M.I., Krimmer, E.C., Doyle, W.J. and Barry, H. 1977 Fluctuating dental asymmetry as an indicator of the stressful prenatal effects of D^9—tetrahydrocannabinol in the laboratory rat. *Toxicol. Appl. Pharmacol.* **42**, 339–344.

Siegel, S. and Castellan, N.J. 1988. *Nonparametric Statistics for the Behavioral Sciences.* 2nd edn. McGraw-Hill, Singapore.

Simmons, L.W. 1994 Correlates of male quality in the field cricket, *Gryllus campestris* L.: age, size, and symmetry determine pairing success in field populations. *Behav. Ecol.* **6**, 376–381.

Simmons, L.W. and Ritchie, M.G. 1996 Symmetry in the songs of crickets. *Proc. R. Soc. Lond.* B **263**, 305–311.

Simmons, L.W., Tomkins, J.L. and Manning, J.T. 1995 Sampling bias and fluctuating asymmetry. *Anim. Behav.* **49**, 1697–1699.

Simpson, G.G. 1944 *Tempo and Mode in Evolution*. Columbia University Press, New York.

Simpson, J.L., Golbus, M.S., Martin, A.O. and Sarto, G.E. 1982 *Genetics in Obstetrics and Gynaecology*. Grune and Stratton, New York.

Singh, S.M. 1982 Enzyme heterozygosity associated with growth at different developmental stages in oysters. *Can. J. Genet. Cytol.* **24**, 451–458.

Singh, S.M. and Zouros, E. 1978 Genetic variation associated with growth rate in the American oyster (*Crassostrea virginica*). *Evolution* **32**, 342–353.

Skinnes, H. and Burås, T. 1987 Developmental stability in wheat to differences in the temperature during seed set and seed development. *Acta Agric. Scand.* **37**, 287–297.

Smith, B.H., Garn, S.M. and Cole, P.M. 1983 Problems of sampling and inference in the study of fluctuating dental asymmetry. *Am. J. Phys. Anthropol.* **58**, 281–289.

Smith, L.D. and Palmer, A.R. 1994 Effects of manipulated diet on size and performance of brachyuran crab claws. *Science* **264**, 710–712.

Smith, M.H., Chesser, R.K., Cothran, E.G. and Johns, P.E. 1982 Genetic variability and antler growth in a natural population of white-tailed deer. In *Antler Development in Cervidae*. R.D. Brown (Ed.), pp. 365–387. Caesar Kleberg Wildlife Research Foundation, Kingsville, Texas.

Smouse, P.E. 1986 The fitness consequences of multiple-locus heterozygosity under the multiplicative overdominance and inbreeding depression models. *Evolution* **40**, 946–957.

Sneddon, L.U. and Swaddle, J.P. 1997 Asymmetry and fighting performance in the shore crab *Carcinus maenas*. MS.

Sniegowski, P.D. and Lenski, R.E. 1995 Mutation and adaptation: the directed mutation controversy in evolutionary perspective. *Ann. Rev. Ecol. Syst.* **26**, 553–578.

Sofaer, J.A. 1979 Human tooth-size asymmetry in cleft lip with or without cleft palate. *Arch. Oral Biol.* **24**, 141–146.

Solberg, E.J. and Sæther, B.-E. 1994 Fluctuating asymmetry in the antlers of moose (*Alces alces*): does it signal male quality? *Proc. R. Soc. Lond. B* **254**, 251–255.

Soulé, M. 1967 Phenetics of natural populations. II. Asymmetry and evolution in a lizard. *Am. Nat.* **101**, 141–160.

Soulé, M.E. 1979 Heterozygosity and developmental stability: another look. *Evolution* **33**, 396–401.

Soulé, M.E. 1982 Allomeric variation. I. The theory and some consequences. *Am. Nat.* **120**, 751–764.

Soulé, M. and Baker, B. 1968 Phenetics of natural populations. IV. The population asymmetry parameter in the butterfly *Coenonympha tullia*. *Heredity* **23**, 611–614.

Soulé, M. and Cuzin-Roudy, J. 1982 Allomeric variation. 2. Developmental instability of extreme phenotypes. *Am. Nat.* **120**, 765–768.

Spedding, G.R. 1987a The wake of a kestrel (*Falco tinnunculus*) in gliding flight. *J. exp. Biol.* **127**, 45–57.

Spedding, G.R. 1987b The wake of a kestrel (*Falco tinnunculus*) in flapping flight. *J. exp. Biol.* **127**, 59–78.

Stamberg, R. and Simchen, G. 1970 Specific effects of temperature on recombination in *Schizophyllum commune*. *Heredity* **25**, 41–52.

Stearns, S.C. 1987 The selection arena hypothesis. In *The Evolution of Sex and its Consequences*. S.C. Stearns (Ed.), pp. 337–349. Birkhäuser, Basle.

Stearns, S.C. 1992 *The Evolution of Life Histories*. Oxford University Press, Oxford.

Stearns, S.C. and Kawecki, T.J. 1994 Fitness sensitivity and the canalization of life-history traits. *Evolution* **48**, 1438–1458.

Stearns, S.C., Kaiser, M. and Kawecki, T.J. 1995 The differential genetic and environmental canalization of fitness components in *Drosophila melanogaster*. *J. evol. Biol.* **8**, 539–557.

Stebbins, G.L. 1950 *Variation and Evolution in Plants*. Columbia University Press, New York.

Stephenson, A.G. 1981 Flower and fruit abortion: proximate causes and ultimate functions. *Ann. Rev. Ecol. Syst.* **12**, 253–279.

Stern, C. 1926 An effect of temperature and age on crossing-over in the first chromosome of *Drosophila melanogaster*. *Proc. Natl. Acad. Sci. USA* **12**, 530–532.

Suchentrunk, F. 1993 Variability of minor tooth traits and allozymic diversity in brown hare *Lepus europaeus* populations. *Acta Theriol.* **38**, Suppl. 2, 59–69.

Sumarsono, S.H., Wilson, T.J., Tymms, M.J., Venter, D.J., Corrick, C.M., Kola, R., Lahoud, M.H., Papas, T.S., Seth, A. and Kola, I. 1996 Down's syndrome-like skeletal abnormalities in ETS2 transgenic mice. *Nature* **379**, 534–537.

Summerbell, D. and Wolpert, L. 1973 Precision of development in chick limb morphogenesis. *Nature* **244**, 228–230.

Suthers, R.A. 1994 Variable asymmetry and resonance in the avian vocal tract: a structural basis for individually distinct vocalizations. *J. Comp. Physiol.* A **175**, 457–466.

Suzuki, D.T. 1963 Interchromosomal effects on crossing over in *Drosophila melanogaster*. II. A re-examination of chromosome inversion effects. *Genetics* **48**, 1605–1617.

Swaddle, J.P. 1996 Reproductive success and symmetry in zebra finches. *Anim. Behav.* **51**, 203–210.

Swaddle, J.P. 1997a On the heritability of developmental stability. *J. evol. Biol.* **10**, 57–61.

Swaddle, J.P. 1997b Developmental stability and predation success in an insect predator-prey system. *Behav. Ecol.* (in press).

Swaddle, J.P. 1997c Within-individual changes in developmental stability affect flight performance. *Behav. Ecol.* (in press).

Swaddle, J.P. and Cuthill, I.C. 1994a Preference for symmetric males by female zebra finches. *Nature* **367**, 165–166.

Swaddle, J.P. and Cuthill, I.C. 1994b Female zebra finches prefer males with symmetrically manipulated chest plumage. *Proc. R. Soc. Lond.* B **258**, 267–271.

Swaddle, J.P. and Cuthill, I.C. 1995 Asymmetry and human facial attractiveness: symmetry may not always be beautiful. *Proc. R. Soc. Lond. B.* **261**, 111–116.

Swaddle, J.P. and Witter, M.S. 1994 Food, feathers and fluctuating asymmetry. *Proc. R. Soc. Lond.* B **255**, 147–152.

Swaddle, J.P. and Witter, M.S. 1995 Chest plumage, dominance and fluctuating asymmetry in female starlings. *Proc. R. Soc. Lond.* B **260**, 219–223.

Swaddle, J.P. and Witter, M.S. 1997a The effects of molt on the flight performance, mass and behaviour of European starlings (*Sturnus vulgaris*): an experimental approach. *Can. J. Zool.* (in press).

Swaddle, J.P. and Witter, M.S. 1997b On the ontogeny of developmental stability in a stabilised trait. *Proc. R. Soc. Lond.* B **264**, 329–334.

Swaddle, J.P., Witter, M.S. and Cuthill, I.C. 1994 The analysis of fluctuating asymmetry. *Anim. Behav.* **48**, 986–989.

Swaddle, J.P., Witter, M.S. and Cuthill, I.C. 1995 Museum studies measure FA. *Anim. Behav.* **49**, 1700–1701.

Swaddle, J.P., Witter, M.S., Cuthill, I.C., Budden, A. and McCowen, P. 1996 Plumage condition affects flight performance in starlings: implications for developmental homeostasis, abrasion and moult. *J. Avian Biol.* **27**, 103–111.

Swain, D.P. 1987 A problem with the use of meristic characters to estimate developmental stability. *Am. Nat.* **129**, 761–768.

Sweet, G.B. 1973 Shedding of reproductive structures in forest trees. In *Shedding of Plant Parts*. T.T. Kozlowski (Ed.), pp. 341–382. Academic Press, New York.

Symons, D. 1979 *The Evolution of Human Sexuality*. Oxford University Press, Oxford.

Szilagyi, P.G. and Baird, J.C. 1977 A quantitative approach to the study of visual symmetry. *Perception Psychophys.* **22**, 287–292.

Taddei, F., Matic, I., Vulic, M., Godelle, B. and Radman, M. 1996 The speed of the Red Queen: mutation rate varies with past and present environment. *Abstracts of the Fifth International Congress of Systematic and Evolutionary Biology, Budapest, August 17–24, 1996*, p. 47.

Tåning, Å. V. 1952 Experimental study of meristic characters in fishes. *Biol. Rev.* **27**, 169–193.

Tarasjev, A. 1995 Relationship between phenotypic plasticity and developmental stability in *Iris pumila* L. *Russian J. Genet.* **31**, 1655–1663.

Tartof, K.D. and Bremer, M. 1990 Mechanisms for the construction and developmental control of heterochromatin formation and imprinted chromosome domains. *Development Suppl.* 35–45.

Tartof, K.D., Bishop, C., Jones, M., Hobbs, C.A. and Locke, J. 1989 Towards an understanding of position effect variegation. *Develop. Genet.* **10**, 162–176.

Teather, K. 1996 Patterns of growth and asymmetry in nestling tree swallows. *J. Avian Biol.* **27,** 302–310.

Tebb, G. and Thoday, J.M. 1954 Stability in development and relational balance of X chromosomes in lab populations of *Drosophila melanogaster. Nature* **174,** 1109.

Teska, W.R., Smith, M.H. and Novak, J.M. 1991 Food quality, heterozygosity, and fitness correlates in *Peromyscus polionotus. Evolution* **44,** 1318–1325.

Thessing, A. and Ekman, J. 1994 Selection on the genetical and environmental components of tarsal growth in juvenile willow tits (*Parus montanus*). *J. evol. Biol.* **7,** 713–726.

Thoday, J.M. 1955 Balance, heterozygosity and developmental stability. *Cold Spring Harbor Sym. Quant. Biol.* **20,** 318–326.

Thoday, J.M. 1958 Homeostasis in a selection experiment. *Heredity* **12,** 401–405.

Thoma, R. 1996 *Developmental Instability, Handedness, and Brain Lateralization: MES and MRI correlates.* MSc thesis, Dept. Psychology, University of New Mexico, Albuquerque, New Mexico.

Thomas A.L.R. 1993a On the aerodynamics of birds' tails. *Phil. Trans. R. Soc. Lond. B* **340,** 361–380.

Thomas, A.L.R. 1993b The aerodynamic costs of asymmetry in the wings and tails of birds: asymmetric birds can't fly round tight corners. *Proc. R. Soc. Lond. B* **254,** 181–189.

Thomas, B.J. and Rothstein, R. 1991 Sex, maps, and imprinting. *Cell* **64,** 1–3.

Thomas, F. and Poulin, R. 1997 Using randomization techniques to analyse fluctuating asymmetry data. *Anim. Behav.* (in press).

Thompson, J.N., Jr. and Woodruff, R.C. 1978 Mutator genes—pacemakers of evolution. *Nature* **274,** 317–321.

Thompson, S.N. 1983 Biochemical and physiological effects of metazoan endoparasites on their host species. *Comp. Biochem. Physiol.* **74B,** 183–211.

Thornhill, R. 1992a Fluctuating asymmetry, interspecific aggression and male mating tactics in two species of Japanese scorpionflies. *Behav. Ecol. Sociobiol.* **30,** 357–363.

Thornhill, R. 1992b Fluctuating asymmetry and the mating system of the Japanese scorpionfly, *Panorpa japonica. Anim. Behav.* **44,** 867–879.

Thornhill, R. and Gangestad, S.W. 1993 Human facial beauty: averageness, symmetry and parasite resistance. *Human Nat.* **4,** 237–269.

Thornhill, R. and Gangestad, S.W. 1994 Human fluctuating asymmetry and sexual behavior. *Psychol. Sci.* **5,** 297–302.

Thornhill, R. and Møller, A.P. 1997 Developmental stability, disease and medicine. *Biol. Rev.* (in press).

Thornhill, R. and Sauer, P. 1992 Genetic sire effects on the fighting ability of sons and daughters and mating success of sons in a scorpionfly. *Anim. Behav.* **43,** 255–264.

Timofeeff-Ressovsky, N.V. 1934 Über der Einfluss des genotypischen Milieus und der Aussenbedingungen auf die Realisation des Genotypes. Genmutation vti bei *Drosophila funebris. Nachr. Acad. Wiss. Göttingen II. Math-Phys.* **1,** 53–106.

Tinbergen, N. 1951 *The Study of Instinct.* Oxford University Press, Oxford.

Todaro, G.J. and Martin, G.M. 1967 Increased susceptibility of Down's syndrome fibroblasts to transformation by SV40. *Proc. Soc. Exp. Biol. Med.* **124,** 1232–1236.

Tomkins, J.L. and Simmons, L.W. 1995 Patterns of fluctuating asymmetry in earwig forceps: no evidence for reliable signalling. *Proc. R. Soc. Lond. B* **259,** 89–96.

Towe, A.M. and Stadler, D.R. 1964 Effects of temperature on crossing over in *Neurospora*. *Genetics* **49**, 577–583.

Tracy, M., Freeman, D.C., Emlen, J.M., Graham, J.H. and Hough, R.A. 1996 Developmental instability as a biomonitor of environmental stress: an illustration using aquatic plants and macroalgae. In *Biomonitors and Biomarkers as Indicators of Environmental Change*. F.M. Butterworth (Ed.), pp.313–338. Plenum Press, New York.

Trivers, R.L. 1985 *Social Evolution*. Benjamin/Cummings, Menlo Park.

Trout, W.E. and Hanson, G.P. 1971 The effect of Los Angeles smog on the longevity of normal and hyperkinetic *Drosophila melanogaster*. *Genetics* **68**, S69.

Tuinstra, E.J., de Jong, G. and Scharloo, W. 1990 Lack of response to family selection for directional asymmetry in *Drosophila melanogaster*: left and right sides are not distinguished in development. *Proc. R. Soc. Lond. B* **241**, 146–152.

Turing, A. 1952 The chemical basis of morphogenesis. *Phil. Trans. R. Soc. Lond. B* **237**, 37–72.

Turkington, R.W. 1971 Hormonal regulation of cell proliferation and differentiation. In *Developmental Aspects of the Cell Cycle*. I.L. Cameron, G.M. Padilla and A.M Zimmerman (Eds.), pp. 315–355. Academic Press, London.

Ueno, H. 1994 Fluctuating asymmetry in relation to two fitness components, adult longevity and male mating success in a ladybird beetle, *Harmonia axyridis* (Coleoptera: Coccinellidae). *Ecol. Entomol.* **19**, 87–88.

Uetz, G.W., McClintock, W.J., Miller, D., Smith, E.I. and Cook, K.K. 1996 Limb regeneration and subsequent asymmetry in a male secondary sexual character influences sexual selection in wolf spiders. *Behav. Ecol. Sociobiol.* **38**, 253–257.

Valentin, J. 1972 The effect of the Curly inversions on meiosis in *Drosophila melanogaster*. III. Interchromosomal effects of homozygous inversions. *Hereditas* **72**, 255–260.

Valentine, D.W. and Soulé, M. 1973 Effect of p,p'-DDT on developmental stability of pectoral fin rays in the grunion, *Leuresthes tenius*. *Fish. Bull.* **71**, 921–926.

Valentine, D.W., Soulé, M.E. and Samollow, P. 1973 Asymmetry analysis in fishes: a possible indicator of environmental stress. *Fish. Bull.* **71**, 357–370.

VanDemark, N.L. and Free, M.J. 1970 Temperature effects. In *The Testis*. A.D. Johnson, W.R. Gomes and N.L. VanDemark (Eds.), Vol. 3, pp. 233–297. Academic Press, New York.

van Noordwijk, A. J. 1984. Quantitative genetics in natural populations of birds illustrated with examples from the great tit *Parus major*. In *Population Biology and Evolution*. K. Wöhrmann and V. Loeschcke (Eds.), pp. 67–79. Springer, Berlin.

van Noordwijk, A., van Balen, J.H. and Scharloo, W. 1988 Heritability of body size in a natural population of the great tit (*Parus major*) and its relation to age and environmental conditions during growth. *Genet. Res., Camb.* **51**, 149–162.

Van Valen, L. 1962 A study of fluctuating asymmetry. *Evolution* **16**, 125–142.

Vogl, C. 1996 Developmental buffering and selection *Evolution* **50**, 1343–1346.

Vøllestad, L.A., Hindar, K. and Møller, A.P. 1997 Patterns of correlation between heterozygosity and fluctuating asymmetry. MS.

von Fersen, L., Manos, C., Goldowsky, B. and Roitblat, H. 1992 Dolphin detection and conceptualization of symmetry. In *Marine Mammal Sensory Systems*. J. Thomas et al. (Eds.), pp. 753–762. Plenum, New York.

Vrijenhoek, R.C. and Lerman, S. 1982 Heterozygosity and developmental stability under sexual and asexual breeding systems. *Evolution* **36**, 768–776.

Waddington, C.H. 1940 *Organisers and Genes*. Cambridge University Press, Cambridge.

Waddington, C.H. 1953 The genetic assimilation of an acquired character. *Evolution* **7**, 118–126.

Waddington, C.H. 1958 *The Strategy of the Genes*. Allen and Unwin, London.

Waddington, C.H. 1960 Experiments on canalizing selection. *Genet. Res., Camb.* **1**, 140–150.

Waddington, C.H. 1961 Genetic assimilation. *Adv. Genet.* **10**, 257–293.

Waddington, C.H. and Robertson, E. 1966 Selection for developmental canalisation. *Genet. Res., Camb.* **7**, 303–312

Wagner, G.P., Booth, G. and Bagheri-Chaichian, H. 1997 A population genetic theory of canalization. *Evolution*, **51**, 329–347.

Wakefield, J., Harris, K. and Markow, T.A. 1993 Parental age and developmental stability in *Drosophila melanogaster*. *Genetica* **89**, 235–244.

Wakelin, D. and Blackwell, J.M. (Eds.) 1988 *Genetics of Resistance to Bacterial and Parasitic Infection*. Taylor and Francis, London.

Wallace, A.R. 1889 *Darwinism*. Macmillan, London.

Wallace, M. 1957 A balanced three-point experiment for linkage group V of the house mouse. *Heredity* **11**, 223–258.

Ward, P.D. 1992 *On Methuselah's Trail: Living Fossils and Great Extinctions*. Freeman, New York.

Ward, P.J. 1994 Parent-offspring regression and extreme environments. *Heredity* **72**, 574–581.

Ward, R.D. and Elliott, N.G. 1993 Heterozygosity and morphological variability in the orange roughy, *Hoplostethus atlanticus* (Teleostei: Trachichthyidae). *Can. J. Fish. Aquat. Sci.* **50**, 1641–1649.

Watkinson, A.R. and White, J. 1986 Some life-history consequences of modular construction in plants. *Phil. Trans. R. Soc. Lond. B* **313**, 31–51.

Watson, P.J. and Thornhill, R. 1994 Fluctuating asymmetry and sexual selection. *Trends Ecol. Evol.* **9**, 21–25.

Watt, W.B. 1979 Adaptation at specific loci. I. Natural selection in phosphoglucose isomerase of *Colias* butterflies: biochemical and population aspects. *Genetics* **87**, 177–194.

Watt, W.B. 1983 Adaptation at specific loci. II. Demographic and biochemical elements in the maintenance of the *Colias* PGI polymorphism. *Genetics* **103**, 691–724.

Watt, W.B., Carter, P.A. and Blower, S.M. 1985 Adaptation at specific loci. IV. Differential mating success among glycolytic allozyme genotypes of *Colias* butterflies. *Genetics* **109**, 157–175.

Watt, W.B., Carter, P.A. and Donohue, K. 1986 Females' choice of 'good genotypes' as mates is promoted by an insect mating system. *Science* **233**, 1187–1190.

Watt, W.B., Cassin, R.C. and Swan, M.S. 1983 Adaptation at specific loci. III. Field behavior and survivorship differences among *Colias* PGI genotypes are predictable from in vitro biochemistry. *Genetics* **103**, 725–739.

Wayne, R.K, Modi, W.S. and O'Brien, S.J. 1986 Morphological variability and asymmetry in the cheetah (*Acinonyx jubatus*), a genetically uniform species. *Evolution* **40**, 78–85.

Wcislo, W.T. 1989 Behavioral environments and evolutionary change. *Ann. Rev. Ecol. Syst.* **20**, 137–169.

Wedemeyer, G.A., Burton, B. and McLeay, D. 1990 Stress and acclimation. In *Methods for Fish Biology*. C. Schreck and P. Moyle (Eds.), pp. 451–490. American Fisheries Association, Bethesda.

Weinstein, R.S., Majumdar, S. and Genat, H.K. 1992 Fractal geometry applied to the architecture of cancellous bone biopsy specimens. *Bone* **13**, A38.

Weir, B.S. and Cockerham, C.C. 1973 Mixed self and random mating at two loci. *Genet. Res., Camb.* **21**, 247–262.

Weitkamp, L.R., van Rood, J.J., Thorsby, E., Bias, W., Fotino, M., Lawler, S.D., Dausset, J., Mayr, W.R., Bodmer, J., Ward, F.E., Seignalet, J., Payne, R., Kissmeyer-Nielsen, F., Gatti, R.A., Sachs, J.A., and Lamm, L.U. 1973 The relation between parental sex and age to recombination in the HL-A system. *Human Heredity* **23**, 197–205.

West, B.J. 1990a *Fractal Physiology and Chaos in Medicine*. World Scientific, Singapore.

West, B.J. 1990b Physiology in fractal dimensions: error tolerance. *Ann. Biomed. Eng.* **18**, 135–149.

West, B.J. and Goldberger, A.L. 1987 Physiology in fractal dimensions. *Am. Sci.* **75**, 354–365.

Weyl, H. 1952 *Symmetry*. Princeton University Press, Princeton.

White, M.J.D. 1978 *Modes of Speciation*. Freeman, San Francisco.

Whiten, A. and Byrne, R. 1988 Tactical deception in primates. *Behav. Brain Sci.* **11**, 233–273.

Whitlock, M. 1993 Lack of correlation between heterozygosity and fitness in forked fungus beetles. *Heredity* **70**, 574–581.

Whitlock, M.C. 1995 Variance-induced peak shifts. *Evolution* **49**, 252–259.

Whitlock, M.C. 1996 The heritability of fluctuating asymmetry and the genetic control of developmental stability. *Proc. R. Soc. Lond.* B **263**, 849–854.

Whitlock, M.C. and Fowler, K. 1997 The instability of studies of instability. *J. evol. Biol.* **10**, 63–67.

Whitt, G.S., Parker, D.P. and Childers, W.F. 1977 Aberrant expression during the development of hybrid sunfishes (Perciformes, Teleostei). *Differentiation* **9**, 97–109.

Wiener, J.G. and Rago, P.J. 1987 A test of fluctuating asymmetry in bluegills (*Lepomis macrochirus* Rafinesque) as a measure of pH-related stress. *Environ. Poll.* **44**, 27–36.

Wiens, D., Calvin, C.L., Wilson, C.A., Davern, C.I., Frank, D. and Seavey, S.R. 1987 Reproductive success, spontaneous embryo abortion, and genetic load in flowering plants. *Oecologia* **71**, 501–509.

Wiggins, D.A. 1989 Food availability, growth, and heritability of body size in nestling tree swallows (*Tachycineta bicolor*). *Can. J. Zool.* **68**, 1292–1296.

Wilber, E., Newell-Morris, L. and Pytkowicz, A. 1993 Dermatoglyphic asymmetry in fetal alcohol syndrome. *Biol. Neonate* **64**, 1–6.

Wildt, D.E., Bush, M., Goodrowe, K.L., Packer, C., Pusey, A.E., Brown, J.L., Joslin, P. and O'Brien, S.J. 1987 Reproductive and genetic consequences of founding isolated lion populations. *Nature* **329**, 328–331.

Williamson, P.G. 1981 Palaeontological documentation of speciation in Cenozoic molluscs from Turkana Basin. *Nature* **293**, 437–443.

Willig, M.R. and Owen, R.D. 1987 Fluctuating asymmetry in the cheetah: methodological and interpretive concerns. *Evolution* **41**, 225–227.

Willson, M.F. and Burley, N. 1983 *Mate Choice in Plants*. Princeton University Press, Princeton.

Wilson, D.M. 1968 Inherent asymmetry and reflex modulation of the locust flight motor pattern. *J. exp. Biol.* **48**, 631–641.

Wilson, J.M. and Manning, J. T. 1996 Fluctuating asymmetry and age in children: evolutionary implications for the control of developmental stability. *J. Human Evol.* **30**, 529–537.

Winberg, G.G. 1936 Temperature optimum of development. *Adv. Modern Biol.* **5**, 560–561.

Witter, M.S. and Cuthill, I.C. 1993 The ecological costs of avian fat storage. *Phil. Trans. R. Soc. Lond. B* **340**, 73–92.

Witter, M.S. and Lee, S.J. 1995 Habitat structure, stress and plumage development. *Proc. R. Soc. Lond. B* **261**, 303–308.

Witter, M.S. and Swaddle, J.P. 1994 Fluctuating asymmetries, competition and dominance. *Proc. R. Soc. Lond. B* **256**, 299–303.

Wodzicki, T.J. and Zajaczkowski, S. 1989 Auxin waves in cambium and morphogenetic information in plants. In *Signals in Plant Development*. J. Krekule and F. Seidová (Eds.), pp. 45–64. Academic Publishing, The Hague.

Wolf, D.P., Byrd, W., Dandekar, P. and Quigley, M.M. 1984 Sperm concentration and the fertilization of human eggs in vitro. *Biol. Reprod.* **31**, 837–848.

Wolf, S.D., Silk, W.K. and Plant, R.E. 1986. Quantitative patterns of leaf expansion: comparison of normal and malformed growth in *Vitis vinifera* cv Ruby Red. *Am. J. Bot.* **73**, 832–846.

Wolff, K. 1987 Genetic analysis of ecologically relevant morphological variability in *Plantago lanceolata* L. 2. Localization and organization of quantitative trait loci. *Theor. Appl. Genet.* **73**, 903–914.

Wood, W.B. 1991 Evidence from reversal of handedness in *C. elegans* embryos for early cell interactions determining cell fates. *Nature* **349**, 536–538.

Woolf, C.M. 1993 Does homozygosity contribute to the asymmetry of common white leg markings in the Arabian horse? *Genetica* **89**, 25–33.

Woolf, C.M. and Gianas, A. 1976 Congenital cleft lip fluctuating dermatoglyphic asymmetry. *Am. J. Human Genet.* **28**, 400–403.

Woolf, C.M. and Gianas, A. 1977 A study of fluctuating dermatoglyphic asymmetry in the sibs and parents of cleft lip propositi. *Am. J. Human Genet.* **28**, 503–507.

Wooten, M.C. and Smith, M.H. 1986 Fluctuating asymmetry and genetic variability in a natural population of *Mus musculus*. *J. Mammal.* **67**, 725–732.

Yablokov, A.V. 1986 *Population Biology. Progress and Problems of Studies on Natural Populations*. Mir, Moscow.

Yampolsky, L.Y. and Scheiner, S.M. 1994 Developmental noise, phenotypic plasticity, and allozyme heterozygosity in *Daphnia*. *Evolution* **48**, 1715–1722.

Yezerinac, S.M., Lougheed, S.C. and Handford, P. 1992 Morphological variability and enzyme heterozygosity: individual and population level correlations. *Evolution* **46**, 1959–1964.

Yokoyama, T., Copeland, N. G., Jenkins, N. A., Montgomery, C. A., Elder, F. F. B. and Overbeek, P. A. 1993 Reversal of left-right asymmetry: a situs inversus mutation. *Science* **260**, 679–682.

Yost, H.J. 1992 Regulation of vertebrate left-right asymmetries by extracellular matrix. *Nature* **357**, 158–161.

Yuge, M. and Yamana, K. 1989 Regulation of the dorsal axial structures in cell-deficient embryos of *Xenopus laevis*. *Develop., Growth, Differentiation* **31**, 315–324.

Zahavi, A. 1975 Mate selection—a selection for a handicap. *J. theor. Biol.* **53**, 205–214.

Zakharov, V.M. 1981 Fluctuating asymmetry as an index of developmental homeostasis. *Genetika* **13**, 241–256.

Zakharov, V.M. 1987 *Asymmetry of Animals*. Nauka, Moscow.

Zakharov, V.M. 1989 Future prospects for population phenogenetics. *Sov. Sci. Rev. F. Physiol. Gen. Biol.* **4**, 1–79.

Zakharov, V.M. and Bakulina, E.D. 1984 Disturbance of developmental stability at crossing different strains of *Drosophila virilis* (variation in the number of ovarioles taken as an example). *Genetika* **20**, 1390–1391.

Zakharov, V.M. and Yablokov, A.V. 1990 Skull asymmetry in the Baltic grey seal: effects of environmental pollution. *Ambio* **19**, 266–269.

Zakharov, V.M., Olsson, M., Yablokov, A.V. and Esipenko, A.G. 1989 Does environmental pollution affect the developmental stability of the Baltic grey seal (*Halichoerus grypus*)? In *Influence of Human Acitivities on the Baltic. Proceedings of a Soviet-Swedish Symposium, Moscow, 14–18 April 1986.* A.V. Yablokov and M. Olsson (Eds.), pp. 96–108. Gidrometeoizdat, Leningrad.

Zakharov, V.M., Pankoski, E., Sheftel, B.I., Peltonen, A. and Hanski, I. 1991 Developmental stability and population dynamics in the common shrew, *Sorex araneus*. *Am. Nat.* **138**, 797–810.

Zar, J.H. 1984 *Biostatistical Analysis*. Prentice Hall, New Jersey.

Zeeman, C. 1989 A new concept of stability. In *Theoretical Biology: Epigenetic and Evolutionary Order from Complex Systems.* B. Goodwin and P. Saunders (Eds.), pp. 8–15. Edinburgh University Press, Edinburgh.

Zeleny, C. 1933 Genetics and embryology. *Science* **77**, 177–181.

Zera, A.J., Koehn, R.K. and Hall, J.G. 1985 Allozymes and biochemical adaptation. In *Comprehensive Insect Physiology, Biochemistry, and Pharmacology*, Vol. 10. G.A. Kerkut and L. Gilbert (Eds.), pp. 633–674. Pergamon Press, New York.

Zink, R.M., Smith, M. and Patton, J.L. 1985 Association between heterozygosity and morphological variance. *J. Hered.* **76**, 415–420.

Zouros, E. and Foltz, D.W. 1987 The use of allelic isozyme variation for the study of heterosis. *Isozymes: Curr. Topics Biol. Med. Res.* **13**, 1–59.

Zouros, E., Singh, S.M. and Miles, H.E. 1980 Growth rate in oysters: An overdominant phenotype and its possible explanation. *Evolution* **34**, 856–867.

Zouros, E., Singh, S.M., Foltz, D.W. and Mallet, A.L. 1983 Post-settlement viability in the American oyster (*Crassostrea virginica*)—an overdominant phenotype. *Genet. Res., Camb.* **41**, 259–270.

Zvereva, E.L., Kozlov, M.V., Niemelä, P. and Haukioja, E. 1996 Delayed induced resistance and increase in leaf fluctuating asymmetry as responses of *Salix borealis* to insect herbivory. *Oecologia* (in press).

Zvereva, E.L., Kozlov, M.V. and Haukioja, E. 1997 Stress responses of *Salix borealis* to pollution and defoliation. *J. Appl. Ecol.* (in press).

Author index

Agladze, K. 46
Agnew, P. 221
Aitken, M. 25, 32
Akam, M. 41
Alados, C.L. 6, 13, 226
Alatalo, R.V. 79
Alekseeva, T.A. 46, 48, 87–8
Alerstam, T. 103
Alexander, C.B. 29
Alexander, R.M. 156
Alibert, P. 128
Allard, R.W. 120
Allen, F.I. 101
Allendorf, F.W. 6, 10, 11, 17, 21, 23, 125, 150
Alley, T.R. 197
Althukov, Y.P. 221
Ames, L.J. 140
Ancel, P. 94
Andersson, M. 195, 201
Aparicio, J.M. 62–3
Arak, A. 183–6, 187
Arcese, P. 226
Arnold, S.J. 211
Arnqvist, G. 113
Asai, R. 43
Astheimer, L.B. 90
Atchley, W. 117, 130, 132
Attneave, F. 187
Auerbach, C. 94–5
Austad, S.N. 117
Ayala, F.J. 121

Bader, R.S. 27, 118
Bagchi, S.K. 210–11
Baillie, D.I. 93
Bailly, F. 48
Baird, D.D. 214
Baird, J.C. 34, 187
Baker, B. 120, 128
Bakulina, E.D. 129
Balmford, A. 25, 27, 163–4, 167, 173–6, 199
Barden, H.S. 125
Barlow, H.B. 186–7
Barraclough, T.G. 105
Barton, N.H. 105
Bateman, K.G. 6, 7
Batterham, P. 40, 91, 94
Bawa, K.S. 194, 216
Beacham, T.D. 8
Beardmore, J.A. 14, 22, 118, 121, 136
Bell, A.E. 101
Belyaev, D.K. 92–3, 97
Bengtsson, B.-E. 140
Benkman, C.W. 21
Bennett, A.T.D. 196
Benson, P. 197
Bergé, P. 58
Berger, E. 120, 125
Bethe, A. 18
Bijlsma, R. 121
Bijlsma-Meeles, E. 121
Birch, M.C. 193

Birney, E.C. 214
Björklund, M. 27, 211
Blackman, R.L. 106
Blackwell, J.M. 94
Blanco, G. 122, 125
Blum, A. 98
Boardman, T.J. 30
Boer, C.H. 214
Bohr, V.A. 103
Booth, C.L. 121
Bornstein, M.H. 186
Borodin, P.M. 92–3, 98
Bottini, E. 120
Botvinev, O.K. 221
Boue, J. 216
Bownes, M. 103
Bradbury, D. 216
Bradley, B.P. 101, 136
Brakefield, P.M. 39, 46
Brandt, M. 137, 145
Bremer, M. 99
Breuker, C.J. 39
Briggs, J.C. 89
Brightwell, L.R. 20
Brown, N.A. 19–20, 56
Brown, R.L. 193
Brueckner, M. 18
Buchholz, J.T. 207, 212
Bullock, S. 186
Bulmer, M.G. 72
Burås, T. 135
Burley, N. 194, 195, 196
Burnaby, P. 32
Burns, J.A. 120
Burton, R.S. 130
Buschang, P.H. 125
Buss, L.W. 42
Byrne, R. 181

Cairns, J. 95
Carmeliet, P. 41
Caro, T.M. 191
Carson, H.L. 189
Carter, P.A. 88
Castellan, N.J. 25, 32

Castets, V. 46
Chakraborty, R. 8, 124
Chamberlain, A.T. 25, 27, 32, 33,
    201–2, 219
Chandley, A.C. 93
Chandra, R.K. 220–1
Charlesworth, B. 78, 120
Charlesworth, D. 120
Chenuil, C. 213
Chippindale, A.K. 61–2
Clarke, G.M. 6, 14, 17, 21, 23, 40,
    114–15, 118, 119, 121, 124–6,
    130–2, 136–7, 140–1, 143, 150,
    209, 210, 220
Clement, D.E. 187
Cliff, D. 186
Clutton–Brock, T.H. 227
Cockerham, C.C. 124
Coelho, J.R. 124
Cook, N.D. 184–5
Corballis, M.C. 186
Córdoba-Aguilar, A. 197, 218
Cornuet, J. 124
Corr, S. 196
Cothran, E.G. 120
Cott, H.B. 22, 190
Coxeter, H.S.M. 12
Coyne, J.A. 22, 86
Crescitelli, F. 87
Cross, S.S. 13, 45
Crow, J.F. 125
Crusio, W.E. 126
Cuervo, J.J. 34, 79–80, 107, 174,
    180, 199–200
Cullis, C.A. 95
Cunningham, M.R. 197
Curio, E. 186
Cuthill, I. 11, 30, 33, 34, 150, 170,
    195, 197–8
Cuzin-Roudy, J. 37, 125–6

Daly, M. 216
Dancik, B.P. 8
Danzmann, R.G. 120
Darwin, C. 77, 104

Davies, A.G. 40
Davis, A.P. 41
Davis, T.A. 18
Dawkins, M.S. 179–80, 185
De Marinis, F. 130
De Pomerai, D. 57
Degn, H. 44, 48
Delius, J.D. 186
Dennis, R. 103
Derr, J.A. 101
Devroetes, P. 48
Dhondt, A.A. 192
Dobzhansky, T. 67, 118, 119, 128, 204
Downhower, J.F. 33, 213
Doyle, W.J. 6, 9, 14, 136, 144–5
Drake, J.W. 95
Dubinin, N.P. 7
Dudash, M.R. 215
Dufour, K. 146, 152
Durrant, A. 95

Eanes, W.F. 6, 8
Ehret, G. 216
Eilbeck, J.C. 46
Eisenman, R. 34
Eisner, F.F. 98
Ekman, J. 102
Eldredge, N. 73, 82, 86, 104, 108, 109
Elliott, N.G. 8
Emlen, J.M. 37, 38, 43–5
Endler, J.A. 86, 100, 105
Enquist, M. 183–6, 187
Eriksson, M. 6, 157, 203, 213, 215–17
Erway, L. 137
Escós, J.M. 6, 13
Evans, M.R. 159, 168, 169, 171, 198, 204

Fahn, A. 42
Falconer, D.S. 78, 97, 98, 102, 115, 117
Farris, M.A. 120

Feder, J. 12
Fedoroff, N.V. 96–7, 103
Felley, J.D. 31, 128–30
Ferguson, M.M. 130
Ferrara, N. 41
Fields, S.J. 27
Fisher, R.A. 116
Fitter, A.H. 103
Fleischer, R.C. 122
Folstad, I. 221, 222, 223
Foltz, D.W. 122
Forbes, M.R. 26, 31, 55, 147, 152
Ford, C.S. 216
Forkman, B. 196
Fowler, K. 113–15
Fox, W. 136, 140
Fraser, A.S. 68, 71, 74–5, 76, 88, 108
Fraser, F.C. 223
Free, M.J. 214
Freeman, D.C. 6, 11, 12, 18, 58, 142, 148, 150
French, V. 46
Frey, K.J. 98, 101
Fujio, Y. 120
Furlow, F.B. 186, 192, 197

Gabriel, M.L. 101
Gajardo, G.M. 121
Gangestad, S.W. 31, 32, 125, 202, 210, 218, 222
García-Vázquez, E. 71
Garland, T. 88
Garner, W.R. 187
Garton, D.W. 120
Gavrilets, S. 70, 72–4, 76
Gebhardt-Heinrich, S.G. 102
Gershwin, M.E. 220–1
Gest, T.R. 136, 145, 211
Gibson, J.B. 101
Gilbert, S.F. 37, 42, 43, 47, 48
Gilliard, E.T. 192
Giurfa, M. 186, 188
Glass, L. 48
Goldberger, A.L. 13

Goldschmidt, R.B. 37, 39, 86, 94, 109, 130–1
Goodwin, B. 47
Gottesman, I.I. 13, 192
Gould, S.J. 73, 82, 86, 104, 108, 109
Govind, C.K. 48, 57
Govindaraju, D.R. 8
Grafen, A. 179–80
Graham, J.H. 6, 7, 11, 12, 13, 14, 15, 17, 21, 23, 31, 37, 38, 43, 47, 48, 49–51, 58, 109, 128–30, 140–1, 151
Grammer, K. 197
Grant, M.C. 120, 125, 209–10
Graubard, M.A. 93
Greene, D.L. 27
Grell, R.F. 93
Griffing, B. 101, 136
Groeneveld, H. 119, 142
Gualtieri, C.T. 142
Guilford, T. 170, 179–80, 185
Gustafsson, L. 102
Guthrie, R.D. 82
Gvozdev, V.A. 109, 131

Habers, G. 186
Hademenos, G.J. 48
Hagen, D.W. 112
Haines, T.A. 140
Hall, B.G. 95, 97
Halldórsdóttir, H. 25, 27, 32, 201
Hallgrímsson, B. 151
Hamilton, W.D. 116
Hanawalt, P.C. 103
Handford, P. 6, 8, 123
Hanson, G.P. 149
Hardin, R.T. 101
Harper, D. 180
Harrington, R.W. 101
Harrison, R.G. 214
Hartl, G.B. 121
Harvey, G.A. 101
Harvey, S. 90–1
Hasegawa, M. 223
Hasson, O. 180

Hastings, A. 70, 72–4, 76
Hastings, H.M. 38
Hatchwell, B.J. 159, 169, 171
Hawkins, A.J.S. 121
Hayman, D.L. 93
Healy, S. 218
Heiskanen, A. 136
Held, L.I. 42
Henderson, J. 214
Hershkovitz, I. 148
Herzog, A. 155
Hilbish, T.J. 120
Ho, L. 103
Hoffmann, A.A. 73, 82, 89, 90, 91, 108, 147
Höglund, J. 30, 33, 79–80, 164, 167, 172–6, 198–9
Holloway, G.J. 98
Horridge, G.A. 186–8
Hostert, E.E. 86, 106
Houle, D. 74, 81, 115–16, 120, 127
Hsieh, C.L. 103
Hubbs, C.L. 15, 101
Hubbs, L.C. 15
Hubert, W.A. 29
Huether, C.A. 101
Huntley, H.E. 12
Huxley, T.H. 86
Hwu, T.-K. 142
Hyde, J.E. 194

Inbau, F. 181
Ivanovics, A.M. 87

Jablonka, E. 95, 106
Jackson, J.F. 128, 130
Jaenisch, R. 103
Jagoe, C.H. 140
Jamison, C.S. 223
Jennions, M.D. 203
Johns, P.E. 120
Johnson, G.R 98, 101
Johnson, S.D. 203
Johnstone, R.A. 180, 183–6
Jokela, P. 149

Jones, K.L. 7
Jones, R.T. 160
Joswiak, G.R. 130
Julesz, B. 186–7

Kacser, H. 120
Karter, A.J. 222
Kat, P.W. 125
Kawecki, T.J. 81
Kerkis, J.J. 94–5
Kieser, J.A. 70, 72–3, 89, 108, 119, 142
Kilias, G. 93
Kimura, M. 125
Kindred, B. 70, 71, 101
King, D.P.F. 8
Kirpichnikov, V.S. 137, 193, 211
Kitchell, J.A. 89
Klein, J. 193
Kobyliansky, E. 122, 125, 130, 213
Koehn, R.K. 8, 46, 120–1
Koella, J. 221
Koji, T. 45
Kondo, S. 43
Kowner, R. 198
Kozlov, M.V. 142
Kozlowski, J. 207, 212
Krakauer, D.C. 52
Kraus, E.J. 216
Krebs, R.A. 91
Kushev, V.A. 93
Kuzin, A.M. 87

Lamb, B.C. 93
Lamb, E. 95, 106
Lamb, T. 130
Lambert, D.M. 105
Lambrecht, M. 192
Lande, R. 189
Lander, A. 56
Langlois, J.H. 197–8
Langridge, J. 101, 136
Larsson, K. 101
Latyszewski, M. 102
Lauder, J.M. 40

Lawrence, M.J. 92
Leamy, L. 26, 27, 28, 114–15, 117, 118, 126
Leary, R.F. 6, 8, 10, 11, 14, 17, 21, 23, 118, 122–6, 128, 136, 144, 150
Lee, S.J. 225
Lee, Y.-H. 193
Leech, S.M. 227–8
Lenski, R.E. 95
Lerman, S. 123, 130
Lerner, I.M. 119, 122
Leroi, A.M. 91
Leung, B. 26, 31, 55, 147, 152
Levin, B.R. 92
Levin, D.A. 89, 104–5, 108
Levitan, D.R. 193
Lewontin, R.C. 37, 115, 120
Lieber, M.R. 103
Lindén, M. 102
Lindgren, D. 94–5
Lindholm, A.K. 21
Lindsey, C.C. 101
Livshits, G. 32, 121, 122, 125, 130, 213
Lobashov, M.E. 94
Locke, J. 99
Lockwood, R. 33, 168, 174
Loeschcke, V. 91
Loidl, J. 93
Loveless, A. 94–5
Lu, B.C. 93
Lubiniecki, A.S. 223
Lucas, A.M. 61
Lucchesi, J.C. 93
Luckey, T. 87
Ludwig, W. 6, 9, 11, 21
Lundström, A. 27

McAndrew, B.J. 8
McBee, K. 147
McClintock, B. 96, 102, 109
McDonald, J.F. 109
Mackay, T.F.C. 101
McKechnie, S.W. 102

McKenzie, J.A. 6, 14, 17, 21, 23, 40, 91, 94, 102, 125, 130–2, 136–7, 143, 210–11, 216, 220
McManus, I.C. 15
McNab, B.K. 103
McNary, H.W. 101
McNelly-Ingle, C.A. 93
Mahsberg, D. 215–16
Makaveev, T. 120, 125
Malina, R.M. 125
Malyon, C. 218
Mandelbrot, B.B. 12, 38
Manning, J.T. 25, 27, 31, 32, 33, 157–8, 197, 201–2, 209, 213, 219
Manwell, C. 120
Markel, A.L. 98
Markow, T.A. 6, 13, 15, 17, 114–15, 192
Markowski, J. 33
Martin, G.M. 223
Mason, L.G. 27
Mason, S.F. 19
Mather, K. 14, 21, 23, 37, 54, 118, 130
Mayhill, J.T. 151
Maynard Smith, J. 22, 23, 47, 108, 133, 179–80, 185
Mayr, E. 73, 105, 109
Meinhardt, H. 42, 46
Menne, M. 186
Merilä, J. 27, 102
Michod, R.E. 92
Midgley, J.J. 203
Mittler, J.E. 95
Mitton, J.B. 6, 8, 14, 46, 120–1, 123–5, 204, 209–10
Møller, A.P. 6, 11, 15, 17, 18, 23, 27, 30, 32, 33, 34, 41, 53, 59–60, 64–5, 66, 72–4, 75, 76, 78, 79, 81, 82, 99, 100, 102, 105, 107, 112–16, 125, 130, 141, 144–6, 150, 152, 156, 157, 158, 159, 161, 167, 168, 170–1, 172–6, 180, 181–2, 183, 185, 189,

192, 193, 194–5, 198–200, 201, 202, 203, 204, 205, 207, 208, 210–13, 214–17, 218–19, 220–3, 224, 226, 227
Montag, B.A. 216
Montag, T.W. 216
Moodie, G.E.E. 202, 213, 225
Moodie, T.E. 202, 213
Mooney, M.P. 137, 145
Moreno, E. 168, 181–2
Moreno, G. 68, 75, 76, 100
Morgan, M.J. 18
Moss, M.L. 144
Mukai, T. 54, 204
Munn, C.A. 181
Murphy, P.A. 101
Murray, J.D. 43

Nagorcka, B.N. 46
Nahon, E. 52
Naugler, C.T. 227–8
Neel, J.V. 92–3
Nei, M. 125
Neville, A.C. 15, 16–17, 18, 20, 22
Newberne, P.M. 220–1
Newell-Morris, L.L. 216
Newton, I. 227
Nielsen, J.T. 224
Nijhout, H.F. 46
Nilsson, J.-Å. 137–8
Norberg, R.Å. 16, 164–6, 172
Norberg, U.M. 160–1
Novak, J.M. 123, 125
Nowak, B. 186
Numerical Algorithms Group 32

O'Brien, S.J. 214
Ockenden, L. 31, 157–8
O'Farrell, K. 40, 125, 210–11, 216
Oldroyd, B.P. 124–5
Olsen, L.F. 44, 48
Osborne, R. 101
Osgood, W. 101
Osorio, D. 22, 186–7, 191
Otronen, M. 196

Otte, D. 86, 105
Owen, R.D. 119, 147
Ozernyuk, N.D. 46, 48, 87, 102

Packard, G.C. 30
Packer, C. 226
Pagel, M. 52–3, 193
Palmer, A.R. 11, 14, 15, 17, 18, 21, 23, 24, 25, 27, 28, 29, 30, 31, 32, 34, 35, 47, 55, 57, 61–2, 112, 114–15, 121, 124
Pankakoski, E. 141, 210–11
Parisi, J. 48
Parker, H.R. 107, 129
Parker, L.T. 126
Parsons, P.A. 6, 33, 73, 86, 88, 89, 90, 91, 92–3, 98, 102, 104, 108, 112, 135–7, 147, 149–52, 156, 213
Pashler, H. 186
Pasteur, L. 19
Patterson, B.D. 121
Patton, J.L. 121
Pearce, J. 48, 57
Peitgen, H. 38
Pennycuick, C.J. 160–1
Perrett, D.I. 197–8
Petersen, C. 193
Petrie, M. 189
Phelan, J.P. 117
Picton, H.D. 33, 137
Pierce, B.A. 120
Piñeiro, R. 71
Plough, H.H. 92–3
Polak, M. 91, 220–1
Pomiankowski, A. 11, 27, 30, 32, 72–4, 75, 76, 78, 81, 82, 99, 114–15, 146, 150, 172–4, 180, 184, 189, 199, 204
Portin, P. 149
Poulin, R. 32
Powers, D.A. 124
Price, T.D. 61, 81, 102
Purnell, D.J. 22
Pusey, A.E. 226

Qazi, Q.H. 142

Radesäter, T. 25, 27, 32, 201
Rago, P.J. 144
Rappaport, J. 34
Rasmuson, M. 6, 7, 8, 84, 130, 143
Ratner, V.A. 96–7
Reeve, E.C.R. 37, 118
Reeves, B.C. 186
Reid, J. 181
Reimchen, T.E. 225
Rendel, J.M. 70, 71, 72
Rensch, B. 186
Rettig, J.E. 144, 226–7
Reznichenko, L.P. 98
Rhodes, F.H.T. 104, 108
Rice, W.R. 34, 84, 105–6, 124
Richards, R.A. 101
Ridsdill–Smith, T.J. 141
Riesenfeld, A. 136, 144
Rifaat, O.M. 93
Rintamäki, P. 197
Ritchie, M.G. 192
Robertson, E. 71, 101
Robertson, J.L. 128
Rodhouse, P.G. 121
Roelofs, W.L. 193
Roff, D.A. 81, 227
Roggman, L.A. 197
Roldán, C.E. 186
Romanoff, A.L. 214
Romashov, D.D. 7
Root, T. 103, 149
Rose, A.M. 93
Ross, K.G. 128
Rothstein, R. 103
Ruban, G.I. 136
Rubio, J. 71
Ryabova, G.D. 121
Ryan, B.F. 25
Ryan, M.J. 183
Ryman, N. 8

Sachs, T. 42
Sæther, B.-E. 32, 33

Sagan, L.A. 87
Saino, N. 223
Sakai, K.-I. 53, 204, 223
Sankaranarayanan, K. 95
Sargent, T.D. 191
Sarvas, R. 216
Sauer, P. 25, 27, 32, 112, 218
Saunders, S.R. 151
Schaeffer, A.A. 23
Scharloo, W. 3, 70, 71, 72
Scheiner, S.M. 8
Schluter, D. 81
Schmalhausen, I.I. 67, 69, 72–3, 75,
    96, 104
Schroeder, M. 43
Schwabl, U. 186
Sciulli, P.W. 137
Seavey, S.R. 194
Selker, E.U. 93, 95
Selye, H. 87, 90–1, 94, 135
Semlitsch, R.D. 101
Seradilla, J.M. 121
Seredenin, S.B. 94–5
Serova, I.A. 94–5
Shackell, N.L. 6, 9, 14
Shami, S.A. 121
Shapiro, B.L. 125, 131
Shapiro, J.A. 97
Shaposhnikov, G.K. 82–4, 106
Sheldon, B.C. 214
Sheldon, B.L. 71
Sheppard, P.M. 191
Shimamoto, Y. 53, 204, 223
Shorey, H.H. 193
Shumway, S.R. 120
Shykoff, J.A. 59–60
Siegel, M.I. 136–7, 144–5
Siegel, P. 140
Siegel, S. 25, 32
Simchen, G. 93
Simmons, L.W. 33, 34, 192, 198,
    201
Simpson, G.G. 82
Simpson, J.L. 214, 216
Singh, S.M. 120

Skinnes, H. 135
Smith, L.D. 57
Smith, M.H. 123
Smookler, H.H. 144
Smouse, P.E. 32, 121, 125
Sneddon, L.U. 218
Sniegowski, P.D. 95
Solberg, E.J. 32, 33
Sondhi, K.C. 22, 23
Sorci, G. 203
Soulé, M. 14, 30, 33, 37, 53, 61,
    123, 125–6, 130, 140–1, 151
Spedding, G.R. 160
Spencer, H.G. 105
Stadler, D.R. 93
Stamberg, R. 93
Stearns, S.C. 81, 208, 212, 214, 227
Stebbins, G.L. 203
Stephenson, A.G. 215
Stern, C. 93
Stettenheim, P.R. 61
Strobeck, C. 11, 14, 15, 17, 18, 23,
    24, 25, 27, 28, 29, 30, 31, 32,
    34, 35, 47, 112, 114–15, 121, 124
Suchentrunk, F. 123, 227
Sugihara, G. 38
Sumarsono, S.H. 41
Summerbell, D. 56
Suthers, R.A. 192
Suzuki, D.T. 93
Swaddle, J.P. 11, 24, 25, 27, 28, 29, 30,
    32, 33, 34, 35, 62–3, 64, 113, 115,
    137–9, 157, 167–70, 173, 195–6,
    197–8, 211–13, 218–19, 224
Swain, D.P. 10
Sweet, G.B. 216
Symons, D. 197
Szathmáry, E. 108
Szilagyi, P.G. 34, 187

Taddei, F. 94, 96
Tåning, Å.V. 100–1
Tarasjev, A. 8
Tartof, K.D. 99
Teather, K. 64

Tebb, G. 67, 128
Teska, W.R. 123
Thessing, A. 102
Thoday, J.M. 67, 112, 128, 130, 136
Thoma, R. 186
Thomas, A.L.R. 160–6, 169, 171–3, 176, 181
Thomas, B.J. 103
Thomas, F. 32
Thompson, J.N. 22, 128
Thompson, S.N.
Thornhill, R. 25, 27, 30, 32, 41, 112–16, 125, 130, 194, 196, 197, 202, 204, 208, 210, 217, 218, 220–1, 222, 223
Thseng, F.-S. 142
Timofeeff-Ressovsky, N.V. 16
Tinbergen, N. 191
Todaro, G.J. 223
Tomkins, J.L. 33, 201
Towe, A.M. 93
Tracy, M. 150
Trivers, R.L. 181, 220
Trout, W.E. 149
Tuinstra, E.J. 112
Turing, A. 42, 47
Turkington, R.W. 45

Ueno, H. 227
Uetz, G.W. 196

Valentin, J. 93
Valentine, D.W. 14, 33, 140–1, 150
van Abeelen, J.H.F. 126
van Noordwijk, A.J. 102
Van Valen, L. 11, 14, 15, 21, 32
VanDemark, N.L. 214
Vaquier, V.D. 193
Vogl, C. 70
Vøllestad, L.A. 123, 126–7, 209
von Fersen, L. 186
Vrijenhoek, R.C. 123, 130

Waddington, C.H. 3–4, 6, 7, 69–70, 71, 101, 108

Wagner, G.P. 81
Wakefield, J. 32, 136
Wakelin, D. 94
Wallace, A.R. 189
Wallace, B. 118
Ward, P.D. 89
Ward, P.J. 100
Ward, R.D. 8
Wassermann, K. 103
Watkinson, A.R. 42
Watson, P.J. 30, 222
Watt, W.B. 121, 125
Wayne, R.K. 33, 118–19
Wcislo, W.T. 90
Weatherhead, P.J. 146, 152
Webb, C.J. 216
Wedemeyer, G.A. 46
Weinstein, R.S. 13
Weir, B.S. 124
West, B.J. 13, 48
White, M.J.D. 42, 106
Whiten A. 181
Whitlock, M.C. 32, 55, 100, 113–16, 121
Whitt, G.S. 130
Wiebe, W.J. 87
Wiener, J.G. 143
Wiens, D. 215–16
Wiggins, D.A. 101
Wilber, E. 32, 142
Wildt, D.E. 214
Williamson, P.G. 73
Willig, M.R. 119
Willson, M.F. 194
Wilson, D.M. 23
Wilson, J.M. 209
Wilson, M. 216
Winberg, G.G. 46, 87
Withler, R.E. 8
Witter, M.S. 11, 34, 62–3, 64, 137–9, 150, 173, 211, 218–19, 225
Wodzicki, T.J. 48
Wolf, D.P. 216
Wolf, S.D. 65
Wolff, K. 122

Wolpert, L. 56
Wood, D. 197
Wood, W.B. 57
Woolf, C.M. 190
Wooster, D. 113
Wooten, M.C. 123

Yablokov, A.V. 6, 10, 140
Yamana, K. 58
Yampolsky, L.Y. 8
Yen, J.L. 21, 136, 143
Yezerinac, S.M. 8
Yokoyama, T. 21
Yost, H.J. 57
Young, R.W. 144

Yuge, M. 58

Zahavi, A. 52, 180
Zajaczkowski, S. 48
Zakharov, V.M. 3, 6, 10, 33, 34, 88,
    121, 126, 129, 136, 140, 142,
    213
Zar, J.H. 27
Zeeman, C. 47
Zeleny, C. 94
Zera, A.J. 127
Zhang, S.W. 186
Zink, R.M. 121
Zouros, E. 120, 122
Zvereva, E.L. 221

Taxonomic index

Accipiter nisus 224
Acheta veletis 15
Acinonyx jubatus 118–19
Acipenser baeri 136
Acipenser stellatus 121
Aedes aegypti 23
Aegopodium podagaria 142
Agelaius phoeniceus 146
Alga, brown 142
Allium ursinum 93
Alphaeus 18
Ambystoma tigrinum 120
Anarhynchus frontalis 16, 22–3
Anisopus fenestralis 23
Ant, fire 128
Anthriscus 82–3
Aphid 82–4, 106
Apis mellifera 119, 124, 126, 187–8
Arabidopsis thaliana 101, 136
Artemia franciscana 121
Avena sativa 101

Bass, largemouth 107, 129, 130
Bear, grizzly 137
Bee, honey 119, 124, 126, 187–8
Beetle 22
Betula 142
Bicyclus anynana 39, 46–7
Bicyclus safitza 46–7
Birch 142
Birgus 20
Black locust tree 141

Blackbird
  European 216, 226
  red-winged 146
Blowfly, Australian sheep 21, 39–40, 126, 131–2, 136–7, 142–3, 210–11, 216
Bopyrus 20
Bos taurus 98, 190
Botryllophilus 20
Bowerbird 192
Branta leucopsis 101
Brassica campestris 101
Brassica napi 101
Bug, milkweed 23
Butterfly 27, 46–7, 121

Caenorhabditis elegans 57, 93
Cancer productus 57
Canis familiaris 190
Capra hircus 190
Capra pyrenaica 13
Capreolus capreolus 155–6
Carassius auratus 7
Carcinus maenas 218
Cat 190
Cattle 98, 190
Cavia porcellus 190
Chaerophyllum bulbosum 82–4, 106
Chaerophyllum maculatum 83–4
Cheetah 118–19
Chicken 101, 144, 156, 158–9, 192, 196, 211

Chrysopa perla 126, 140
Coconut 18
Cocos nucifera 18
Colaptes auratus auratus 128
Colaptes auratus cafer 128
Colias 121
Columba livia 156
Convolvulus arvensis 141
Coprinus lagopus 93
Cormorant, double-crested 140
Cottus bairdi 213
Crab
   fiddler 16
   hermit 20
   horseshoe 89
   robber 20
   shore 218
Crassostrea virginica 120
Cress, thale 101, 136
Cricket 15
   field 15, 192
Crossbill 17, 21

Dama dama 218
Deer
   fallow 218
   roe 155–6
   white-tailed 120
Dendrobates tinctorius 22
Dictyostelium 48
Dog 190
Drosophila melanogaster 4, 7, 8, 21,
   22, 40, 56, 69–71, 75, 76, 81,
   84, 88, 91, 92–3, 96–7, 101–2,
   113, 118, 121, 128, 129, 136,
   137, 140,143,147, 149, 189, 213
Drosophila nigrospiracula 220–1
Drosophila simulans 101
Dryomyza anilis 196
Dysaphis anthrisci majkopica 82–4, 106
Dysaphis chaerophyllina 106
Dysdercus bimaculatus 101

Earwig 201
Elm, broad-leaved 65–6, 211, 220–1

Enneacanthus gloriosus 128
Enneacanthus obesus 128
Epilobium angustifolium 142, 203,
   213, 215–17
Equus caballus 157–8, 190
Escherichia coli 94, 95–6

Falco tinnunculus 62–3, 160
Felis catus 190
Ficedula albicollis 102
Finch, zebra 195–6, 212–13, 218–19
Fireweed 142, 203, 213, 215–17
Fish 22, 101
Flicker
   red-shafted 128
   yellow-shafted 128
Fly
   Australian bush 126, 141
   dung 224
   house 220, 224
Forficula auricularia 201
Frog 22, 57, 58
Fruitfly 4, 7, 8, 21, 22, 40, 56, 69–
   71, 75, 76, 81, 84, 88, 91, 92–3,
   96–7, 101–2, 113, 118, 121, 128,
   129, 136, 137, 140, 143, 147,
   149, 189, 213
Fucus furcatus latifrons 142
Fundulus heteroclitus 101, 124

Galleria mellonella 87
Gallus domesticus 101, 144, 156,
   158–9, 192, 196, 211
Gasterosteus aculeatus 191, 202, 213,
   224–5
Gazella dorca 12
Gazella granti 191
Gazelle
   dorcas 12
   Grant's 191
Gecko 22
Gemsbok 212–13, 218
Geospiza fortis 102
Giraffa camelopardalis 189
Giraffe 189

Glycine max 148
Goat 190
Goldfish 7
Grape 65
Grasshopper 192
Grouse, black 197
Grunion 140–1
Gryllus bimaculatus 15, 192
Guineapig 190

Halichoerus grypus 10, 140
Hamster 190
Hare, brown 123, 226–7
Harmonia axyridis 227
Hemigrapsus nudus 61–2
Hetaerina americana 197, 218
Hetaerina cruentata 197, 218
Hirundo rustica 18, 59–60, 64–5,
  100, 102, 141, 145–6, 161, 164,
  166, 168, 170–1, 181–2, 194–5,
  201–2, 204–5, 211, 212–14, 218–
  19, 220–3, 224
Homarus americanus 57
Homo sapiens 27, 34, 41, 113, 125,
  131, 142, 186, 187, 192, 196,
  197–8, 202, 209, 210, 212–13,
  215–16, 218, 220–3
Hordeum vulgare 101
Horse 157–8, 190
Human 27, 34, 41, 113, 125, 131,
  142, 186, 187, 192, 196, 197–8,
  202, 209, 210, 212–13, 215–16,
  218, 220–3
Hyla cinerea 130
Hyla gratinosa 130
Hyla langsdorfii 22

Ibex, Spanish 13

Jynx torquilla 189

Kestrel, European 62–3, 160
Killifish 101, 124

Lacerta 136

Lacerta agilis 87–8
Lacerta vivipara 213
Lepomis cyanellus 130
Lepomis macrochirus 143
Lepus europaeus 123, 226–7
Leuresthes tenuis 140–1
Limulus 89
Linanthus androsaceus 101
Lingula 89
Linum catharticum 101
Lion 226
Lithinus nigrocristatus 22
Loach 102
Lobster, American 57
Locust 23
Loxia curvirostra 17, 21
Lucilia cuprina 21, 39–40, 126, 131–
  2, 136–7, 142–3, 210–11, 216
Lychnis viscaria 213, 216–17
Lyrurus tetrix 197

Macaca nemestrina 216
Maize 96–7
Malacosoma disstria 227–8
Malus silvestris 216
Man-o-war, Portuguese 16
Mesocricetus auratus 190
Micropterus salmonides 107, 129,
  130
Microtus 82
Misgurnus fossilis 102
Mosquito 23
Moth
  forest tent caterpillar 227–8
  wax 87
Mouse
  deer 190
  house 21, 40–1, 56, 71, 93, 97–8,
    102, 118, 128–9, 131–2, 137,
    147, 190, 216
Mus musculus 21, 40–1, 56, 71, 93,
  97–8, 102, 118, 128–9, 131–2,
  137, 147, 190, 216
Musca domestica 220, 224
Musca vetustissima 126, 141

Muskrat 210–11
Mussel, blue 121
Myrmeleotettix maculatus 192
Mytilus edulis 121

Natrix fasciata 22, 101
Neurospora crassa 93
Nicotiana tabacum 223
Nuthatch, European 137–8

Odocoileus virginianus 120
Oenanthe leucura 192
Oilbird 192
Oncopeltus 23
Oncorhynchus clarki 29–30
Oncorhynchus mykiss 136, 144
Ondatra zibethicus 210–11
Oribi 226
Ornithonyssus bursa 222
Oryctolagus cuniculus 190
Oryx gazella 212–13, 218
Oryza sativa 54, 142
Ourebia ourebi 226
Ovis aries 155, 190
Oyster, American 120

Pandinus imperator 215–16
Panorpa nipponensis 217
Panorpa ochraceopennis 217
Panorpa vulgaris 218
Panthera leo 226
Parus caeruleus 102, 211
Parus major 102, 192
Parus montanus 102
Pavo cristatus 189, 216, 223
Peacock, blue 189, 216, 223
Peromyscus floridanus 136, 145
Peromyscus gossypinus 136, 145
Peromyscus maniculatus 190
Phalacrocorax auritus 140
Phaseolus vulgaris 101
Phylloscopus inornatus 61
Physalia physalis 16
Pig 190
Pigeon, domestic 156

Pinus radiata 216
Pinus silvestris 216
Plover, wry-billed 16, 22–3
Poeciliopsis lucida 130
Poeciliopsis monacha 130
Poplar 144, 226–7
Populus euramericana 144, 226–7
Prunus cerasus 216
Pseudacris ornata 101
Python regius 22

Rabbit 190
Rangifer tarandus 221, 223
Rat 117–18, 136–7, 140, 144–5, 190,
    211
Rattus norvegicus 117–18, 136–7,
    140, 144–5, 190, 211
Reindeer 221, 223
Rhynchaenus rufus 220–1
Rice 54, 142
Rivulus marmoratus 101
Robinia pseudoacacia 141

Sail-by-the-wind, jack 16
Salamander, tiger 120
Salamandra 22
Salix borealis 221
Salmo gairdneri 118
Salmo salar 122
Salmon, Atlantic 122
Salvelinus confluentus 21, 128
Salvelinus frontinalis 128
Scatophaga stercoraria 224
Sceloporus undulatus 130
Sceloporus woodi 130
Schistocera gregaria 23
Scomber scombrus 22
Scorpion 215–16
Seal, grey 10, 140
Sheep 155, 190
Shizophyllum commune 93
Shrew, common 141, 142, 213
Shrimp
    brine 121
    snapping 18

Sitophilus oryzae 98
Sitta europaea 137–8
Snake, garter 136, 211
Solenopsis invicta 128
Solenopsis richteri 128
Sordaria brevicolis 93
Sorex araneus 141, 142, 213
Soybean 148
Sparrowhawk, European 224
Spider 196
Starling 27, 28–30, 63–4, 137–9,
    168–70, 211, 218–19, 225
Steatornis caripensis 192
Stickleback, three-spined 191, 202,
    213, 224–5
Streamertail, red-billed 171–2
Sturgeon
  Siberian 136
  stellate 121
Sturnus vulgaris 27, 28–30, 63–4,
    137–9, 168–70, 211, 218–19, 225
Sunfish
  banded 128
  blue-spotted 128
  bluegill 143
  green 130
Sus scrofa 190
Swallow
  barn 18, 59–60, 64–5, 100, 102,
    141, 145–6, 161, 164, 166, 168,
    170–1, 181–2, 194–5, 201–2,
    204–5, 211, 212–14, 218–19,
    220–3, 224
  tree 64, 101
Syrphus viridiceps 23

Tachycineta bicolor 64, 101
Taeniopygia guttata 195–6, 212–13,
    218–19

Tamanovalva limax 15
Teak 210–11
Tectonia grandis 210–11
Thamnophis elegans 136, 211
Tigriopus californicus 130
Tisbe holothuriae 118
Tit, great 192
Tobacco 223
Tribolium castaneum 101
Triticum aestivum 101, 135–6
Trochilus polytmus 171–2
Trout
  brook 128
  bull 21, 128
  cutthroat 29–30
  rainbow 136, 144
Turdus merula 216, 226

Uca musica 16
Ulmus glabra 65–6, 211, 220–1
Uroplates fimbriatus 22
Ursus arctos horribilis 137

Vanellus vanellus 22
Vellela vellela 16
Vipera berus 22
Vitis vinifera 65

Warbler, yellow-browed 61
Weevil, rice 98
Wheat 101, 135–6
Wheatear, black 192
Willow 221
Wryneck 189

Xenopus laevis 57, 58

Zea mays 96–7

Subject index

Adenylate energy charge 86–7
Aerodynamics 159–77
  avian tails 160–1, 163–8, 170–2
  avian wings 161–3, 167–70
  functional importance of
    traits 172–5
Alleles, deleterious recessive 125
Antisymmetry 16–20
Asymmetry, directional 15–20
Asymmetry, fluctuating
  acidification 140
  aerodynamics 159–77
  alcohol 142
  analysis 23–35
  audiogenic stress 144–5
  chemical factors 138, 140–2
  competition 226–7
  cost of 21–3
  DDT 140
  definition 9–11
  developmental selection 212,
    214–17
  dominance, social 217–20
  fecundity 210–14
  gametes 214
  growth 209–11
  heavy metals 141
  heritability 22, 112–17
  heterozygosity 122–7
  hybridisation 128–30
  immune defence 222
  in meristic characters 10

inbreeding 117–19
individual property 11
intraspecific competition 142–4
longevity 226–8
loss of symmetry 189–90
maintenance 116–17, 175–7, 203–5
maternal age 136
mutation 131–3
nutritional stress 137–9
ontogeny 60–6
parasitism 220–3
performance 155–9
population density 142–4
population statistic 11
predation 223–5
radiation 145–6, 204–5
repeatability 116
size dependence 30–1
smoking 142
survival 226–8
temperature 135–7
testosterone 223
transitions between types of
  asymmetry 20–1
water temperature 136

Canalisation 3, 68–72
  selection experiments 69–71
Cell–cell communication 52–3, 193–4
Chaos theory 42–52
Coefficient of variation, additive
  genetic 78, 81

Coefficient of variation,
    phenotypic 7–8, 79–80
    under adverse conditions 100–2
Competition
    interspecific 226–7
    intraspecific 142–4, 226–7
Corticosterone 87, 90

Distribution of species under
    adverse conditions 103–4
DNA repair 87
Dominance, social 217–20

Epistasis 75
Error, measurement 27–30
Evolution, gradual 82

Feathers, development of 61
Fecundity, developmental
    stability 210–14
Feedback, negative 44–6, 57–8
Fisher's fundamental theorem 116
Fractals 12–13, 38
Free-radical scavengers 87

Gait 158–9
Gametes, developmental stability 214
Growth
    developmental stability 209–11
    non-linear 43–52

Handedness, development 55–6
Herbivory, adaptations to 82–4
    developmental stability 220–1,
    223
Heritability 97
    of developmental stability 112–17
Heterosis, multiple-locus 125
Heterozygosity
    advantage 125
    developmental instability 122–7
    fitness 119–22
    hybridisation 128–30
Homeostasis, developmental 3
Homeotherms, advantage of

heterozygosity 123, 126–7
Hormesis 87
hox genes 41
Hybridisation 107–8, 128–31
    disruption of coadapted
    genome 107–8, 128–31

Immune defence 222
Inbreeding, developmental
    instability 117–19
Indices
    frequency 5–9
    repeated-formation 5–6, 9–14
Individual asymmetry
    parameter 53–5
Innovations, major
    evolutionary 108–9
Insecticide resistance 39–40

Landscape, epigenetic 3
Life-history theory 81

Meta-analysis 113–14, 126–7, 194
Metabolic efficiency 86–90, 120–1,
    149, 209–10
Modifiers, genetic 73, 99
Modular organisms 42–3
Monitoring 150–2
Morphogen
    oscillations 47–8
    standing waves 46–7
Mutagen 94–6
Mutation
    biased 204–5
    developmental stability 131–3,
    204–5
    directed 95–6
    under adverse conditions 94–6

Noise, developmental 4

Overdominance 125

Palaeontology 82
Parakeratosis 155–6

Parasitism 220–3
Performance 155–9
Phenocopy 94
Phenodeviant 5–7
Phyllotaxis 12
Pleiotropy, antagonistic 91–2
Poikilotherms, advantage of
    heterozygosity 123, 126–7
Population asymmetry
    parameter 53–5
Predation 223–35
Punctuated equilibria 82

Randomisation tests 32
Rashevsky–Turing model 48–52
Reaction norms of asymmetry 58–
    60
Reaction–diffusion model 48–52
Recombination, under adverse
    conditions 92–3
Repeatability, of fluctuating
    asymmetry 116
Resistance, stress 90–2

Selection, developmental 212, 214–
    17
Selection, directional 76–7
    under adverse conditions 100, 102
Selection, disruptive 76–7
Selection, modes of
    genetic variance 72–6
    phenotypic variance 72–6
Selection, response to 97
Selection, sexual 77–80, 194–203
    asymmetry, fluctuating 79–80,
        194–203
    coefficient of variation, additive
        genetic 78
    coefficient of variation,
        phenotypic 79–80

Selection, stabilising 76
Signalling,
    aposematic coloration 191
    behaviour 191–3
    cellular signals 193–4
    colour patterns 189–91
    electric signals 193
    major histocompatibility
        complex 193–4
    neural networks 183–6
    predator–prey signals 190–1
    receiver psychology 186–8
    sensory bias 183
    sexual selection 194–203
        facial attractiveness 197–8
        in plants 202–3
        male–male competition 197,
            200–1
        mate choice 194–7, 198–200
    tactile signals 193
    theory 52–3, 179–82
Speciation
    adverse conditions 104–8
    allopatric 105
    peripatric 105
    sympatric 105–6
Stability, developmental 3–4
    behavioural traits 13, 191–3
Stress, developmental 4–5, 85, 148–
    9
    specific response 147–8
    susceptibility, trait-specific 145–7
Survival, developmental
    stability 226–8

Transposable elements 96–7, 109
Transposition 96–7

Variation, additive genetic, under
    adverse conditions 97–100